MW01701900

INDIA, PAKISTAN AND THE SECRET JIHAD

This book explores the history of jihadist groups in Jammu and Kashmir, documenting the course of their activities and their changing character from 1947 to 2004. Drawing on new material, including classified Indian intelligence dossiers and records, Praveen Swami shows that jihadist violence was not, as is widely assumed, a phenomenon that manifested itself in Indian-administered Jammu and Kashmir only after 1988. Rather, a welter of jihadist groups waged a sustained campaign against Indian rule in Jammu and Kashmir from the outset, after the Partition of India.

This book first analyses the ideology and practice of Islamist terrorism as it changed and evolved from 1947–1948 onwards. It subsequently discusses the impact of the secret jihad on Indian policy-making on Jammu and Kashmir, as well as its influence on political life within the state. Finally, looking at some of the reasons why the jihad in Jammu and Kashmir acquired such intensity in 1990, the author suggests that the answers lie in the transfiguration of the strategic environment in South Asia by the nuclear weapons programmes of India and Pakistan.

As such, the book argues, the violent conflict which exploded in these two regions after 1990 was not a historical discontinuity: it was, instead, an escalated form of what was by then a five-decade old secret war.

This new work will be of much interest to students of the India–Pakistan conflict, South Asian politics and security studies in general.

Praveen Swami is New Delhi Chief of Bureau for *Frontline* magazine, and writes on security- and intelligence-related issues. He has reported on the crisis in Jammu and Kashmir, as well as other low-intensity conflicts in India, for over a decade.

ASIAN SECURITY STUDIES

Edited by Sumit Ganguly
Indiana University, Bloomington and Andrew Scobell,
US Army War College

Few regions of the world are fraught with as many security questions as Asia. Within this region it is possible to study great power rivalries, irredentist conflicts, nuclear and ballistic missile proliferation, secessionist movements, ethnoreligious conflicts and inter-state wars. This new book series will publish the best possible scholarship on the security issues affecting the region, and will include detailed empirical studies, theoretically oriented case studies and policy-relevant analyses as well as more general works.

CHINA AND INTERNATIONAL INSTITUTIONS
Alternate paths to global power
Marc Lanteigne

CHINA'S RISING SEA POWER
The PLA navy's submarine challenge
Peter Howarth

IF CHINA ATTACKS TAIWAN
Military strategy, politics and economics
Steve Tsang (Ed.)

CHINESE CIVIL-MILITARY RELATIONS
The transformation of the People's Liberation Army
Nan Li (Ed.)

THE CHINESE ARMY TODAY
Tradition and transformation for the 21st century
Dennis J. Blasko

TAIWAN'S SECURITY
History and prospects
Bernard D. Cole

RELIGION AND CONFLICT IN SOUTH AND SOUTHEAST ASIA
Disrupting violence
Linell E. Cady and Sheldon W. Simon (Eds.)

POLITICAL ISLAM AND VIOLENCE IN INDONESIA
Zachary Abuza

US-INDIAN STRATEGIC COOPERATION INTO THE 21ST CENTURY
More than words
Sumit Ganguly, Brian Shoup and Andrew Scobell (Eds.)

INDIA, PAKISTAN AND THE SECRET JIHAD
The covert war in Kashmir, 1947–2004
Praveen Swami

INDIA, PAKISTAN AND THE SECRET JIHAD

The covert war in Kashmir, 1947–2004

Praveen Swami

Routledge
Taylor & Francis Group
LONDON AND NEW YORK

First published 2007
by Routledge
2 Park Square, Milton Park, Abingdon, Oxon OX14 4RN

Simultaneously published in the USA and Canada
by Routledge
270 Madison Ave, New York, NY 10016

Routledge is an imprint of the Taylor & Francis Group, an informa business

© 2007 Praveen Swami

Typeset in Times New Roman by
Integra Software Services Pvt. Ltd, Pondicherry, India
Printed and bound in Great Britain by
Biddles Ltd, King's Lynn

All rights reserved. No part of this book may be reprinted or reproduced or utilised in any form or by any electronic, mechanical, or other means, now known or hereafter invented, including photocopying and recording, or in any information storage or retrieval system, without permission in writing from the publishers.

British Library Cataloguing in Publication Data
A catalogue record for this book is available from the British Library

Library of Congress Cataloging in Publication Data
Swami, Praveen.
India, Pakistan and the secret jihad : the covert war in Kashmir, 1947–2004 / Praveen Swami. -- 1st ed.
p. cm. -- (Asian security studies)
"Simultaneously published in the USA and Canada."
Includes bibliographical references and index.
ISBN 0–415–40459–2 (hardback)
1. Jammu and Kashmir (India)--Politics and government.
2. India--Foreign relations--Pakistan. 3. Pakistan--Foreign relations--India. 4. Jihad. I. Title. II. Series.
DS485.K27S93 2006
954′.604--dc22
2006014509

ISBN10: 0–415–40459–2 (hbk)
ISBN10: 0–203–96840–9 (ebk)

ISBN13: 978–0–415–40459–4 (hbk)
ISBN13: 978–0–203–96840–6 (ebk)

For my daughter Tarsha, the light of my eyes
When she is old enough to read this book, I hope she will at least understand – if not forgive – my sudden and sometimes prolonged disappearances

CONTENTS

ACKNOWLEDGEMENTS

I could not have written this book without the generous support of several individuals, some of whom I cannot name.

Amongst those I can, Sumit Ganguly must come first: his persistent prodding and encouragement led me to begin work on this book, and he generously gave of both his time and energies in the course of my writing it.

To N. Ram, the Editor-in-Chief of *Frontline* magazine, I will always be indebted. He entrusted a young and untested reporter with covering India's most sensitive conflict, and stood by me during difficulties with both the establishment and non-state actors. *Frontline* has often been complimented for both the breadth and the depth of its reportage on the conflict in Jammu and Kashmir. This is not because of any exceptional abilities I possess, but because of the magazine's unparalleled willingness to commit financial and organizational resources to real news – something that is, sadly, a diminishing feature of Indian journalism.

Finally, without the support of the United States Institute of Peace (USIP), which awarded me a Senior Fellowship in 2004–2005, I could not have actually written this book. Their decision to grant a prestigious and generous fellowship to a relatively obscure journalist amazing – and I will always be grateful for the opportunity they provided.

To the individuals at USIP who worked with me as the book was being written, I also wish to record my gratitude: John Crist, my programme officer, was saint-like in his patience and concern; Kiran Sharma provided library support of a quality I had not imagined could exist; C. Christine Fair was a source of not only great intellectual stimulation but personal strength. To Razili Datta I owe special thanks; I was blessed to have a research assistance of such intelligence and integrity. To my friends in Washington, DC, Murtaza Solangi, Ayesha Siddiqa, and to Hannah Bloch, Claudia Kinkela and Laurent Murawiec I owe special thanks for their love and patience.

I am indebted, too, to Ajai Sahni, A.K. Doval, Anupam Gupta, A.M. Watali, Dinkar Gupta, Farooq Khan, Farooq Ahmad, Gurbachan Jagat, K.P.S. Gill, K. Rajendra, Mohammad Yusuf Tarigami, N.N. Vohra, Pushpa Iyer, Sumedh Saini and Usman Majid for the time they gave to discussing issues directly related either to this book or to helping develop, over the years, some understanding of covert warfare.

Above all, though, I am grateful to J., J., A., K., M., R., R., G., N., and D., the people who made it possible to gather the material needed to write this book and allowed me to peek into the secret world they inhabit, at not inconsiderable risk to their careers and, in some cases, their lives.

China
Taj.
Afghanistan
Territory ceded by Pakistan to China in 1963; claimed by India
Northern Areas
North-West Frontier Province
Indian claim
Siachen Glacier
Aksai Chin
Fed. Admin. Tribal Areas
Line of Control
Kargil
Tarbela Reservoir
Indus
Wular Lake
Line of Actual Control
Azad Kashmir
Srinagar
Vale of Kashmir
Jammu and Kashmir
ISLAMABAD
Islamabad Capital Territory
Pakistan
Jammu
Jhelum
Himachal Pradesh
Punjab
China
India
Chenab
Sutlej
Govind Sagar

1

A HOUSE ON A HILL

> The last argument is the sword.
> Nurul Amin,
> Chief Minister of East Pakistan
> and President of the East Pakistan
> Muslim League, 1950[1]

Perched above the Dal Lake in Srinagar, on the slopes of the Shankaracharya mountain, is one of Jammu and Kashmir's least known monuments: a modest two-storey house that sits under the shade of a magnificent *Chinar* tree. It shows signs of neglect – the hand-carved walnut-wood ceilings have been painted over with hospital-white enamel, and a spectacularly unappealing concrete office block has come up on its right flank – but even the considerable efforts of the Public Works Department have not succeeded in obscuring its beauty. Prime Minister Indira Gandhi, India's third Prime Minister and the architect of its most decisive military victory over Pakistan, is reputed to have spent a part of her honeymoon in this house. While the story might be apocryphal, few who have seen the house would disagree that she would have been well advised to do so.

On all sides of the house on the hill are other landmarks of Jammu and Kashmir's recent political history. A few hundred metres to the left is the house of Sheikh Mohammad Abdullah, the leader who was central in crafting the state's independence from monarchical rule and was without dispute its most important political figure. The Lieutenant-General of Indian Army's XV Corps, who commands the military forces that defended what is now Indian-administered Jammu and Kashmir from Pakistani assault in 1947–1948 and have held it ever since, lives and works across the road. Other centres of power are scattered all around: the home of the Indian state's supreme representative in Jammu and Kashmir; that of the head of its external intelligence service, the Research and Analysis Wing (RAW) and the Pari Mahal, or the Palace of Fairies, which has served both as a home to many of Srinagar's most powerful bureaucrats and as a top-secret interrogation centre.

Students of the spatial geography of power might find the location on the house of the hill significant: from there, it is but a short downhill stroll to any of these places. *Their* inhabitants, on the other hand, must march *up* the slope if they wish to visit the home of the Assistant Director of India's Intelligence Bureau, the covert service responsible for the nation's domestic security and counter-intelligence. This book is a history of the secret storm that swirled around the house on the hill: the long jihad that has been fought in Jammu and Kashmir from 1947–1948 to the present day. In the first half of this introductory chapter, I shall provide a brief overview of my arguments and a discussion of their significance, an introduction to the sources and documents I have used, as well as some conceptual questions. The second half of this chapter provides an overview of the strategic significance and pre-Independence politics of Jammu and Kashmir, leading up to the long jihad that began in 1947–1948.

The Jihad in Kashmir

To my mind, the course of the long jihad in Jammu and Kashmir raises questions that far transcend the spatial stage on which it was and is still being enacted: questions that may give both students of conflict and policy-makers cause for reflection on our understanding of and responses to jihadist forces.

In a world transfigured by the tragic events of September 11, 2001, and their still-developing fallout, we have become accustomed to placing Afghanistan and the events that transpired there after its invasion by the Soviet Union at the centre of our understanding of global war unleashed by the Islamist far-Right. Much of the literature on modern Islamist groups sees them as children of the unhappy histories of post-colonial regimes, notably their poor administrative structures, poverty and repressive character, and the consequent expression of popular wrath through primordial religious identities. Islamist terrorism is, in this reading of history, the consequence of failures of nationhood; the politics which underpin it are seen as narratives that lie outside of and in opposition to the system of modern nation-states.

The jihad in Jammu and Kashmir, however, defies almost all of these conclusions. Contrary to much received wisdom on the subject, this jihad did not arise as a consequence of the dramatic events which transfigured what is sometimes called the Muslim World in the third quarter of the twentieth century. It commenced, instead, in 1947, within weeks of the birth of India and Pakistan and decades before the Iranian revolution or the movement of Soviet forces across the Amu Darya into Afghanistan. It was, as we shall see in coming chapters, the child of the very *creation* of two modern nation-states, India and Pakistan. Its extraordinary longevity is, albeit perversely, evidence of their *success*. Most of the protagonists of the long jihad were deeply entwined with the state-structures of both India and Pakistan: the military and covert apparatus of the two states, their geo-strategic fortunes, their ideological concerns and their existential anxieties. If we are to engage in a serious search for peace in Jammu and Kashmir, where the conflict has claimed well over 40,000 lives of both civilians and combatants in the last decade alone, we must first revisit the conflict and examine again our premises about its causes and course.

Ever since 1990, when the conflict in Jammu and Kashmir began to engage the world's attention, a number of excellent books have sought to examine the political causes behind the violence. Victoria Schofield's magisterial *Kashmir In Conflict*, for example, provides a fine overview of the roots of India–Pakistan contestation in the region, and the origins of the political forces that for a time threatened to evict India from the part of the state it controls.[2] Sumit Ganguly's *The Crisis in Kashmir* provides an incisive account of the post-Independence collision between new social forces and flawed institutions that underpinned the explosion of violence from 1990.[3] Navnita Chadha-Behera has, in turn, carefully examined the conflicts of ethnicity and regional identity in Jammu and Kashmir.[4] All of these are but a very small sampling of the great volume of high-quality work on the origins and structures of the ongoing conflict. Most of these accounts see the violence in Jammu and Kashmir as a phenomenon of relatively recent genesis. Mohammad Amir Rana has, typically, asserted that the modern Pakistani "culture of jihad was strengthened by the revolution in Iran, nurtured by the Americans via Operation Cyclone, nourished by the extremist views and money of Osama bin Laden and came to fruition in the acts of the Taliban".[5]

While this argument has some merit – the intensity and character of the jihad as well as its popular legitimacy was without dispute transformed by the cataclysmic changes in South, Central and West Asia witnessed in the 1970s and 1980s – it is far from a complete rendering. The jihad in Jammu and Kashmir had in fact raged on ever since Jammu and Kashmir acceded to the Union of India in 1947, and Indian troops landed in Srinagar to defend the state against Pakistani irregulars. Although much scholarship contains references to the existence of terrorist activity at various points between Kashmir's accession to India and the events of 1989–1990, there has so far been little history of what actually transpired in the interim period. Both India and Pakistan have different reasons for having maintained a discreet silence on the question. India has not wished to acknowledge the intensity and depth of resistance to its control of Jammu and Kashmir from the very outset. Pakistan, for its part, has had no desire to admit its sustained support for terrorism over almost six decades, conduct perhaps unprecedented in the relationships between two nation-states anywhere and at any time. Given the singular lack of archival material, the history of the long jihad in Jammu and Kashmir has existed only, so to speak, in the footnotes.

In essence, this book presents a new map of Jammu and Kashmir, a re-drawing of events with the jihad and its authors at the centre. It traces the long jihad through several distinct phases. The first began with the war of 1947–1948 and continued until the early 1960s, waged by small groups of Pakistan-backed covert operatives, whose principal objective was to bring pressure to bear on political processes in Indian-administered Kashmir. Nehru, appropriately, described this as "an informal war".[6] From the early 1960s to the mid-1960s, the informal war acquired greater momentum and structure with the emergence of the second phase of the jihad. Led by what came to be known as the Master Cell and its subsidiary covert organizations, this second phase of activity was intended to create the conditions for a mass rebellion in Jammu and Kashmir, and was informed by both expressly Islamist notions of jihad and emerging Pakistani military doctrines on sub-conventional warfare. After Indian counter-intelligence eliminated the Master Cell, the lessons learned from its failure were applied to a third phase of the jihad and manifested in al-Fatah, a group that thrived from the late 1960s until the war of 1971.

Al-Fatah, like its predecessors, failed to achieve its objective. Again, like its predecessors, it helped create a corpus of committed personnel, trained in covert warfare tactics and techniques, who would play a considerable role as mentors and inspirational figures in the future. In the immediate aftermath of its destruction and the territorial vivisection of Pakistan during the war of 1971, the task of rebuilding the jihad fell into the hands of the National Liberation Front. Suspect in the eyes of Pakistan's covert services, the National Liberation Front nonetheless succeeded in running the fourth phase of the long jihad with some successes. It was, however, to gain little institutional support from Pakistan until geo-strategic circumstances in the late 1970s, and the subsequent successes of that country's nuclear weapons programme, afforded the opportunity to escalate the long jihad to unprecedented levels. Much of the fifth phase of the jihad was fought outside of Jammu and Kashmir itself – a conflict I have described as a "war of many fronts". It was this fifth phase of the jihad that led to the events witnessed after 1989–1990: a sub-conventional war fought under the cover of Pakistan's nuclear weapons.

Several commentaries on the India–Pakistan conflict have metaphorically described it as unending or unceasing. This book suggests, among other things, that this description is also accurate in a very literal manner. The four India–Pakistan wars – of 1947–1948, 1965, 1971 and 1999 – and the multiple crises that nearly led to more wars can be visualized as the great pillars that hold up the multiple spans of a bridge. Each of them constituted a defining historical moment that both shaped the course of the long jihad and was in turn influenced by its unfolding. Between these wars, however, there was no peace. While the jihad was fought by small numbers of covert operatives, the scale of whose armed activities was by today's standards trivial, they had an enormous impact on both political life and policy-making. As shall become evident in this book, some of the participants in the early phases of the long jihad were to have a direct role in the events after 1989–1990. Social, economic, political and ideological forces far larger than the jihad itself indisputably contributed to that cataclysmic event: my effort here is to highlight an ignored narrative thread in the history of those events.

Documents and sources

My interest in writing this book was triggered, in part, by an event that took place in the house on the hill.

In the late 1990s, when terrorist violence in Jammu and Kashmir was at its peak, an elderly post office clerk showed up at the house, asking to meet with an officer who held some savings in the post office. It emerged that the clerk had no real interest in the officer or in the matter of his savings. In the mid-1950s, the clerk said he had been assigned to a remote post office near the Banihal Pass, the high Himalayan passage that was then the sole land route from the Indian plains into the Kashmir valley. One brutal winter evening, a message arrived informing him that a man who would soon make his way across the snow was to be given immediate access to a telephone upon his arrival. Despite the blizzard raging outside, the man did indeed arrive. He exchanged a few terse words with whoever was at the other end of the telephone line, and then asked for a personal favour – that his servant be asked to ready his home for his arrival and prepare water for a hot bath.

With little else to do, the clerk made a few inquiries over the next few days. It turned out that the man on the other end of the line had been Jammu and Kashmir's second Prime Minister, Bakshi Ghulam Mohammad. The man who had appeared through the snow was Colonel Hasan Walia, the first official inhabitant of the house on the hill. Ever since then, the clerk had wished to see the inside of the home.

True or otherwise, the story of the postal clerk piqued my curiosity. India's first spymaster in Jammu and Kashmir – a confidant of both India's first Prime Minister, Jawaharlal Nehru and its first Home Minister, Sardar Vallabhbhai Patel – Hasan Walia is little remembered today. To his contemporaries in Jammu and Kashmir, though, he was a central presence. Sheikh Abdullah saw the spymaster as an emblem of what he believed were New Delhi's intrigues and machinations in Kashmir, and a driving force behind his alienation from Nehru.[7] Abdullah's strenuous efforts to have Hasan Walia removed from Jammu and Kashmir, however, yielded nothing. Hasan Walia's professional skills may have won the argument, for India's covert services secured considerable successes in their battle against their Pakistani counterparts through the 1950s. At the outset of my research, I had hoped to excavate the course of Hasan Walia's war and the political battle that had raged around it. In this enterprise, I had little success.

My search for material on the early phases of India–Pakistan covert warfare in Jammu and Kashmir did, however, lead me in the direction of a considerable volume of material, much of it classified, that had been generated during the long jihad. Notable among this collection were two large volumes of investigation records authored by the head of Jammu and Kashmir's own counter-espionage service, Surendra Nath, a police officer who played a key role in Indian counter-terrorist policy-making and execution until his death in an air crash in 1993. In addition, I succeeded in exhuming some diaries maintained by participants, as well as posters, photographs and propaganda material. Much of this material is now part of the collection of Indiana University at Bloomington, whose staff has rescued these documents from the near-hopeless condition in which I found them. In addition to this material, my work as a journalist gave me considerable access to participants in the conflict. Where possible, I have cited specific documents and sources. I have not, however, referred to conversations with sources who I cannot name, for I believe that unverifiable citations serve no purpose.

It is worth acknowledging, at this stage, the fairly obvious limitations of my sources: to underline the fact that I have only succeeded in peeking through the window into the secrets that might be contained in the house on the hill. First, neither India nor Pakistan declassify intelligence-related documentation; neither, indeed, have a legislative mechanism that would enable them to do so. What I have obtained was made available by sources I had access to as a journalist. It is entirely possible, even likely, that new material could emerge in the future that would challenge all or significant proportions of my conclusions or, at the very least, my emphasis. One reason that I find the documents I have used to be credible is because their authors never intended for them to be made public. Nonetheless, like all official and non-official documentation, they do recount history from particular points of view. Many key individuals, who I would have liked to have spoken with had passed away before my work even began, including Surendra Nath himself. Many others in the covert world, both officials and their adversaries, were unwilling to talk. I had no access, most importantly, to the many Pakistani nationals whose stories, should they tell them, may lead to a reassessment of many of my conclusions.

If this book nudges some of the many individuals who authored the events I describe to reveal their stories, or to the official disclosure of greater amounts of archival material from the covert services of India and Pakistan, I believe the effort that has gone into writing it would be worthwhile.

Jihad and terrorism

In the course of my book, the terms "jihad" and "terrorism" shall appear with some frequency. Writing in 2005, at a time when the United States of America's campaigns in Iraq and Afghanistan have led both of these terms to be deployed as polemical abuse, my decision to utilize them requires some explanation. It is not my purpose here to attempt a scholarly discussion of the meanings of these words; that is not a project I, a journalist rather than a social scientist, am competent to engage in. My objective is, rather, to make explicit the position from which I see them, and the meaning I vest in them.

Of these two terms, the use of "jihad" is perhaps easier to address. I have chosen to describe the groups who have waged the long war in Jammu and Kashmir as "jihadist" principally because they themselves defined their project in this fashion. It is not, however, intended as a judgement on the legitimacy of the use of this term. Scholars and theologians have energetically disputed the degree of sanction Islam gives to the use of force, and indeed the very meaning of the term jihad itself. I have no competence to discuss these questions. My use of the term merely draws on its deployment by the groups whose history I trace in this book and the state which sponsored them. Thus, this book is at its core a history of the use of terrorism by the forces of the jihad in Jammu and Kashmir, and the relationship of these forces with the two nation-states, India and Pakistan.

Jihad was not, of course, a form of warfare invented by Pakistan. Leaving aside its place in the history of Islam, jihad had in the course of the nineteenth century developed into a part of modern warfare in South, West and Central Asia. Calls for holy wars were frequently made through the age of imperial expansion, as the great powers of the time jostled for influence. Covert operatives of Great Britain, Germany and Russia each vied to have their local allies declare the other power an enemy of Islam and thus give strength and legitimacy to their own cause. Great Britain's covert services, for example, encouraged a rebellion by right-wing clerics in 1924 and 1928 to destabilize attempts by the Afghan regime of Aman Allah to encourage democratization and ensure the education of women.[8] Opponents of imperial expansion, in turn, sought to mobilize the religious beliefs of their subjects and supporters to resist the growing influence of the great powers. Bar the superb scholarship of the historian of Islam, Yoginder Sikand, and some others, relatively little work is available on the intellectual development of the Islamist far Right in India. None, moreover, touches on the development of the jihad as a modern way of warfare and its evolution across space and time.

My use of "terrorism" may arouse greater contention, particularly since it has in recent years come to be used in an indiscriminate and abusive fashion. My decision to use this term, rather than several possible alternatives, is underpinned by two reasons, one personal and the other theoretical. It is perhaps best to state the first of these up-front. Indian journalists who reported on the struggle for the creation of a separate Sikh state, Khalistan, had traditionally used the terms "extremists" or "terrorists" to describe the character of the groups engaged in this enterprise. Khalistan groups subsequently imposed a set of codes on civil society in general, and on the media in particular, which among other things deemed the use of these terms impermissible.[9] Known as the Panthic Codes, these rules of reportage were imposed upon the media at gunpoint. The term "militant", now widely used in the Indian press to describe armed opponents of the State, was the product of this coercion. As a journalist who worked through that period, and because the term "militant" conflates non-violent political radicalism with specific forms of armed activity, I find its use unacceptable.

More important than my personal preferences, however, is the fact that the word "terrorism" describes a particular form of armed activity with considerable accuracy. Part of the current confusion caused by the term is the result of the fact that it is used, loosely and inaccurately, to describe the ideological character of armed actors, rather than their military tactics. Underpinning this confusion is the assumption that there is such a thing called "the terrorist", rather than a covert or overt military operative who uses terrorism to achieve certain political ends; one who may, furthermore, use other means to achieve these ends at other times and most likely has other identities, for example as a cricket fan or a loving parent. Terrorism may be used by actors of the ideological left or of the right, a fact which historians have long recognized. Despite his considerable admiration for the politics of the revolutionary Semeno Azharkovich Ter-Petrossian, Eric Hobsbawm without cant or squeamishness described him as "a brave and tough Armenian terrorist".[10] States use terrorism just as non-state actors do, and the intensity of their terrorist acts can be considerably greater. Nazi Germany's bombing of Stalingrad has been succinctly described as "a pure terror raid; its purpose to kill as many civilians as possible, overload all the services, sow panic and demoralization".[11] There is no shortage of similar examples either in World War II or in any other conflict of any scale.

What is this tactic, though, that I am referring to as "terrorism"? Regular militaries fight wars which may use both regular and irregular means, including terror, to achieve victory. Insurgent groups differ from regular armies in the character of their organization, their ideological underpinnings, their structures of control and so on, but in one central way they are similar: they seek to overthrow the state and its apparatus, coercive and non-coercive, through military engagement. Terrorist groups, on the other hand, acknowledge that their military capabilities are not such that they may have any realistic prospect of bringing down the state, or even imposing significant costs upon its coercive apparatus. Rather, terrorist groups seek to undermine the foundations of the state through selective actions, in the main targeting civilians and civilian authority-figures, in order to undermine the state's ability to command compliance. Such groups may see such actions as an end in themselves, or as a means to generate the conditions in which an insurgency or even a revolution may emerge.

In his recent work on Pakistan, Stephen Cohen has offered a lucid conceptualization of the practice of terrorism. It is not for naught, Cohen notes, that terrorism has often been compared to theatre, with "bystanders", or civil society, as its principal audience. He argues:

> The goal of the terrorist is to use an extreme act to change the way in which this group [civil society] sees reality. Thus, the terrorist is literally a bad actor, a bit player in a drama that seeks to change reality by a theatrical performance of increasingly unimaginable horror. As in the case of violence in literature and films, the level of horror has to increase over time to attract the attention of bystanders, who have their own mechanism of coping with the awful.[12]

For much of the course of the long jihad, the groups who waged war on the Indian state in Kashmir were organizations of limited military resources: they well understood it could not be defeated in battle. Barring perhaps a brief period in the early 1990s, jihadist groups cannot in any meaningful way be described as insurgents; certainly, their principal victims during year after year of the ongoing, bloody phase of the conflict have been civilians, in the main Kashmiri Muslims in whose name this war has been waged. To borrow from Hobsbawm, again, the warriors of the long jihad for the most part existed on "the cloak-and-dagger fringe of illegal revolutionary movements, where the smugglers, terrorists, forgers, spies and 'expropriators' operate".[13] Critics of the use of the term "terrorism" often use some variation or elaboration of the claim that one man's terrorist is another man's freedom-fighter, a debate which is remarkable not only for its conceptual sterility but because of its presentation of an opposition between things which are not opposed. To my mind, at least, a freedom fighter may among other things be a terrorist, smuggler or forger, or a terrorist, forger or smuggler may, among other things, be a freedom fighter. The *objectives* of terrorism do not change its military character in the least, nor serve to either add or take away from the legitimacy of the tactic.

Kashmir and the great game

Accustomed as we are to seeing Jammu and Kashmir as a metaphor for paradise – an image built among others, by the poet Thomas Moore, who described it as home to "roses the brightest earth ever gave", despite never having visited the place – it is hard to think of it as a battlefield.[14] For the last six decades, it has been exactly that – a site of contestation for the national identities and wills of India and Pakistan. These two nation-states, however, were not the first to comprehend the strategic significance of Jammu and Kashmir. Long before anyone had even conceived of the creation of the modern states of India and Pakistan, Jammu and Kashmir had become a key piece in the infinitely complex covert battle waged by imperial Great Britain and Russia for control of Asia: a contestation immortalized by Rudyard Kipling as the Great Game. In this sense, the conflict in Jammu and Kashmir long pre-dates the existence of either of its two principal protagonists.

In March 1846, a welter of sub-Himalayan and Himalayan territories was bound together into a new kingdom through the instrument of the Treaty of Amritsar, an agreement signed between Gulab Singh, the monarch of the minor kingdom of Jammu, and the East India Company. Gulab Singh had been a vassal of the Maharaja Ranjit Singh, who had ruled the Sikh Empire from the city of Lahore. Both the East India Company and Ranjit Singh had, for a variety of reasons, curbed their considerable territorial ambitions and respected a frontier along the Satluj River in Punjab. However, Maharaja Ranjit Singh's death created considerable strife within the Lahore court, and enabled the British to go to war against the Sikh forces in 1845. Gulab Singh wisely chose to back the East India Company in the conflict and, when their joint forces prevailed, he was rewarded with the title and holdings of a Maharaja. Gulab Singh became the founder of the Dogra dynasty, which would hold power for a century and a year. The origin of Jammu and Kashmir, as this new kingdom was known, was thus organically linked with the arrival and growth of British imperial rule in India.

Imperial Britain had good reason to give birth to this new state. Among Britain's concerns in South Asia was the prospect of southward Russian expansion, which it was feared would one day lead to the forces of the Czar pushing their way through to the waters of the Indian Ocean. Any such thrust would have had to come through Central Asia, and then Afghanistan or Tibet, regions little known to the East India Company well into the second half of the nineteenth century. Perched at the junction of these strategic regions, Jammu and Kashmir was of critical military importance. By the time of his death at Bukhara in 1825, the veterinary surgeon William Moorcroft had surveyed large swathes of the western Himalayas and central Asia. Although his expeditions had no official sanction, they generated great amounts of material of military-intelligence value for the East India Company, not the least within which was an exhaustive description of Kashmir's famed shawl industry. Moorcroft relied not only on associates from the Indian plains for information and aid, but also, notably, on the great commercial networks of traders in Jammu and Kashmir, which stretched as far west as China and what is now Myanmar.[15]

Moorcroft's discovery of two European-bred and European-trained dogs near the Kailash summit in Tibet led him to believe that Russian agents had passed that way in search of routes into India.[16] While there is little reason to believe that this speculation was correct, the fact is that Russian operatives were indeed headed in that direction. In 1819, the Russian military officer Nicolai Muraviev was dispatched on a mission to the kingdom of Khiva, deep in central Asia, to propose a commercial and strategic relationship with Moscow. Such efforts accelerated as the century progressed. By the mid-nineteenth century, the British had set up an intelligence service known as the Pundits to carry out covert military cartography in Tibet, Afghanistan and central Asia. Trained to march in precise steps, and thus taking measurements that could then be turned into functional maps, the Pundits often travelled disguised as mendicants, using ritual beads to record distances and a log-book concealed inside the copper tube of a prayer wheel.[17] Indian *munshis*, or secretaries, also played a key role in British expeditions into the inner Himalayas. One of the most famous of these, Mohan Lal, came from a Kashmiri family with a long tradition of service to the British crown.[18]

To some in both Russia and Britain, these covert activities were precursors to an inevitable full-scale war of imperial domination between the two great rivals. The Soviet revolution ensured that this war never took place, but another military enterprise was among the proximate causes of the creation of Kashmir. Between 1839 and 1842, the East India Company dispatched an expeditionary force into Afghanistan, in an ill-advised effort to strengthen their control over this strategically crucial area. The first Anglo-Afghan war ended in the decimation of almost 40,000 East India Company troops, because of both logistical problems and an intelligence failure – a lack of understanding of the complex web of tribal affiliations that underpinned the position of the Afghan monarchy. Though the British, fearful of the impact of these events on their possessions in the plains, sought to suppress news of their defeat, the merchants at Delhi soon received reports of these events. One firm even arranged for the ransoming of women and children held captive in Kabul, a sign of the advanced development of commercial networks in these regions.[19] Two lessons that were to shape imperial policy in Jammu and Kashmir emerged from this experience. First, the British emerged from the Anglo-Afghan war with no desire to commit troops in the high Himalayas. The second, related lesson was that Indian allies and collaborators could do the job better. Gulab Singh and his heirs fit the bill perfectly.

The making of a battlefield

Despite their strategic need for each other, relations between the British and the Dogras were not always smooth. Suspicions periodically arose that Jammu and Kashmir's rulers were attempting to strike independent deals with Russia, or were otherwise seeking freedom from their role as vassals. Britain used the many instances of Dogra misrule, notably the appalling conditions of the mass of the peasantry in Kashmir and frequent outbreaks of famine, to push for a greater direct role. Such arguments not only were disingenuous – large-scale and horrific deprivation was commonplace both in other princely states where the British had a greater role and in those parts of India it directly administered – but were nonetheless successful instruments to secure leverage. By 1885, a British Resident, or administrative overseer, had been appointed for Jammu and Kashmir, as a symbol of the growing influence of imperial officials in running the state. Successive Dogra rulers resented this imposition, but their own internal position was too precarious to offer serious resistance to it.

Popular opposition to Dogra rule became evident almost immediately after it began. Less than a year after the Treaty of Amritsar was signed, Dogra tax policies on artisans led to a large-scale outflow of shawl workers to the Punjab plains, a development which decimated the economically vital industry. Widespread famine aggravated the situation and, in April 1865, the shawl workers of Srinagar rose in protest against the regime. "It was", F.M. Hassnain has argued, "perhaps the first organized demands day in the history of class struggle in India".[20] The shawl workers' revolt was brutally suppressed –28 protestors were believed to have been killed by Dogra forces, and arrests and punitive fines were imposed on their leaders. Despite some fitful efforts at administrative reform, working class protests broke out with regularity in the coming decades, mirroring trends in popular struggle across South Asia. The economic depression that followed the Great War of 1914–1918 further heightened these tensions. In 1924, for example, workers of the state-owned Silk Factory initiated a large-scale wage struggle, which again had to be suppressed by the use of brute force.

By the time of the Silk Factory strike, both the Dogra *durbar* in Jammu and Kashmir and the British *raj* were under siege from an emerging social class: Indians trained in the new universities and colleges that had sprung up under imperial rule, who were increasingly conscious of their exclusion from power and the inequities of imperial rule. Among the leading figures of this new class was Sheikh Mohammad Abdullah, the son of a peasant family who would have a central role in shaping Jammu and Kashmir's political future. Like many of the new class of politicized young men emerging from Jammu and Kashmir, Sheikh Abdullah turned to Lahore for his education. There, he developed a great sensitivity to the political and economic deprivation of his people, as well as an engagement with the major currents of anti-imperialist thought in British-ruled India. Both secular-nationalist ideas and religious revivalism shaped the ideological filters through which this new class viewed imperialism. Competing Hindu and Islamic chauvinist movements were sweeping much of the region, led by emerging elites who sought mass legitimacy. Pan-Islamists and Hindu revivalists competed for space, along with socialists, communists and nationalists of various hues.

Jammu and Kashmir, with its Hindu ruler and mainly Muslim population, was inevitably a key focal point for such revivalist contestation: these structures, and the processes of mobilization, would ensure that "religion and politics became inextricably intertwined".[21] Just before the Silk Factory strike, a group of Muslim clerical and business leaders submitted a memorandum charging the Dogra monarchy with systematically excluding Muslims from governance and with obstructing their practice of their faith. Interestingly, this leadership seems to have had little to do with the long tradition of working class struggle – the nine-point memorandum did not mention among its demands the need for better wages and working conditions – but did provide articulation to the widespread feeling that the monarchy could not and would not offer its Muslim subjects either equity or progress. It has been pointed out that Memorialists, as the authors of the 1924 memorandum were known, had overstated their case: Muslims did indeed occupy some positions of considerable influence, both in the coercive apparatus of the Dogra state and in its administrative organs, and there had been a considerable growth in the educational facilities available to them.[22] Nonetheless, the fact remains that the traditional feudal elites, in the main the Hindu Pandit community and also some Rajput-caste notables of Jammu, both Hindu and Muslim, were grossly over-represented.

Abdullah, like many of his class and generation, developed his activist skills while working with the political organizations that emerged from the Memoralists. Among the most important of these was the Young Men's Muslim Association. Backed both by the new, educated aspirants for government jobs, like Abdullah, and by influential clerics such as Maulvi Mohammad Yusaf Shah, the head of the powerful hereditary order of the Mirwaiz of Srinagar, the Young Men's Muslim Association consisted of two major tendencies. One, represented by Abdullah, sought to use democratic and constitutional means to pressure Hari

Singh to bring about reforms for the greater social and economic advancement of Muslims. The other, a minority led by Ghulam Nabi Gilkar, sought to lead a revolution against Dogra rule, which they saw as an instrument for the subjugation of Islam, intended to lead to the installation of a Muslim Sultan. In April 1931, an incident occurred that tipped the balance of power among the Young Men's Muslim Association in favour of the religious right. During Eid prayers that month, a Hindu police official in Jammu was alleged to have desecrated a copy of the Quran. Gilkar pushed Abdullah, by some accounts against his better judgement, to deliver a speech condemning the incident from a mosque in Srinagar. It was the first mass political gathering of its kind: both the religious themes on which it was based and the fact that Abdullah's speech was delivered at a mosque were of considerable significance.

Religion offered a means of mass mobilization to Jammu and Kashmir's emerging leadership – but, as events were to establish, it was a blunt tool that resulted in considerable collateral damage. In June 1931, the Young Men's Muslim Association met to elect its leaders. Towards the end of the meeting, Abdul Qadeer Ghazi Khan, a member of a clerical family with long-standing links to the Islamist ideologue Jamal-ud-Din Afghani Astarabadi, delivered a speech demanding an uprising against Hari Singh. Khan represented the radical right of the clerical establishment in India. Having arrived in Jammu and Kashmir disguised as the servant of a British tourist, Khan had hoped to influence the course of events there. His mentor, Jamal-ud-Din Afghani, had repeatedly called for jihad which would free the world from British imperial rule; Khan hoped to bring about something of the kind within Jammu and Kashmir. To a greater degree than he perhaps anticipated, Khan succeeded in drawing Jammu and Kashmir into the brewing Hindu–Muslim conflagration in northern India. Incarcerated for his seditious speech, Khan became a focal point for anti-Dogra sentiment in the Kashmir valley. Abdullah claimed that the cleric was being persecuted "for the cause of Islam and for the Muslim masses", and called for his supporters to be "prepared to be sacrificed for the sake of Islam".[23] On the day of Khan's trial, July 13, 1931, a fight broke out between protestors and policemen outside the jail in Srinagar. What started as a minor scuffle rapidly escalated, and 28 protestors were killed in the showdown that followed.

For the first time, events in Jammu and Kashmir generated a major pan-India political response. Muslim leaders from across India met at Shimla to express their outrage at the jail massacre, and decided to call for a day of action against the monarchy. On September 22, the day chosen for this protest, thousands of people gathered at the Jamia Masjid in downtown Srinagar demanding the release of Sheikh Abdullah and Mufti Jalal-ud-Din, who had been incarcerated for their role in the violence that followed Khan's trial. Another massacre followed. As the protestors shouted "Islam *zindabad* [long live Islam]", the Maharaja's troops opened fire, killing at least 25 people.[24] After this second massacre, the Islamist idiom of the protest sharpened. Rioting directed at Hindu-owned businesses in urban Kashmir grew in scale.[25] Mirwaiz Yousaf Shah called for a jihad, leading thousands of his supporters to mass at the shrine of Dastagir Sahib in Srinagar, armed with knives, swords and guns.[26] Soon, however, it became clear the cleric had no desire to allow events to spiral out of hand. Using the services of several Muslim notables loyal to the throne, notably Nawab Khusro Jang, Abdul Rahman Afandi, Khwaja Salaam Shah and Khwaja Noor Shah, Maharaja Hari Singh was able to defuse this second phase of protests, and arrive at an accommodation with the Srinagar clerical establishment and its supporters.

By September 1931, Hari Singh had succeeded in strengthening his accommodation with Yusaf Shah. It marked the end of the alliance between new and old elites in Kashmir, and the activists of both groups were soon at war with each other. Abdullah and his supporters now formed a new organization, the All Jammu and Kashmir Muslim Conference. Yusuf Shah regularly charged this organization with being a front for the Ahmadiyya sect, deemed heretic by orthodox Sunni Muslims; Abdullah, in turn, claimed the cleric had sold out to the Dogra monarchy. Over the coming decade, the distance between the two groups steadily expanded. Abdullah's linkages with the all-India anti-imperial movement grew steadily, as did his attraction to the socialist ideas of the man who would become India's first Prime Minister, Jawaharlal Nehru. In June 1939, shortly after he attended a conference of the All-India National Congress, Sheikh Abdullah changed the name of the Muslim Conference, dropping the word that denoted its communal affiliation. It was now called the Jammu and Kashmir National Conference, signalling its commitment to represent all of the peoples of the state, irrespective of their faith. Over the next six years, the National Conference would increasingly align itself with the Indian National Congress; Abdullah's opponents, by contrast, would turn to the Muslim League, which was fighting for the creation of Pakistan, a new state to be carved out of the Muslim-majority areas of British India.

How radical a break with the past was the creation of the National Conference? Abdullah, as we shall see, was entirely willing to make opportunistic use of chauvinist forces to discredit his opponents. Despite his stated commitment to secular and socialist principles, he never severed his connections with the politics of the mosque. Yet, the fact remains that the National Conference represented a radically new form of political organization in Jammu and Kashmir, one severed, at least formally, from the communal bases of its predecessors. Indeed, Abdullah's decision seems to have been motivated by the realization that the growing communal fissures in Jammu and Kashmir served no one's interests other than those of the Dogra state and his clerical enemies. In 1937, for example, a disputation on the right of Muslims to slaughter cows gave legitimacy to both Hindu and Muslim revivalists, enabling the Dogra monarchy to represent itself as the sole credible arbiter of their differences. In the build-up to Indian Independence, the battles between the National Conference and its rivals accelerated. Using the cadre of Mirwaiz Yousaf Shah as a nucleus, the Muslim Conference was revived in May 1940. The intensity of the competition between the born-again Muslim Conference and the National Conference is evident from the fact both organizations used quasi-military tactics; street battles between the heirs of these two groups were, indeed, to remain common until the late 1980s. While the National Conference succeeded in beating off the threat to its position within the Kashmir valley, the Muslim Conference expanded in those regions where Hindu–Muslim tensions were the most fragile, capitalizing on growing communal fissures.

Towards war

India and Pakistan were born in August, 1947.

Like other princely states, Jammu and Kashmir had been asked to choose between joining one of the two new states. Hari Singh, suspicious of Nehru's socialist leanings, was reluctant to join India; he was equally skeptical of his own future in Muslim-majority Pakistan. The Maharaja thus played for the time, hoping against hope that he could secure a deal which would give him at least some degree of independence. Abdullah, too, was unsure of whether he would be able to command state-wide support for a decision to join India. Time, however, was running out. In the Muslim-majority areas of Jammu, Hindu chauvinist forces, with no small degree of backing from the Dogra state, initiated a series of large-scale massacres. By one account, 200,000 Muslims were slaughtered in the Hindu-majority regions of Jammu, and another 300,000 were turned into refugees.[27] It is likely that similar numbers of Hindus suffered at the hands of rioters in the Muslim-majority parts of Jammu, who operated with the support of demobilized Muslim soldiers of the British–Indian Army. Muslim troops rebelled in Gilgit, the Maharaja's northern-most territory, and proclaimed their accession to Pakistan. The Kashmir valley itself remained peaceful, but the fact was the Dogra state had started to fall apart.

As India and Pakistan celebrated their independence, Jammu and Kashmir teetered on the fulcrum of its destiny. To Pakistan's military and political establishment, it seemed in that bloody summer of 1947 that the smallest shove would settle the issue. Irregulars, backed by the Pakistani state and its military, made a daring thrust towards Srinagar. Jammu and Kashmir's forces were swept aside. Panicked, Hari Singh turned to India for assistance. The Maharaja was informed that India could commit troops to the defence of Jammu and Kashmir only if he first agreed to the kingdom becoming a part of the new country. With Pakistani forces just a few kilometres from his capital, Hari Singh caved in. The terms of that accession remain bitterly disputed. Accounts sympathetic to Pakistan's position, notably that of the British historian Alistair Lamb, have asserted that Hari Singh's decision to accede was flawed, and that India's claims to Jammu and Kashmir were therefore illegitimate.[28] Many Indian accounts, by contrast, assert that Hari Singh's decision to accede to India was not only a legitimate and legally sound response to Pakistani aggression, but had the support of the largest political formations in Jammu and Kashmir.[29] I shall not address the complex legal debate over the accession of Jammu and Kashmir to the Union of India in this book, but its existence should be borne in mind as my story unfolds.

Pakistan's military push, as we shall see in the next chapter of this book, settled nothing whatsoever, and served only to start a war which continues to this day. While this book deals principally with the covert aspects of India–Pakistan military contestation over Jammu and Kashmir, it also touches on the many other wars with which this conflict was organically entwined. These include the struggles between those of Jammu and Kashmir's residents who believe that their future lies in independence or with Pakistan, and those who see their destiny linked to India; between Islamists, secularists and Hindu nationalists; between modernity and those who reject it; between peasants and elites; between cities and country and between competing visions of Islam itself. No one book, of course, can tell the tale of all these and all the other struggles which together constitute the Jammu and Kashmir conflict. As the anthropologist Jack Weatherford has noted "history cannot be neatly tucked between the covers of a book, and filed away like so many pressed botanical specimens".[30] This book traces just one thread of a complex weave. It is, however, a thread that few have paid attention to and the careful study of which, I believe, yields considerable insight and understanding about the making of the whole.

2

THE INFORMAL WAR

> The liberation of Kashmir is a cardinal belief of every Pakistani. It is an integral part of Pakistan, and Pakistan would remain incomplete until the whole of Kashmir is liberated.
>
> Khawja Nazimuddin, Governor-General of Pakistan, September 19, 1950[1]

In October 1947, Abdul Rashid Butt took the last bus from Baramulla to Muzaffarabad: the very last commercial transport which would traverse that route for almost six decades.

To understand the cultural climate in Jammu and Kashmir as Butt boarded the bus from Baramulla, we may turn to literature: no historian has described it with anything like the evocative precision as Salman Rushdie did in his epic novel, *Midnight's Children.*[2] Jammu and Kashmir sat on the edge of a cataclysmic change, change that had been brewing through the century of Dogra and British Imperial rule and that would reach its climax in the murderous Partition of India. While parts of the province of Jammu were to witness horrific communal violence, most of Jammu and Kashmir was spared this carnage. All of the state, however, was to feel the impact of the storm that had ripped apart South Asia. Represented in Rushdie's novel through the characters of Aadam Aziz and Naseem, the competing influences of an anxiety-suffused modernity and reactionary traditionalism; of local identities and new nationhoods; of syncretic cultures and communal solidarities: all these were to underpin the transfiguration of the small mountain state into a battlefield on which India and Pakistan would assert the legitimacy of the ideological project each represented.

Butt was just 16 when he took the last bus to Muzaffarabad: he, with two cousins and two servants, left never having considered the possibility that he would not be able to return home. His family had dispatched him to enquire about the fate of a cargo of fruits and spices they had sent to Kohala, the small town that marked Jammu and Kashmir's border with the new country that had come into being two months earlier, Pakistan. Normal public transport had come to a standstill, since Pakistan had blocked fuel supplies to Jammu and Kashmir, claiming it had none to spare. Butt and his cousins succeeded in using their contacts to board a bus chartered by a Lahore-based leader of the Muslim League to evacuate his British wife and three children who were on vacation in Srinagar. By Butt's account, he spent the night at a hotel in Muzaffarabad, and then took a horse-drawn cart to Kohala the next morning. There, he learned the goods that his family had sent had disappeared – and that Pakistani irregulars had entered Jammu and Kashmir

Without cash or hope of returning home, Butt stayed on in Muzaffarabad, and eventually made a life for himself in the city. He obtained a degree, set up a business and married. Across the Line of Control (LoC) in Srinagar, his parents and four siblings went about their own lives, and passed away. In 2005, when Butt's son, Saqib Butt, told him that India and Pakistan bus service was about to be started linking with Muzaffarabad, the journalist Tariq Naqash recorded that he "started crying like a child".[3] This chapter tells the tale of the first phase of the long jihad which ensured Abdul Rashid Butt missed his bus home.

The war of 1947–1948

Independence had brought enormous upheaval not just to Jammu and Kashmir, but to all India and Pakistan. Both countries faced enormous political, institutional and economic challenges. In addition, the horrific violence that accompanied Partition, which was to claim over 1,000,000 lives, had generated a crippling flow of refugees across the borders of the two Dominions.

On the eve of Partition, Maharaja Hari Singh found himself trapped between two difficult options. Prime Minister Jawaharlal Nehru's support for the National Conference had, quite obviously, led the Maharaja to fear the prospects for maintenance of his power if he acceded to India. On the other hand, the Hindu ruler had little reason to be enthused by the prospect of a future in Pakistan, either. Independence appeared the most attractive option before the Dogra monarchy, but that option was unviable from the optic of both India and Pakistan. Sensing an approaching crisis, Prime Minister Jawaharlal Nehru pressed the Maharaja to release political prisoners under a general amnesty and asked that Sheikh Mohammad Abdullah be brought into government in order to ensure popular legitimacy for the regime. The Maharaja did release the National Conference leader, but continued to hold out on the issue of the organization's role in governance. The political impasse continued through the end of October 1947.[4]

Meanwhile, time was running out for Hari Singh. Civil unrest had broken out in Poonch that July, in part in protest to the Maharaja's taxation policies. As Independence approached, however, the protests were reshaped by the Hindu–Muslim communal conflagration sweeping northern India. Muslims in the region, who had been disarmed by the Maharaja's troops, turned for help, and the acquisition of weapons, across the north-west frontier.[5] Hindus in the region, by some accounts, turned to the Maharaja's forces for help and communal warfare in Poonch continued to escalate over the following months. In an official telegram to the authorities in Kashmir, dispatched on October 12, 1947, Pakistan's Government described the sight of "large numbers of villages that can be seen burning from the Muree hills", and hinted that some of its troops were considering intervention.[6] Later that month, Pakistan again insisted that a "reign of terror" had been let loose in Poonch.[7] Kashmir authorities responded that the violence was the work of "invaders" who had arrived from "across [the] border [of] Rawalpindi and Hazara District".[8] This last missive included specific reports on atrocities directed at Hindus and Sikhs appended to it, including a massacre at Baral, near Palandari, which had claimed the lives of 36 people.

Indian signals intelligence personnel first detected Pakistani military activity in the Poonch area in early October 1947. While the Army's own signals apparatus was overburdened with traffic on the communal violence in northern India, equipment from the Air Force and Navy was used to monitor Pakistani communications. One of the first intercepted messages of relevance indicated that Gorkha troops from an unknown formation were still holding out at a location called Sensa. India's Military Intelligence Directorate could determine nothing about either the location or troops referenced in this communication. Staff at the Directorate had no compendium of place names, nor a map of Jammu and Kashmir – indeed, departing British officials had destroyed all sensitive files other than a single document about a defunct clandestine organization that had operated in Myanmar during Second World War. A similar sack had been conducted of other intelligence services: even documentation on major political formations of the religious right, notably the Hindu Mahasabha and the Muslim League, had been removed to ensure the new regime would have no access to imperial Britain's relationships with these groups. Indian signals intelligence, however, was soon able to resolve the mystery. Some days after the Sensa traffic was monitored, the Directorate obtained a second intercept, this time referring to the capture of a position code-named Owen. This second batch of intercepts contained a reference to Poonch, enabling the Directorate to at last determine that Gorkha troops in the Jammu and Kashmir State Forces were under attack.[9] In subsequent attacks on the Bhimber garrison of the State Forces, evidence emerged that Pakistan had used light tanks after assaults by irregulars failed to break through.[10]

Pakistan had in fact determined to take a drastic course of action long before sending out its October 12, 1947 telegram – or initiating its military actions in Poonch. Major-General Akbar Khan, then Military Advisor to Pakistan's Prime Minister Liaqat Ali Khan, has provided a cogent account of the events that precipitated Pakistan's decision to launch its first war on Jammu and Kashmir. Until a few days before Partition, Major-General Khan has recorded, Mohammad Ali Jinnah remained convinced that neither Hari Singh nor Sheikh Abdullah would be able to impede Jammu and Kashmir's accession to Pakistan, for simple reasons of geography. Pakistan's leadership believed that the Punjab-province district of Gurdaspur, which had a Muslim majority, would form part of the territory of the new country. Since the sole usable road from the Indian plains to Jammu and Kashmir traversed Gurdaspur, India would then have been left without a usable route to the state. As such, Jinnah and his advisors believed, the Maharaja would have no option other than to accede to Pakistan. To their dismay, however, the Radcliffe Award of August 8, 1947, which drew the borders of India and Pakistan, handed over the district of Gurdaspur to India. Although the route through Gurdaspur to Jammu, and on across the Pir Panjal mountains to Srinagar, was only usable in the dry summer months and for a part of the winter, India now had a frontier with Jammu and Kashmir, and a road that could with a little effort be turned into an all-weather logistical channel.

Several Pakistani commentators, in both scholarly and polemical accounts, have characterized the decision to give Gurdaspur to India as perfidious: the consequence, variously, of the close relationship of the British Viceroy, Lord Louis Mountbatten, with Nehru; an alleged affair between Edwina Mountbatten, the Viceroy's wife, and India's first Prime Minister and the Viceroy's visceral dislike of Jinnah. Recent scholarly investigation of the issue by the historian Shereen Ilahi has undermined this long-standing consensus. Ilahi has noted that the basic unit used to divide Punjab was the *tehsil*, or sub-district, rather than the district itself. The Radcliffe Award gave three of Gurdaspur's four *tehsil*s to India, two of them Muslim-majority, for a variety of reasons to do with security concerns in Punjab. However, Ilahi has pointed out, even if the two Muslim-majority *tehsil*s had gone to Pakistan, the fact that the *tehsil* of Pathankot was Hindu-majority would have left India with control of the land route to Jammu and Kashmir. As such, the frequently reiterated charge that the award of Gurdaspur to India was part of a conspiracy to ensure that Jammu and Kashmir became part of that country has no real foundation in fact. "There is no evidence", Ilahi concluded, "to imply that anyone gerrymandered the boundary because of its implications for the princely state".[11]

To Pakistan's fledgling strategic establishment, however, the Radcliffe Award posed a threat not only to "the safety and wishes of our brethren in Kashmir", but also to "our own safety and welfare".[12] In Major-General Khan's view "Kashmir's accession to Pakistan was not simply a matter of desirability but of absolute necessity for our separate existence."[13] He wrote:

> One glance at the map was enough to show that Pakistan's military security would be seriously jeopardized if Indian troops came to be stationed along Pakistan's western border. Once India got the chance, she could establish such stations anywhere within a few miles of the 180 miles long vital road and rail route between Lahore and Pindi. In the event of war, these stations would be a dangerous threat to our most important civil and military lines of communication. . . . From an economic point of view the position was equally perilous. Our agricultural economy was dependent particularly upon the rivers coming out

> of Kashmir. The Mangla Headworks were actually in Kashmir and the Marala Headworks were within a mile or so of the border. What then would be our position if Kashmir was in Indian hands?[14]

Due to Hari Singh's dithering, Pakistan came to believe the Maharaja was tilting towards India, something that "created the gravest suspicions and uneasiness".[15] Akbar Khan was now tasked with organizing a covert invasion of Jammu and Kashmir, in an effort to settle the issue by force. His efforts resulted in a written plan, *Armed Revolt Inside Kashmir*, which was approved with some variations after a conference with top political leaders in Pakistan, including Prime Minister Liaqat Ali Khan. The assault plans envisaged the use of irregulars, former soldiers and a small core of serving personnel, armed with 4,000 Punjab Police rifles and Army ordnance secretly diverted for use in Jammu and Kashmir.[16]

Although the Muslim League leader Mian Iftikharuddin had told Khan in September 1947, that "any action was to be of an unofficial nature, and no Pakistani troops or officers were to take an active part in it", the reality was very different.[17] Officers of the Pakistan Air Force provided winter clothing, ammunition and weapons, for example, while the senior civil servant Khawja Abdul Rahim collected funds, volunteers and weapons.[18] All of this was fairly evident to contemporary observers. Writing in the London *Observer* of November 2, 1947, Alan Morehead reported that recruitment was underway "not only in tribal territory, but in Pakistan itself".[19] The eminent photographer and journalist Margaret Bourke-White has recorded that local offices of the Muslim League were used to pass arms to the assault groups – an account that tallies with Akbar Khan's insider story, and sheds light on the complicity of official Pakistan in the enterprise.[20] Jinnah's own personal secretary Khurshid Ahmed was arrested in Srinagar on November 2, 1947, in the possession of maps and documents suggesting that he had hoped to organize an uprising against the administration.[21]

Events in Poonch provided a gateway for regular and irregular Pakistani forces to assault the main prize – the Kashmir valley and, the state's capital, Srinagar. On October 22, 1947, troops of the Muzaffarabad-based 4 Jammu and Kashmir Infantry, a unit raised from the Poonch area, mutinied, killing their comrades and their commanding officer, Lieutenant-Colonel Narain Singh. Ironically, Narain Singh had previously rejected efforts by his force headquarters to replace the Poonch Muslim troops under his command, scoffing at claims that they might prove disloyal.[22] Shortly afterwards, the rebels of the 4 Jammu and Kashmir Infantry were joined by Pakistani irregular forces that had been waiting across the border. The invaders were soon in Uri, where they brushed off resistance by outnumbered State Forces, and Baramulla by October 27. Srinagar was now just a few miles away, for all practical purposes, undefended.

At this point, however, the invading force made a critical error. The tribesmen – Mohmand and Afridi, Wazir and Mahsud – of the north-west were superb fighters, but saw little distinction between military conquest and war for pillage. For three entire days, Baramulla was subjected to an orgy of destruction, rape and loot. "Sometimes their help to their brother Muslims", Bourke-White acidly observed,

> was accomplished so quickly that the trucks and buses would come back within a day or two bursting with loot, only to return to Kashmir with more tribesmen, to repeat their indiscriminate "liberating" – and terrorizing of Hindu, Sikh and Muslim villager alike.[23]

And, it would seem Christians. The nuns of the Baramulla Mission Hospital, the most modern in the Kashmir valley, were butchered *en masse* when they attempted to intercede on behalf of their patients.[24]

Perhaps the worst fate of all was reserved for the National Conference leader Mir Maqbool Sherwani, who led covert operations by the Kashmir Militia against the tribal irregulars, and organized popular resistance. Captured by the raiders, Sherwani was asked to make an announcement of support for Pakistan. He refused. Bourke-White has recorded what happened:

> It was a curious thing that the tribals did next. I don't know why these savage nomads should have thought of such a thing unless their sight of the sacred figures in St. Joseph's Chapel on the hill just above had suggested it to them. They drove nails through the palms of Sherwani's hands. On his forehead they pressed a jagged piece of tin and wrote on it: "The punishment of a traitor is death."[25]

Sherwani was finally executed by a firing squad, but his death was not in vain. In retrospect, those three days of carnage in Baramulla were to cost Pakistan all of Jammu and Kashmir. Confronted with the enemy at his gates, Hari Singh finally acted. He dropped his resistance to Abdullah's involvement in the government, and signed documents acceding to India, thus enabling direct military intervention. Arrangements for assistance "proceeded at breakneck speed", and Indian troops were flown in to secure Srinagar airport.[26] Akbar Khan, personally present on the field of battle, gave a valiant speech about "what the issue of Kashmir meant to us as Muslims and Pakistanis".[27] He adopted the *nom de guerre* "General Tariq", in honor of the legendary Moor commander reputed to have burnt his boats behind him upon landing for the conquest of Spain in the seventh century.[28] General Khan's use of Islamic myth would inspire those who planned a subsequent military campaign, the war of 1965, but did little to energize the forces he commanded to prevent Indian troops from beating back the invasion to Uri. A similar process of reversal took place in Jammu. By the end of November 1947, Indian forces succeeded in relieving much of the province.

Efforts to secure a political settlement continued throughout this military campaign. Although he initially called for a compromise on both sides, the rejection of his pleas by Pakistan led Mountbatten to press for "unilateral concessions from the party over which he had real influence – India".[29] Nehru, by contrast, increasingly believed that an aggressive military response was the only option.[30]

This belief was further cemented by fresh raids into the Mirpur region and other areas across the Poonch border in the first week of December 1947. News of the fall of the town of Jhangar on December 24 created an elevated sense of urgency in Delhi, and Nehru concluded that military preparations should speed up and be completed by mid-January. Intervention by Mountbatten and the British Prime Minister, Clement Attlee, however, "finessed" Nehru into choosing a two-track course of action: a reference to the UN, and contingency planning for attacking the invaders' bases in Pakistan. While he proceeded in good faith on the first, Mountbatten "held back on the second point".[31]

What motivations might have underpinned British partisanship at this stage? At the United Nations in early 1948, British diplomats, advised by the United Kingdom's Foreign Office, continually took positions and proffered proposals that favoured Pakistan. This policy emerged from the developing crisis in Palestine, concern about an "Arab backlash", and the sense that "Arab opinion might be further aggravated if British policy on Kashmir were to be seen as being unfriendly to a Muslim state".[32] All of this was a piece with evolving British policy in the region. As early as 1944, we know from the memoirs of Sir Francis Tucker, the last General Officer Commanding of the British Indian Eastern-Command, Imperial strategists had supported the case for Pakistan, seeing it as a buffer against efforts by the Soviet Union to expand its influence into South Asia. Convinced that the Hindu faith, which was in Tucker's view "to a great extent one of superstition and formalism", would be displaced by "a material philosophy such as Communism", British strategists believed it "very necessary to place Islam between Russian Communism and Hindustan".[33] Tucker, like many British strategists, believed that:

> There was much therefore to be said for the introduction of a new Muslim power supported by the science of Britain. If such a power could be produced and if we could orient the Muslim strip from North Africa through Islamia Deserta [*sic*], Persia and Afghanistan to the Himalayas, upon such a Muslim power in Northern India, then it had some chance of halting the filtration of Russia towards the Persian Gulf. These Islamic countries, even including Turkey, were not a very great strength in themselves. But with a northern Indian Islamic state of several millions, it would be reasonable to expect that Russia would not care to provoke them too much.[34]

Pakistan, thus, in Britain's imagination served a purpose that far transcended India: it was planned as part of a mosaic of interlocking pieces intended to contain the Soviet Union long before the Partition of India became a certainty. Given the central role of Afghanistan in containing the Soviet Union, and the fact that Kashmir was placed next to it, Britain's interest in ensuring that Pakistan controlled the levers of power in Jammu and Kashmir is self-evident. When the United Nations did act on Jammu and Kashmir, the terms of its intervention fell well short of Indian expectations that it would unequivocally condemn the invasion. In April 1948, the Security Council issued a resolution establishing the UN Commission on India and Pakistan which was instructed to proceed immediately to the region for mediation between the warring countries. The Security Council also recommended that Pakistan secure the withdrawal of the raiders, after which India would reduce forces in Kashmir to the minimum level required. Finally, the resolution asked India to ensure that the State government was expanded by the inclusion of all major political groups in a new Plebiscite Administration, with assistance from the United Nations.[35]

United Nations intervention came to little. Of more immediate consequence was the fact that the military campaign had begun to reach an impasse. An Indian offensive in the spring of 1948 achieved little, while Pakistan's efforts to open a new front into the Ladakh region of eastern Jammu and Kashmir were repulsed. In the six months before a ceasefire eventually went into effect on New Year's Eve in 1948, both sides had reached a position where "with the mountainous terrain favouring defence and opposing forces equally balanced, a major attack would probably result in minor gains being achieved at a relatively high price in casualties".[36] Military strategists in both India and Pakistan believed that the commitments of greater numbers of troops would have helped achieve decisive victory. Lieutenant-General L.P. Sen, who commanded the 161 Brigade during the war of 1947–1948, expressed frustration at being denied additional offensive formations, and speculated that some at Military Headquarters, still staffed by British officers, in fact sought a situation from "where only a stalemate could result".[37] Major-General Akbar Khan, for his part, insisted that delaying the ceasefire could have helped Pakistan secure further territorial gains along the Chakoti–Baramulla axis, and in Jammu.[38]

Ceasefire or not – and the disappointment of Generals notwithstanding – the end of the war was not in fact in sight. Like the several jihadi enterprises that were to follow, then, the war of 1947–1948 was a crusade initiated as an instrument of state policy, not an outbreak of religious anger or communal passion. Pakistan had existential interests in Jammu and Kashmir, which it would pursue. Nehru understood this clearly. "The invasion of Kashmir is not an accidental affair resulting from the fanaticism or exuberance of the tribesmen", he noted, "but a well organized business with the backing of the State . . . we have in effect to deal with a State carrying out an informal war, but nevertheless a war".[39]

The informal war

Surendra Nath's twin-engine aircraft ploughed into the side of a mountain near Kulu, in the Indian state of Himachal Pradesh, on June 9, 1994. There were no survivors. Newspapers mourned the death of the Governor of the Indian state of Punjab; of a man who played in the crushing of a movement of the Sikh religious right that for a time ranked among the most brutal conflicts in the world, and would have a direct bearing on Jammu and Kashmir's fortunes.

His secret life in Jammu and Kashmir received no notice: few even knew of it.

Twenty-eight years before his death, Nath, then a senior police official, had authored a classified history of India's counter-intelligence campaign against terrorist groups in Jammu and Kashmir from 1948 onwards. The still-secret *Report on Pakistani Organized Subversion, Sabotage and Infiltration in Jammu and Kashmir* is perhaps the sole history of the Informal War, and provides encyclopaedic account of its course. Writing two decades after Nehru proclaimed the existence of the Informal War, Nath affirmed many of its conclusions. The "cease-fire which came into effect on the 1st of January, 1949", Nath wrote "was merely a prelude to the Pakistani efforts to grab Kashmir by other means".[40] "The object" of Pakistani sub-conventional efforts, he argued,

> was to create conditions in which the Government established by law in this State could not function, to arouse communal passions, to assassinate important nationalist leaders and ultimately overthrow the Government and capture power either through their agents or by direct intervention.

Pakistan was uniquely well poised to launch such a covert campaign, for reasons which are little understood. Despite its ostensible military superiority, India's intelligence apparatus was in ruins after Independence. Its Intelligence Bureau, staffed mainly by police officers, was charged with reporting on events in Jammu and Kashmir. At Independence, however, the Intelligence Bureau was in what one participant has described as "a tragic-comic state of helplessness".[41] Qurban Ali, the senior-most Indian in the organization's last months as an institution of British India, was to choose Pakistani citizenship – and had used his offices to transfer every file of importance to that country. India's intelligence personnel were left with "the office furniture, empty racks and cupboards, and a few innocuous files dealing with office routine".[42] India's Military Intelligence Directorate had the capability to monitor events in Jammu and Kashmir, but the chaos across much of northern India meant the Army had neither the time nor resources to do so.[43] Put simply, Pakistan's covert warriors were the only team on the field.

Low-level covert activity mirrored Pakistani conventional military responses to India's spring counter-offensive of 1948. That year, the Jammu and Kashmir Police recovered 643 crude bombs, 666 hand-grenades and 83 tins of fuses in raids, which led to 22 arrests. Authorities claimed that these explosives had been brought from Pakistan by a Srinagar resident working for Pakistani intelligence, Salim Jehangir Khan. What little published material is available suggests that Pakistan's intelligence services, and powerful elements in its political establishment, used such tactics fairly widely. One remarkably candid admission has come from Lieutenant-General Gul Hasan Khan, who served as the last commander-in-chief of the Pakistani armed forces. General Khan's memoirs record that an unnamed "elder statesman" in Pakistan organized covert supplies of weapons to the princely state of Hyderabad in 1948, which was using armed force to resist accession to the Indian Union. According to General Khan, the "elder statesman" organized at least one shipment of .22 pistols on a DC-3 aircraft.[44] Indian intelligence reports that Hyderabad was negotiating a large weapons deal with a Czechoslovak firm had caused considerable disquiet, but no one seems to have picked up on the flow of weapons from Pakistan.[45] The weapons appear to have been used by two ideologically distinct but tactically allied forces, the Islamist gangs of Kasim Rizvi and the Communist Party. In an April 1948 press conference, using rhetoric that anticipated jihadist polemic of the 1990s, Rizvi announced that he would plant the flag of the Muslim monarch of Hyderabad "on the Red Fort in Delhi".[46]

Pakistan's support of such covert activity served a wholly rational purpose: acquiring Jammu and Kashmir remained a real possibility despite the end of the war of 1947–1948, and sub-conventional warfare in support of its political objectives was the sole leverage it had in the parts of the state India now controlled. A significant escalation in this form of warfare became evident from 1951. The year is of some significance. Elections to Jammu and Kashmir's Constituent Assembly were scheduled for later that year. Pakistan clearly understood that events were heading towards a full-scale integration of Jammu and Kashmir with the Indian Union, provoking a renewed crossing of swords at the United Nations. In March, Pakistan's representative at the UN, Sir Zafarullah Khan, charged India with having orchestrated "a deep conspiracy and with long preparation for sending troops to Kashmir"; Prime Minister Nehru responded by insisting this was "cent per cent false".[47] The aggression was not just polemical. Between the spring of 1951 and April 1952, a full-scale mobilization was ordered by Pakistan in anticipation of the possibility that the sub-conventional war could escalate into full-blown hostilities.[48]

Amidst this near-crisis, Pakistani intelligence planners set in place the first post-war covert initiative. On August 20, 1951, the Government Rest House at Akar, on the Srinagar–Pahalgam road, was set on fire, along with a nearby bridge. Jammu and Kashmir Police investigators discovered that the group responsible for the Government Rest House arson had also carried out five earlier, unexplained attacks, the burning of the Kangan, Sagipora and Singhpora Bridges on roads leading out of Srinagar, the destruction of a Forest Department Hut in Nagranag and the cutting of a military telephone line from Srinagar to Gulmarg. The attacks preceded the elections to the Jammu and Kashmir Constituent Assembly and, in the state government's view, were intended to disrupt the democratic process.

Fourteen conspirators were eventually tried for the attacks, of whom nine were convicted. Significantly, another 18 conspirators charged with the attacks could never be tried, as Indian criminal law does not allow for the prosecution of suspects *in absentia*. These suspects included several Pakistani nationals, notably Abbas Ali Shah, the Superintendent of Police in charge of the Criminal

Investigations Department in Rawalpindi, Major Asghar Ali Shah, a security officer at Hillan, in Pakistan-administered Kashmir and Taj Din, Inspector of Police in Nowkote, again in Pakistan-administered Kashmir. As far as I have been able to determine, these charges were the first legally supported allegations by Indian authorities of Pakistani covert sponsorship of terrorism in India.[49] Over the coming decades, dozens of similar charges would be leveled against Pakistan, as the secret war was ratcheted up to ever-higher levels.

In addition to this kind of low-grade terrorism, the India–Pakistan Cease-Fire Line (CFL) in Jammu and Kashmir itself remained unstable, much as it was to be after the rise of jihadi terror in the 1990s. On May 7, 1955, for example, an armed patrol of the Pakistani Border Police crossed on to the Indian side of the Sialkot–Jammu frontier and opened fire on a party made up of both soldiers and civilian employees of the State Agricultural Farm. As a result, an officer, five other ranks and six civilians were killed. The UN Military Observer Group subsequently carried out an investigation, which deemed the incident a violation by Pakistani troops.[50] Such incidents could not have but fuelled Indian concerns about the direction that events in Jammu and Kashmir were headed.

Critics of Indian policy during the period have often asserted, correctly, that it did not apply the same standards of democracy to Jammu and Kashmir at this time as were applied to other States. Sumit Ganguly has noted that "the national political leadership, from Jawaharlal Nehru onwards, adopted a singularly peculiar stand on the internal politics of Jammu and Kashmir: as long as the local political bosses avoided raising the secessionist bogey, the government in New Delhi overlooked the locals' political practices, corrupt or otherwise".[51] Nehru himself candidly admitted that it was "true that political liberty does not exist there in the same measure as in the rest of India".[52] The fact remains, however, that Jammu and Kashmir, unlike any other State, was on the theatre of a continuing sub-conventional conflict which could have at any stage escalated into an India–Pakistan war.

Politics after the war

The threat of war – and the fact that Jammu and Kashmir's accession to India was being contested through covert warfare by Pakistan – shaped the course of politics in Jammu and Kashmir.

Sheikh Mohammad Abdullah took power as Prime Minister of Jammu and Kashmir in 1948, having earlier served as head of the war-time Emergency Administration. The pressures on him were immense. He had little experience in administration, and, from the outset, his relationship with Maharaja Hari Singh was strained. Hari Singh complained that Abdullah was communal; Abdullah in turn complained that the Maharaja was sponsoring violence by Hindu fundamentalist groups in Jammu. Hindu elites opposed the National Conference because it threatened to strip them of their land and privileges; Muslim elites because of Sheikh Abdullah's stubborn opposition to Pakistan.[53]

Yet, on one thing Abdullah seemed unequivocal. "We have decided to work with and die for India", he said soon after taking office.[54] This statement exemplifies what Abdullah had repeatedly said since Independence. In June 1948, he insisted that the accession of Jammu and Kashmir to India was not merely a consequence of the Pakistani invasion. Abdullah argued that

> the people of Jammu and Kashmir have thrown our lot with Indian people not in the heat of passion or a moment of despair, but by deliberate choice. The union of our people has been fused by the community of ideals and common sufferings in the cause of freedom.[55]

Again, on March 7, 1949, he insisted that "we have decided to work with and die for India".[56]

The new Jammu and Kashmir Prime Minister's argument for accession to India was simple, and it was one designed to address the peasants who formed the core of the National Conference constituency. Sheikh Abdullah's reforms were to strip the landholding aristocracy in Jammu and Kashmir of their privileges without compensation, and enable the state to expropriate all estates larger than 22.5 acres to the tenants who worked them. The peasantry supported India, one commentator has argued, "because they understood – and Abdullah told them – that such reform would never be possible in a Pakistan which protected feudalism and landlordism".[57]

By 1949, however, signs of discontent were evident, even among the National Conference's core constituency. Village elites and revenue officials were able to subvert the new land laws, enabling the relatively rich to take control of a large share of the fields which ought to have become available. Then, two successive crop failures in 1949–1950 and 1950–1951 caused enormous hardship. The Government of India pitched in and dispatched large amounts of emergency aid. Much of this, however, was reported to have been misappropriated with the connivance of the distributing organization, the Food Control Department. Cooperative stores, the major distributors of key commodities like rice, salt and cloth, functioned poorly; an official inquiry held in 1952 was to eventually hold their managements guilty of "corruption and malpractices".[58] "The people's patience was exhausted", one observer has noted of the time, "and when the Food Minister, Mirza Afzal Beg, rose to address a meeting of the National Conference workers, he was heckled and made to leave the conference room".[59]

At about the same time, Abdullah began shifting away from his earlier commitment that Jammu and Kashmir's future lay firmly within the Indian Union. In January 1948, Abdullah spoke at the United Nations, defending India's claims to Jammu and Kashmir. His private position, however, was very different. In a meeting with the United States diplomat Warren Austin, he expressed support for independence, saying it could then seek "American and British aid for the development of the country" – an odd position for a politician who marketed himself as a socialist at home, and was suspected of communist sympathies by some in New Delhi. Abdullah then proceeded to lobby Pakistani officials for support, claiming an independent Kashmir would "naturally be closer to Pakistan, firstly because of a common religion and secondly because Lahore is near and Delhi is far off".[60]

By 1949, he was saying the same things publicly. In an interview with the journalist Michael Davidson, he asserted that

> accession to either side cannot bring peace. We want to live in friendship with both the Dominions [India and Pakistan]. Perhaps a middle path between them with economic cooperation with each will be the only way of doing it. However, an independent Kashmir must be guaranteed not only by India and Pakistan, but also by Great Britain.[61]

A perceptive contemporary media report noted that the object of Abdullah turning towards independence was to "strengthen his personal support, which now seems to be falling off".[62]

From 1949, matters began to come to a head. One key line of confrontation was with the Hindu right-wing in Jammu, representing both Dogra feudal privilege and ultra-nationalist opposition to the special status of Jammu and Kashmir. Politically, these forces coalesced into the Praja Parishad, set up with the backing of the Hindu-nationalist leader Shyama Prasad Mookerjee. Abdullah sought to suppress the movement by incarcerating Mookerjee, who died while in custody. This provoked further fury – and further repression. In elections to the Jammu and Kashmir Constituent Assembly, held in 1951, 45 out of 49 Praja Parishad candidates were disqualified on flimsy technical grounds.[63] Sheikh Abdullah's National Conference won 75 seats in the state, 73 of them uncontested.

> Questionable political practices of wholesale rejection of the opposition candidates' nomination papers and unopposed seats set an unhealthy precedent in the state, a precedent which would remain firmly established for a long time.[64]

Abdullah had got what he wanted – and again switched sides. Speaking at the inauguration of the Constituent Assembly, Abdullah made a powerful argument for accession. India's Constitution, he said, "had set before the country the goal of a secular democracy, based upon justice, freedom and modern democracy". By contrast, Pakistan was "a feudal state in which a clique is trying to maintain itself in power". Abdullah firmly ruled out independence. From "August 15 to October 22, 1947", he pointed out, "our State was independent and the result was that our weakness was exploited by the neighbour with invasion [*sic*]".[65] In July 1952, Abdullah finally arrived at a concord – what is now called the Delhi Agreement – with New Delhi on the content of Jammu and Kashmir's special status.

By this time, however, chauvinist movements in Jammu – and in Ladakh, where land reforms had stripped the Buddhist clergy of their estates – were gathering momentum. These movements generated counter-pressures in the Kashmir valley, where Muslim religious anxieties were fuelled by the mobilizations in the south and east. Torn by these competing stresses, Abdullah again turned towards the independence option. In May 1953, he confided in the United States diplomat Adlai Stevenson that he supported independence. That July he told the audience at a public gathering that it was not necessary to be an appendage of either India or Pakistan.[66] Meanwhile, Abdullah dragged his feet on implementing the Delhi Agreement, provoking Nehru to bitterly complain "that a settlement arrived at between us should be by-passed or repudiated, regardless of the merits".[67]

Lurches of this kind fuelled the suspicions of Sardar Vallabhbhai Patel, India's then Home Minister and the leading figure of the right-wing within the Congress Party. Nehru and Patel had long differed over Kashmir policy; the Union Home Minister had, in December 1947, even offered his resignation on the issue. According to the then head of India's intelligence services B.N. Mullik, Patel on one occasion warned that "Sheikh Abdullah would ultimately let down India and Jawaharlal Nehru and would come [out] in his real colours".[68] There were feuds over the appointment of Colonel Hasan Walia as the head of the Intelligence Bureau station in Srinagar, with Abdullah arguing that the officer had been sent to spy on him, not Pakistani agents.[69] Nehru defended Abdullah in several of these battles, but as the Sher-i-Kashmir's pro-independence position grew stronger, it became harder to do so. "It is always painful to part company after long years of comradeship", he said in the face of Abdullah's stalling of the Delhi Agreement, "but if our conscience so tells us, or in our view an overriding national interest requires, there is no help for it".[70]

Nehru had served notice. On August 8, 1953, two of Sheikh Abdullah's most trusted associates, Bakshi Ghulam Mohammad and Ghulam Mohammad Sadiq, staged a party coup which removed him from office. He was then jailed on charges of endangering India's security. What had gone wrong? Not all accepted that the stated reasons for Abdullah's incarceration were correct. Y.D. Gundevia, who worked with Nehru, attributed the break to a powerful group in New Delhi who wished to stall land reform. According to this theory, the group, with the backing of the Union Home Ministry used "right wing propaganda, a whispering campaign and political backbiting" to bring about the coup.[71] Abdullah himself attributed his removal from office to the classes most hit by the land reform movement. "Our plan affected both Hindu and Muslim capitalists and *zamindars* [landlords] equally", he told one interviewer, "but the Hindus had direct lines to Delhi":

> I was in turn called a British agent, a communist agent, and an American agent. My enemies even undermined the loyalty that my associates had had for me. I wanted to take action against these persons and asked for the permission of Pandit Nehru to do so – but instead of giving me permission to prosecute them, he dismissed me and interned me.[72]

Any of these explanations for the breakdown of the relationship between Nehru and Abdullah may indeed have been partially true. There was, however, another piece to the puzzle: the flow of intelligence reports on the sustained Pakistani covert warfare in Jammu and Kashmir. Although communications between the Intelligence Bureau and the Union Ministry of Home Affairs may never become public knowledge, the documentation that became available in the course of research for this book gives us unprecedented insight into the perceptions of the intelligence community, and their understanding of Abdullah's conduct. Sheikh Abdullah himself, however, would have no direct role in events for the next several years. The stewardship of his cause would remain in the hands of the only lieutenant who had stayed loyal to him, Mirza Afzal Beg. In 1955, Beg would set up the Plebiscite Front, the group that would proceed to spearhead the struggle for an independent Kashmir, and at once provide a pool of cadre for Pakistan's covert campaign against India.

The builder's regime

From the optic of Indian intelligence, the low-level covert activity that had pock-marked the first years after Independence was, in essence, the armed form of the continuing political contestation of Jammu and Kashmir's accession to India. The covert war that continued on from the war of 1947 was intended to sabotage the process of Jammu and Kashmir's integration within the Indian union. It was on this political process that the energies of both India and Pakistan were to be focused over the coming years.

If Pakistan had learned one lesson from the war of 1947, it was that a military victory against India was not possible without external help. Its search for allies was to have momentous consequences for Pakistan itself. In 1952, the Pakistani Foreign Minister, Zafarullah Khan, visited several West Asian states, attempting to sell the idea of an Islamic Bloc stretching from the Indian Ocean to the Mediterranean. The idea was greeted with hostility in Egypt, Syria and Lebanon, where it was seen, with good reason, as a British-sponsored enterprise to create a reactionary Islamist counterweight to Arab nationalism. Choudhury Khaliquzaman, a prominent Muslim League politician who was to become Governor of East Pakistan in April 1953, had no greater success. During a meeting with Egypt's King Farouq, he was bluntly told that "Islam was not born on August 14, 1947", the date of Pakistan's independence.[73]

By the time Pakistan declared itself an Islamic Republic on October 31, 1953, the dream of a global Islamic bloc seemed far-fetched. The new designation was driven by domestic compulsions, not overseas ambitions. In February that year, a group of Islamist parties led a violent agitation, demanding that members of the heterodox Ahmadi sect be declared infidels. Within a month, the government of Prime Minister Khwaja Nazimuddin was forced to admit that it had lost control of law and order, and asked the Army to take charge of Lahore. General Azam Khan, appointed as martial law administrator in the city, soon succeeded in restoring order, but the affair also laid the foundations for the Pakistani military's subsequent domination of the country's political life.[74] In the wake of the events of 1953, newly Islamic Pakistan turned to the Western powers for support. In February 1955, the United Kingdom, Iraq, Iran and Turkey signed a treaty known as the Baghdad Pact, which allowed them to "co-operate for their security and defence".[75]

Although the ostensible purpose of the Baghdad Pact was to defend West and South Asia from communism, the implications of increased Western military aid to Pakistan were not lost on New Delhi. Securing a stable legal relationship between New Delhi and Srinagar became a renewed imperative. On January 26, 1957, the Constituent Assembly of Jammu and Kashmir approved the state's Constitution. Jammu and Kashmir's destiny was now irrevocably linked to the constitutional structure of the Indian Union. This new legal order provoked angry protests from Abdullah, and from the United Nations, but with little effect. In March, just two months after the Constitution was ratified, elections were held for the Jammu and Kashmir Legislative Assembly. Bakshi Ghulam Mohammad was elected Prime Minister, backed by a crushing majority of 68 seats in the new legislature. Over the coming years, Bakshi – called "The Builder" by his party faithful – would develop an unsavory reputation as an authoritarian leader who used state patronage both to ensure control of the National Conference cadre apparatus and to enrich party elites.[76]

Bakshi's ability to retain power lay in positioning; in his placement of his office at the fulcrum of the contestation between ethnic Kashmiri chauvinists in the Valley and Dogra chauvinists in Jammu. As a consequence, the real threat to Bakshi came from secular political organizations in both the provinces of Kashmir and Jammu – organizations that could challenge his authority without fuelling chauvinist anxieties among ethnic Kashmiris and Dogras. Balraj Puri, a perceptive contemporary observer himself active in Jammu and Kashmir politics, has noted:

> Bakshi was able to demonstrate in the Valley that he was more successful in containing the [Praja] Parishad threat. And in Jammu, he sought to create the impression that he alone could curb the 'anti-national' activities of Abdullah. However, Bakshi had a vested interest in ensuring a minimum strength for the Parishad and for Abdullah so that Bakshi could arouse fears against them in Kashmiris and Dogras, respectively, and emerge as their protector.[77]

In the build-up to the 1957 elections, the kinds of secular and progressive forces that could pose a real challenge to Bakshi began to emerge. The Democratic National Conference, made up of Ghulam Mohammad Sadiq, Mir Qasim and D.P. Dhar (who was later, as Home Minister, to commission Surendra Nath's reports), broke rank with Bakshi. It demanded the extension of key safeguards in the Indian Constitution to Jammu and Kashmir, where the Constitution did not have a chapter on fundamental rights. These included the right of appeal to India's Supreme Court, and poll oversight by the Election Commission. The

Democratic National Conference also sought the release of Sheikh Abdullah – who, ironically enough, would most likely have opposed their Constitutional demands. At once, the Praja Socialist Party, a secular formation, gathered strength in Jammu and called for a dialogue with Sheikh Abdullah. Bakshi bitterly attacked both organizations, and at times used what one observer described as "totalitarian methods" to contain them.[78]

In January 1958, with Bakshi firmly ensconced in office and Jammu and Kashmir's constitutional relationship with India on relatively firm ground, Sheikh Abdullah was released from jail. Pandit Jawaharlal Nehru's abiding friendship with his old political colleague may have been one major factor in bringing about the release; so too may have been the desire to ensure the reincorporation of Abdullah in mainstream Jammu and Kashmir politics. Behera has suggested, following Puri, that the release (and, indeed, the fact that Mirza Afzal Beg was released from jail and allowed to mobilize a pro-Abdullah campaign) was something of a conspiratorial performance, intended to undermine the Democratic National Conference – Praja Socialist Party mobilization. "After achieving that objective", she has suggested, "they could be quietly incarcerated in the name of safeguarding India's national interests in Kashmir".[79]

Though this school of thought has considerable currency in Jammu and Kashmir, it does appear somewhat simplistic. Sheikh Abdullah was after all only released *after* the 1957 elections, not before them, when he would have most hurt the Democratic National Conference's prospects. It is also worth noting that a third conspiratorial explanation exists for Abdullah's release, inferred from a statement by the then chief of the Intelligence Bureau, B.N. Mullik, India's first spymaster: "we were sure that he would indulge in such activities as would enable us to get further direct evidence against him".[80] In other words, Abdullah was released in order to secure the evidence to arrest him yet again.

Whatever the truth, Mullik's observation points to the near-predestination of what was to follow. Struggling for political space, Sheikh Abdullah was in no position to seek accommodation with New Delhi. Out of jail after four years, he did nothing to alleviate Nehru's concerns. In a statement to the press, Sheikh Abdullah proclaimed that the

> expression of the will of the people through a plebiscite is the one formula which has been agreed upon by the parties concerned and in a mass of disagreement about details, this common denominator has held the field so far.

Lest someone should have read this to mean that the field might shift in the future, Sheikh Abdullah underlined his position. Bakshi could "shout from the top of the Banihal pass" that the accession of Jammu and Kashmir was "final and irrevocable", but such claims had no legitimacy for the government was made up of "*goondas* [thugs], opportunists and thieves".[81]

Soon after, political violence broke out, culminating in the murder of a ruling-party affiliated politician Mohiuddin Bandey in a clash with National Conference cadre at Hazratbal. In April 1958, Sheikh Abdullah was again detained. This time the charges were deadly serious: if India's intelligence establishment had indeed been engaged in laying a trap, the prey had bitten.

Unending war

For at least one key military official in Pakistan, the war of 1947 had never ended. Soon after the end of the war, Major-General Akbar Khan wrote a tract demanding armed assistance for a rebellion in Jammu and Kashmir. In essence, he argued that even if such assistance was to lead to war, it would be "so much the better as this would tend to threaten the existing international peace and only then would there be reason for the United Nations again to take note of the [Kashmir] problem".[82] Khan's tract, of which 5,000 copies were published, excited the interest of India's politicians and press.

Pakistan's politicians were also listening. Khan was now asked to brief Pakistan's then President, Iskandar Mirza, on just how a renewed military enterprise to take Jammu and Kashmir might be run. Khan laid out plans for a covert war, involving just 500 men trained in guerilla warfare and sabotage.

> My emphasis was on the use of lesser and lesser numbers, so that a pair of men would have at least a clear mile to operate in. Thus, they would be almost impossible to detect and they would have no difficulty in going across the cease-fire line which was open in so many places to such an extent that unauthorized traffic of men and animals was constantly going on across it regularly. They would have to be preferably locals, or at least in local clothes, armed only with some dynamite for blowing up bridges and pliers for wire cutting. For their own protection each could have a knife or a small local made pistol. They would not need to fight against police or troops. Their target would be unguarded bridges, isolated wires and unprotected transport.[83]

Mirza, Khan has recorded, also discussed plans to defend Lahore, in case such a sub-conventional operation provoked a full-scale Indian military response. President Mirza, according to the Major-General, "said that having satisfied himself about the security of Lahore, he had advised General Ayub Khan to proceed with that scheme for occupied Kashmir".[84] Major-General Khan was never told precisely what Ayub Khan had been tasked to do, but does record that "in those very days, some small explosions took place in occupied Kashmir".[85] Subsequently, Malik Feroze Khan Noon, soon after taking charge as Pakistan's Prime Minister in 1957, whispered in Major-General Khan's ear, "we have started it".[86] Responsibility for the new operations, Noon told Major-General Khan, had been entrusted to a policeman – Deputy Inspector-General Mian Anwar Ali.

Indian counter-intelligence officials read the new wave of covert war as integrally linked with the political chaos initiated by the palace coup of 1953. Sheikh Abdullah's summary removal from office, Nath wrote, had:

> started a new and more intensive phase of Pakistani inspired subversive activity. The relations and close associates of Sheikh Mohammad Abdullah who now felt frustrated in their designs, took increasingly to under-ground subversive activity. Pakistani authorities also pumped in money, subversive literature, arms, ammunition and explosives on an ever increasing scale into the Valley through such disgruntled elements and paid agents. A so-called "war council" was constituted to organize all these activities.[87]

Investigation of this affair, conducted by officials of the Indian Intelligence Bureau and the Jammu and Kashmir Police officer Ghulam Qadir Ganderbali, was to emerge as what is known as the Kashmir Conspiracy Case – perhaps the most controversial chapter in the troubled relationship between New Delhi and Sheikh Abdullah. Charges of criminal conspiracy, under Section 121 of the Ranbir Penal Code, were filed on May 21, 1958 against 25 people, including Mirza Afzal Beg and five serving Pakistani Intelligence Bureau personnel. Sheikh Abdullah's name was added to the list of those accused on October 23, 1958. I shall return to the several lucid critiques of the criminal charges against Abdullah, and to the political circumstances that led to the prosecution eventually being dropped. This part of my narrative, however, is concerned with events as they were understood by counter-intelligence officials within Jammu and Kashmir.

Mirza Afzal Beg was released from jail in 1954. It was, the *Report on Pakistani Organized Subversion, Sabotage and Infiltration in Jammu and Kashmir* baldly informs us, a political decision, intended to ensure those close to Sheikh Abdullah would at least by proxy have the opportunity "to participate in the deliberations of the Constituent Assembly".[88] As I have pointed out earlier, others have offered more sinister motives, notably the claim that Beg's release was intended to inflame Hindu chauvinist fears, and thus legitimize Bakshi's rule among the Dogras of Jammu. Whatever the truth, he formed the Jammu and Kashmir Plebiscite Front on August 9, 1955, with the avowed purpose of compelling India to hold a referendum.

Anti-India pamphlets and posters began streaming into Jammu and Kashmir across the Cease-Fire Line. So, according to Indian intelligence, did four groups of covert operatives, all acting under the command of Khan Mohammad Khan, a Deputy Superintendent of Police in the Pakistan Intelligence Bureau. Three operatives, Bagh Ali, Ismail and Rahim, were tasked with operating in the Jammu province; Abdul Rahim and Jamal-ud-Din were to operate in Poonch. Two cells were prepared for operations in the Valley, one made up of Jehangir Khan, Akbar Mirpuri and Mohib Ullah Beg; the other under the command of Aziz Parwana.[89]

Indian counter-intelligence investigation of these groups broadly tallied with that of Major-General Khan in attributing responsibility for the new phase of covert warfare to Pakistan's Intelligence Bureau, a police organization, rather than to its military. Indian intelligence operatives' description of the training of these covert operatives is remarkably similar to those available of the military training provided to terrorists trained in jihadi camps in the 1990s, a fact of no small importance. Instructors taught covert operatives the critical guerilla skill of operating in small groups of three or four individuals, a scheme remarkably similar to what Major-General Khan had advocated. Apart from basic weapons training, the course content included training in hand-to-hand combat, laying ambushes, jungle warfare skills like the maintenance of hideouts and tactics for assaults on defensive posts.[90]

Religion formed a major part of the curriculum, much as would decades later among the jihadi groups that operated in Jammu and Kashmir in the 1990s:

> During training hours early in the morning, all the companies were required to get together in the parade ground for a short while. A Maulvi (Muslim cleric) and one or two high army commanders spoke to them on religious teachings and recited verses from [the] Quran Sharif. It would be emphasized that warfare was recognised [as legitimate] in [the] Quran and, as such, it was the duty of every person of the Mujahid force to obey and die on the command of their officers and be loyal to and fight for their country and defend its borders. Further it was exhorted that they had to fight against India to liberate the Kashmiri Muslims whose religious activities were curbed and who were not free to offer their Nimaz [Namaz, *sic*] prayers or even call (give *azan*) the faithful to prayer. It was further alleged that Islamic festivals, etc. were not allowed to be celebrated in India.[91]

Later in this book, I shall return to a more detailed analysis of the question of whether the growing influence of jihadi groups in the 1990s, as the eminent scholar Yoginder Sikand has argued, in fact demonstrated "a marked transformation in the terms of discourse with which the Kashmiri liberation struggle has sought to express itself".[92] Worth noting, however, are the striking similarities between what operatives were being told by their Pakistani official instructors in the 1950s and the indoctrination of jihadi cadre decades later. Sikand's account of the Markaz Dawa wal'Irshad, the parent political organization of the Lashkar-e-Taiba terrorist group, is illuminative:

> The Markaz sees the conflict in Kashmir not as a territorial dispute between India and Pakistan, nor even as a clash between cultures, but

> as nothing less than a war between two different and mutually opposed ideologies: Islam, on the one hand, and disbelief (kufr), on the other. This is portrayed as only one chapter in a long struggle between the two that is seen as having characterized the history of the last 1,400 years ever since the advent of the Prophet Muhammad. The roots of the Kashmir problem are seen in its Muslim rulers having been replaced, first by the Sikhs and then by the Hindu Dogras through British assistance. With India (i.e., the "Hindus") having taken over Kashmir in 1947, a long and protracted reign of bloody terror is seen to have been unleashed on the Kashmiri Muslims. This is seen as a direct and logical consequence of Hinduism itself, because, it is alleged, 'the Hindus have no compassion in their religion.' Hence, it is the duty of Muslims to wage jihad against the 'Hindu oppressors.' All Hindus are tarred with the same brush. Thus, Hafiz Muhammad Sa'eed declares: "In fact, the Hindu is a mean enemy and the proper way to deal with him is the one adopted by our forefathers . . . who crushed them by force".[93]

Language of this kind would have been familiar to the covert groups who fought the Informal War. Trained in Sialkot, a small town in Pakistan-administered Kashmir, Bagh Ali's group was the first to strike. Their first operation was the bombing of a shop in Jammu's main shopping boulevard, Residency Road, on June 18, 1957. By the time of their arrest in November, 1957, the group had executed four other strikes on targets of potential military value – two bombings of bridges, one targeted at the United Nations Military Observer Group's office in Jammu and another directed at an Indian Airlines Corporation van.

One action, however, was expressly terrorist in character, an attack on a temple in Dharmkhoh on October 1, 1957. The intention was, presumably, to incite Hindu–Muslim violence. Abdul Rahim's Poonch group came close to this objective by assassinating a prominent Hindu, Roopay Shah, and injuring several others, in a bomb attack on July 23, 1957. Mohib Ullah Beg's group used similar tactics in the course of its work in the Valley. While it carried out attacks on legitimate targets – notably the bombing of two small culverts across streams in Pulwama and Budgam – it also sought to incite communal violence. On September 8, 1957, the unit planted two booby-trapped explosive devices inside a mosque in Maisuma, one of the most politically violent neighborhoods of Srinagar. A local resident, Abdul Ahad Bakaya lost both his sons when one of the mosque bombs went off, and his wife and daughter were injured.[94]

A variety of factors seem to have motivated members of these covert units, not all of them expressly political. Aziz Parwana, for example, began his career as a Pakistani covert operative through the startling medium of falling in love with a Srinagar woman, Attiqa Bano. Parwana was told, however, that he would first have to obtain the permission of her father, Ahmad Ullah Bhat, who had migrated to Rawlakote in Pakistani Kashmir. Desperate to marry Attiqa Bano, Parwana made the perilous journey across the CFL. Having left Indian-held territory, he made contact with Sajawal Khan, a Pakistani Security Officer posted at the Mori Maidan pass. The enterprising Khan prevailed on Ahmad Ullah Bhat the necessity of offering his daughter's hand in marriage only if Parwana first agreed to work as a Pakistani agent. Acting under Khan's supervision, Parwana eventually organized a cell of 14 members, who set off bombs at Srinagar's Palladium Cinema and under a bridge in Allochi Bagh. Four more bombs were recovered after Parwana's interrogation, two planted in the Hind Kashmir Hotel and two under bridges around Srinagar.[95]

By the accounts of Indian counter-intelligence operatives, Sajawal Khan was instrumental in giving political content to these covert operations. Khan was claimed, by prosecutors of the Kashmir Conspiracy Case, to have first met with key members of the Plebiscite Front, including Mirza Afzal Beg, Mohammad Akbar Sofi and Peer Maqbool Vilgami, in 1955. The meeting was held at the home of another activist, Ghulam Mohammad Bhat. Sajawal Khan paid a second visit to Kashmir in 1956, travelling across the CFL through the densely forested, high-altitude Yus Maidan pass with Jehangir Khan, this time carrying explosives. On a third visit, again in 1956, Sajawal Khan again met Jehangir Khan – evidence, as Nath saw it, of "the active and direct manner in which Pak [Pakistan] authorities controlled and guided the activities of the conspirators".[96]

So far, however, Pakistan's Intelligence Bureau had registered little success. Occasional bomb blasts had not brought the Indian state to its knees, and there had been no popular uprising against Sheikh Abdullah's incarceration. Hidden away in some archives in Islamabad, there must be accounts of how Pakistan's intelligence community understood this reversal, and how they planned the next phase of the covert war. What evidence is already available, however, allows us to attempt at least an initial answer to the most crucial question of all: what in fact was the relationship of the covert war with the political war?

The contested truths of the Kashmir conspiracy case

On the face of it, there ought to be little reason for surprise at the proposition that Sheikh Abdullah's circle *may* have turned to Pakistan for assistance. As early as 1952, Victoria Schofield has noted, Abdullah had "become disillusioned with India's secularism". And, although "he remained opposed to the two-nation theory, contrary to his earlier expectations, Pakistan was proving viable".[97] Faced with a hostile government in New Delhi, it would have been only natural for his supporters to have taken what help they could from Pakistan.

Jammu and Kashmir police authorities filed their charges against Abdullah and his supporters on May 21, 1958. In a legal document presented to a Srinagar magistrate, D.W. Mehra, the then Inspector General of Police, alleged on behalf of the State that Begum Abdullah and the circle around her:

> approached Pakistan agencies and officials to devise plans to overthrow the Government of the State and to make it a part of Pakistan.

> Furthermore contact, direct and indirect, was maintained by the accused with Pakistan agencies and intelligence officers; meetings were held with Pakistani officers at Srinagar and other places in the state and outside; weapons and explosives were obtained from Pakistan agencies; volunteers were raised and organized under the name of 'Karkuns'; communal feelings were inflamed amongst the Muslims, who were exhorted by appeal to religion to work against the Government of the State and in favour of Pakistan; false and malicious propaganda was published against the Government of the State; enmity and hatred were attempted to be created between different classes; hatred, contempt and disaffection were sought to be spread towards the Governments of Jammu and Kashmir and India.[98]

Indian intelligence operatives were in no doubt about the integrity of their evidence against Sheikh Abdullah. "When he went back to jail", Mullik smugly wrote, "we were quite satisfied that he had built up an unassailable case against himself".[99] Yet, the case itself dragged on interminably, with all concerned blaming each other for the courtroom delays. For four years, the Kashmir Conspiracy Case was stalled in a Magistrate's Court, before finally being committed to a Court of Sessions in April 1962. Bureaucrat Y.D. Gundevia, at Nehru's side during the time, has provided this account:

> In the Sessions the case went even slower, so much so that the Intelligence Bureau was put on the defensive. Nehru was again losing his patience. Charges and counter-charges were made by the Intelligence Bureau against the accused and the accused against the Intelligence Bureau, on who was responsible for the delay in the proceedings before the Sessions Judge. The furor became serious enough for the Intelligence Bureau and the Home Ministry to bring out a pamphlet (for free circulation, of course) to explain who was delaying the case.[100]

The problems in court were not, however, only procedural. One key complication for the prosecution case was that Begum Akbar Jehan Abdullah, Sheikh Abdullah's wife, was alleged to have issued receipts for funds received from Pakistani diplomat S. Mohammad Raza to the agents who carried them.[101] The official examiner of the documents in question certified that the handwriting on the receipts, bearing pseudonyms Zeenat-ul-Islam, Alif Din and Hamshera, was indeed that of Begum Abdullah; three of these documents are appended to the *Report on Pakistani Organized Subversion, Sabotage and Infiltration in Jammu and Kashmir*. She was neither arrested nor charged. "One would have thought", wrote a commentator otherwise sympathetic to both Nehru and Abdullah,

> that she would be the principal accused, with hundreds of letters alleged to be in the hands of the prosecution 'proving' that she was the principal go-between for the receipt of Pakistani funds. For this serious gap in the

prosecution case the only explanation possible, I suppose, is – Nehru's chivalry.[102]

Yet, Nehru's chivalry in itself would not have been adequate to explain the prolonged delays in the prosecution. Even if Begum Abdullah *had* been charged, it is far from clear that convictions could ever have been secured. The *Report on Pakistani Organized Subversion, Sabotage and Infiltration in Jammu and Kashmir* offers fascinating new insight into the primary evidence on the Kashmir Conspiracy Case, much of it so far unavailable to scholars studying the period. Notably, the material in the *Report* suggests that Mullik's claim that the case was "unassailable" was at some distance from the truth.

For one, each of the alleged conspirators used a complex web of codes, designed to make their communication seem innocuous. Money was referred to, variously, as penicillin, books, clothes or eggs; the Indian Army as *Malech* or untouchable. Sheikh Abdullah was called "Assadullah"; Mirza Afzal Beg "Israil"; "Barader Akram" or "Nizamuddin."[103] Allegedly communicating with Pakistani intelligence operative Major Asghar Ali, for example, Begum Abdullah wrote to the "Contractor Khwaja Firm." Her letter stated that the "bearer is a trusted man and can be taken to your GM [General Manager] and after discussing the matter we shall be eager to get the contract going on a firm footing". Another letter written by her thanked "Ibrahim Sahib," the alleged code for the Pakistani intelligence operative Khan Mohammad Khan, for Rs 500 sent through a courier called "Behari Ji," allegedly Abdullah's political associate Pir Sonaullah Shah. In yet another case, this time a letter from Mirza Afzal Beg, a reference to "gardener" was taken to mean an assassin, charged with "pruning the branches".[104] Prosecutors could quite easily have proved that the "Contractor Khwaja Firm" did not exist. It would have been less easy, though, to prove the term was code for a Pakistani official. Since officials of the Pakistan High Commission would not, quite obviously, testify, and Begum Abdullah or her couriers would have been unlikely to incriminate themselves, it is unlikely the charges would have met the rigorous legal standards of evidence required for a criminal conviction.

What direct evidence the *Report on Pakistani Organized Subversion, Sabotage and Infiltration in Jammu and Kashmir* does make available on Sheikh Abdullah's personal role is also ambiguous. A letter from a Pakistani intelligence official Taj-ud-Din, to one of his Kashmir-based operatives, stated that "the war Council [the covert groups] had merged with [the] Plebiscite Front under the direction of Sher-i-Kashmir [Abdullah] and under the leadership of Beg Sahib and Begum Sher-i-Kashmir". Yet, any competent defence lawyer would have pointed out, it was unclear why the operative needed to be provided with this sensitive information in writing – and that too in what intelligence personnel call plain text, as opposed to code. Significantly, Abdullah himself issued no missives that could, by any stretch of imagination, be called incriminating. One circular he issued, which was intercepted by Indian intelligence officials, called for a war on unbelievers. It was liberally drawing on inflammatory religious idiom, but stopping well short of a proclamation of insurrection:

> During the battle of Uhod, some opponents set afloat the rumour that the Prophet [Mohammad] had been killed . . . Addressing the Muslims [the] Prophet said, "supposing the rumour set afloat during the battle of Uhod were true, would it mean that with the death of [the] Prophet, your devotion to God would also die"? Thus, we should remember that whether the person who raises his voice is among us, or is forcibly removed from amongst his own people (a reference to his own detention), his presence or absence makes no difference. Oh believers, if you obey the infidels, they will make you retrace your steps back to faithlessness.[105]

We shall never know which way the case might have gone: precisely what evidence the prosecution might have produced from its files; what defences may have been mounted and what legal twists and turns may have changed the course of history in Jammu and Kashmir. Nor have the protagonists in the affair left behind anything resembling a full account of its covert aspects. In the multi-volume reports on the trial published by Sheikh Abdullah's legal defence team, there is only very occasional reference to the charges that the Plebiscite Front was linked to Pakistan's covert services. On one of the few occasions where Sheikh Abdullah addressed the issue in court, he was dismissive about the charges. Abdullah argued that it would have been legal for him to demand accession to Pakistan, and pointed out that under these circumstances he was unlikely to have used "methods which are not only against the law of the country but are hateful against my creed and principle".[106] He made free with religious metaphor, describing his trial as being of similar epic character to "a similar situation in the sandy plains of Karbala, where the mighty forces of Yazid were ranged against Imam Hussain and a handful of members of his family, who were helpless and devoid of all material resources".[107] Both Abdullah and Beg devoted the vast bulk of their testimony to the larger issue of the legality of the accession of Jammu and Kashmir to India, and the political legitimacy of New Delhi's decision to prosecute them.

History, however, would be shaped by events that had nothing to do with courtroom disputation or the mechanics of the Informal War. The clouds of a new crisis were building fast outside the courtroom – and Sheikh Abdullah would soon be swept out of the jail cell by the storm.

The lion uncaged

Hazratbal stands gleaming over the east bank of the Dal Lake, guarding the water approaches to old-city Srinagar. The shrine was a key centre of National Conference authority, a counterweight to the Jamia Masjid mosque commanded by Maulvi Mohammad Farooq and the *Bakras*. It was home to perhaps the most revered holy relic in Jammu and Kashmir, the *moe-e-muqaddas*, a hair reputed to have belonged to the beard of the Prophet Mohammad.

On December 27, 1963, the relic disappeared from the shrine. Jammu and Kashmir was plunged into chaos. For the next seven days, a cross-party alliance of opposition figures known as the Action Committee emerged as the *de facto* administration of Jammu and Kashmir. Chaired by Maulana Mohammad Sayeed Masoodi, the Action Committee consisted of both National Conference figures, notably Sheikh Abdullah's son Farooq Abdullah, and Islamists like Mirwaiz Mohammad Farooq. As mobs attacked properties owned by the Bakshi family, and the state government retreated behind well-guarded doors, the Action Committee ran "an unauthorized parallel administration, controlling traffic, prices and commerce".[108]

Mullik, the Intelligence Bureau chief, has provided one of the few detailed accounts of the disappearance of the *moe-e-muqaddas* and its mysterious reappearance; even the spymaster, however, shied away from spelling out the details.[109] One popular but empirically unfounded, version of events is that the disappearance was engineered by Bakshi himself. The Chief Minister had resigned from office that October as part of a reorganization of the party apparatus, to be replaced by a relative lightweight, Khwaja Shamsuddin. Thirty-eight charges of corruption were eventually brought against Bakshi by a judicial investigator, of which 15 were proven.[110] In the pop-political rendition of events, Bakshi hoped to use the chaos to establish his indispensability to the Indian state. My own interviews of contemporaries in the Indian intelligence establishment from the time set forth suggest that Mullik had retrieved the relic from a prominent Srinagar family. There was, however, no great consistency to the accounts offered by informants.

One consequence of the cloak-and-dagger retrieval of the *moe-e-muqaddas* was that few on the streets of Srinagar were willing to believe that the hair authorities had produced was in fact the genuine relic. Agitators demanded that a *deedar*, a special exhibition of the relic sanctioned by custom, be held to establish its authenticity. Nehru, by Gundevia's account, personally interceded and over-rode senior officials in the Union Ministry of Home Affairs who opposed the holding of a *deedar*. Maulana Masoodi declared the relic to be genuine at the *deedar*, defusing the crisis. Things could well have gone the other way, Gundevia recalled:

> As we went back to our aircraft to fly back to Delhi that afternoon, after a long silence [Nehru's Cabinet colleague and successor as Prime Minister, Lal Bahadur] Shastriji said to me, half musing to himself: "Gundevia, what would have happened if the Maulana Saheb had declared, at that moment, that the *bal* [hair] wasn't genuine"? "Don't think of it, for God's sake," I said, "it is all over"![111]

In fact, events had just begun to unfold. Nehru, for one, understood the need to "reconsider the basic premise and structure of the Kashmir policy", the Indian state had so far pursued.[112] At an emergency sub-committee meeting of India's Cabinet, he asserted that

> if Kashmir is so destabilized that an ordinary incident of the theft of a relic provokes the people to the extent of trying to overthrow the government, it is time to adopt a new approach and to bring about a revolutionary change in our viewpoint.[113]

This perception was shared by Ghulam Mohammad Sadiq, who had broken ranks with Bakshi by demanding greater democracy. One of the beliefs which had underpinned Indian policy, he argued, "is that the influence of Pakistan on the Kashmiri Muslims is fairly wide and firmly rooted".[114] From this assumption, the government had developed "a primordial fear of the people". The *moe-e-muqaddas* agitation had torn apart the assumption that New Delhi could rule through "interested and self-seeking individuals".

Many of the questions confronting Indian policy-makers during the *moe-e-muqaddas* crisis were to weigh on the minds of officials who faced similar situations twice during the decade of terror, the 1990s. I shall return to these issues later in the course of this book, but it is worth noting that then, as later, it took violence and a collapse of state authority to bring about a reconsideration of policy. Perhaps the most important outcome of the *moe-e-muqaddas* crisis was that it paved the way for the release of Sheikh Abdullah from jail. Nehru, for one, "realized that Sheikh Abdullah remained an important political force . . . [and that] it was necessary to release him to restore public confidence and reach a political accord".[115] Sadiq, who was chosen to succeed the effete Shamsuddin as Chief Minister in the wake of the crisis, supported this line of action, which he saw as part of a general "policy of liberalization".[116]

Nehru, by some accounts, appears never to have been very persuaded by the prosecution's case in the first place, and saw Abdullah's release as the symbolic righting of an historic wrong. Gundevia has described the arrest and prosecution as a "coup", saying Nehru was never "convinced, at any stage, that Sheikh Abdullah was a communalist and was conspiring against India in league with the pro-Pakistan elements in Kashmir".

> [W]hen I had the privilege of working very close to him, as Commonwealth Secretary and, later, as Foreign Secretary, I never heard Jawaharlal Nehru utter one unkind word against Sheikh, not to me and not in my hearing to anyone else,

Gundevia recorded. He asserted that Nehru "never once maligned Sheikh Abdullah and never said one word against him".[117] If Nehru did have doubts about Abdullah's integrity, it would appear that they were ideological rather than criminal in nature.

On April 8, 1964, Sheikh Abdullah became a free man. The Kashmir Conspiracy Case was withdrawn.[118] The prolonged delays had become an embarrassment; one mainstream Indian newspaper had even proclaimed that while "Sheikh Abdullah [was] on trial, India [was] in the Dock".[119] Nath evidently saw little merit in this view, or for that matter in Nehru's conviction's. He wrote, unable to mask his ire at a course of action endorsed by the Chief Minister and Home Minister he served:

> the case was withdrawn by the Government as a measure of normalisation and liberalisation of the State['s] politics. It was hoped that this gesture would divert the attention and energy of a misguided section of people from clandestine and subversive activity to healthy political channels. The manner in which this hope was frustrated by the continued, rather intensified, Pakistani interference in the internal affairs of this State would [soon become] obvious. . . . [120]

Nath, it could be argued, was being unfair. To Nehru, the *moe-e-muqaddas* affair would have shown the limitations of coercive measures. Counter-intelligence personnel in Jammu and Kashmir had succeeded in terminating the covert units active in the 1950s – whether or not Abdullah and Beg were involved in them or not – but this, in itself, did little to bring stability to the state. The release of Sheikh Abdullah at least offered the prospect of some resumption of normal political life, even if it at once contained the potential threat of giving pro-Pakistan forces a platform from which they could pursue their ends. Indian policy-makers have struggled with similar problem in the decades since. No easy answers were available then; none appear evident now.

Towards a new phase

To arrive at an appraisal of the course and impact of the Informal War, two questions need to be answered. First, what were its theoretical and ideological underpinnings? And second, what did its architects hope to achieve?

Major-General Khan's memoirs offer not a little insight into the first of these questions. Pakistan's use of irregular soldiers during the war of 1947–1948 was in essence pragmatic, driven by its lack of conventional military resources and the multiple crises confronting its institutions of state in the wake of Independence. The tribal irregulars dispatched against Srinagar were the sole credible offensive resource available to Pakistan at the time. Pakistani military planners quite clearly understood the limitations of this form of warfare, once India had made the decision to commit its forces to Jammu and Kashmir:

> One Mahsud tribesman aptly described to me their tactics as being like that of the hawk. The hawk flies high in the sky, out of danger; he flies round and round until he sees his prey and then he swoops down on it for one mighty strike and when he has got his prey, he does not wait around, he flies off at once to some far off quiet place where he can enjoy what he has got. The tribesman is indeed very similar – he

> must have mobility, he must have the freedom to choose his own time and target, and he must have security to return to. This is why he is not willing to accept long drawn out actions which tie him down. And clearly, the task that awaited him around Srinagar was just such an action, now that regular Indian troops were arriving.[121]

Once India had consolidated its military position in Jammu and Kashmir, new means were clearly needed if Pakistan's objectives in the state were to be realized. Seen from the point of view of Pakistan's strategic establishment, in particular their concerns about the new country's water resources, it was necessary that pressure be brought to bear upon India, by all means available. Major-General Khan was convinced that, had Pakistan committed greater numbers of troops in 1947–1948, events might have turned out differently. "However", he wrote, "now the time for such action was gone and we could not again break the cease-fire, but the longer we waited the stronger India would become, and so the only course left open for us was to help the people of occupied Kashmir internally with weapons, money and propaganda so that in due course they would be enabled to rise and fight for themselves".[122] The activities of the covert groups of Mohib Ullah Beg, Abdul Rahim or Aziz Parwana were intended to bring about precisely this outcome, drawing on the political capital made available by Sheikh Abdullah's growing anxieties about his own position in Jammu and Kashmir, and that of the state within the Union of India.

Although the covert groups who fought the Informal War had an expressly Islamist character, there was little new or radical in their actual activities. Their operations were, in essence, a small-scale version of the activities of the irregulars unleashed against Srinagar in 1947. All of this was of a piece with the history of sub-conventional warfare. As the historian Eric Hobsbawm has observed,

> until the ultimate phase of guerilla war, when the guerilla force becomes an army, and may actually face and defeat its adversaries in open battle, as at Dien Bien Phu, there is nothing in the purely military pages of Mao, Nguyen Giap, Che Guevara or other manuals of guerilla warfare which a traditional *guerrillero* or band leader would regard as other than simple common sense.[123]

Pakistani strategists, including Major-General Khan, did pay close attention to the doctrines of Mao and Ngyuen Giap, and sought to understand how the lessons of the Vietnam war might be applied to their offensive in Jammu and Kashmir. All this, however, was some distance in the future. For much of the course of the Informal War, Pakistani strategists seem to have been convinced that even a small covert push would serve the purpose of evicting India from Jammu and Kashmir.

Why was this so? Part of the answer might be that Pakistani strategists, like Tucker, saw India, as it emerged from Partition, as a doomed national project, condemned by what they perceived to be the essential characteristics of its people and their principal religious affiliation. The Pakistan Intelligence Bureau's first director, Qurban Ali, elevated this system of belief to an institutional credo that suffused Pakistan's covert services and military. "The Indian masses", Major General Khan wrote, "are ridden by the caste system, superstition, religious intolerance, racial animosities, poverty, malnutrition, physical debility – and the habit of submission and servility".[124] In his reading, this servility was the consequence of the fact that "we", that is Pakistan's Muslims, "have ruled them for eight centuries – matters which cannot be wiped off the memory of the masses overnight". As for India's greater economic resources, and the larger size of its military:

> In the remotest of our villages, the humblest of our people possess a self-confidence and ready willingness to march forward into India – a spirit the equivalent of which cannot be found on the other side. It takes many generations to create such a spirit. In addition, our Frontier tribesmen, no less than 300,000 armed men who have for centuries found India an attractive hunting ground, can still be unleashed against the enemy borders.[125]

Essentialist myths of this kind, as we shall see in subsequent chapters, would be mercilessly exposed: and yet, they would have a remarkable ability to survive adversity, shaping phase after phase of covert combat as well as outright war. Jammu and Kashmir, to the minds of the architects of the Informal War, was just the first prize in a larger war against Hindu India that both history and their efforts would bring to its knees. In some key senses, then, the Informal War that ran from 1947 onwards set the tone for India–Pakistan engagement on Jammu and Kashmir over the coming years, far more so, perhaps, than the first India–Pakistan war. The truth is that Jammu and Kashmir's accession to India was challenged not only politically and diplomatically, although the available literature focuses almost exclusively on these fronts, but also militarily. Jammu and Kashmir was thus a zone of continued warfare – low grade warfare, it is true, but warfare none the less. This is the prism through which policy-makers would have seen political events in the state, events which in other circumstances may have been allowed to play themselves out at leisure.

History, then, must be understood against the backdrop of one stark fact: that both sides perceived their existence as being at stake in Jammu and Kashmir, and were willing to use what means were necessary to secure their survival. Patel's attitude to Sheikh Abdullah, and the slow breakdown in the relationship between Abdullah and Nehru, was shaped not just by his ideological proclivities, but also by the fact that Jammu and Kashmir was at a constant state of war during the period. Many scholarly accounts have been sharply critical of Patel's conduct of India's Jammu and Kashmir policy. Patel may indeed, like most politicians in any time and place, have been guilty of wearing ideological blinkers and of an almost reflexive hostility to those he saw as opponents. None the less, as Union Home Minister, he would on a daily basis have received intelligence reports on the regular acts of sub-conventional warfare carried out by personnel trained by Pakistan, and of the links these personnel had with elements close to Abdullah. It would have been surprising, indeed, had Patel been able to segregate his policy perceptions from this flow of information; indeed, it may have been grossly irresponsible.

As Prime Minister, Nehru would have been in much the same position. As late as 1952, his position was not dissimilar to that taken by Abdullah. Kashmir, the legal scholar A.G. Noorani has pointed out, was still "an open question".[126] In a speech delivered to India's Lok Sabha on June 26, 1952, Nehru argued that:

> We have not got a clean slate to write upon, we are limited, inhibited by our commitments to the United Nations by this, by that. But, nevertheless, the basic fact remains that we have declared and even if we had not declared the fact would remain – that it is the people of Kashmir who must decide. And I say with all respects to our Constitution that it just does not matter what your Constitution says, if the people of Kashmir do not want it, it will not go there . . . Let us suppose there was a proper plebiscite there – and the people of Jammu and Kashmir said, 'We do not want to be with India,' well, we are committed to it, we would accept it. It might pain us, but we would not send an army against them. . . .[127]

What led Nehru to shift his position? We sadly have no archival material on Nehru's responses to what Indian intelligence personnel were telling him about Abdullah's activities. It seems reasonable to speculate, however, that the information gathered by India's counter-intelligence apparatus must have played some part in the Prime Minister's decision to go along with the filing of the Kashmir Conspiracy Case. Had Abdullah's actions been purely political in their significance and impact – had the Informal War not contained within it the potential to create circumstances by which Pakistan could again attempt to take Jammu and Kashmir by force – Nehru's responses to the challenge may well have been different. If nothing else, the existence of the Informal War helps explain the urgency and sense of crisis which permeated decision-making during this period. As with Patel, reports on Pakistan's covert campaign in Jammu and Kashmir would have been on his table each morning. Substantiation of the conclusions of the spies who generated them would not have been hard to come by; both the government apparatus and the newspapers regularly reported on the bomb blasts and arson attacks that marked the progress of the covert campaign.

Paradoxically, then, the first phase of Pakistan's covert war may have failed *because* of its success. Agents were infiltrated, acts of sabotage and terror carried out and politics influenced – and yet, this served only to harden Indian attitudes. Nehru after 1953 would never again consider what Nehru before 1953 had seemed willing to do. Within Jammu and Kashmir, too, there was no sign that covert warfare would tip the balance against India. No uprising had taken place after the arrest of Sheikh Abdullah; even Beg's demand for a lebiscite had done little to destabilize the Bakshi regime.

A more decisive push was needed. Even before Sheikh Abdullah had made his way out of jail, Pakistan's intelligence apparatus had set its mind to considering what could now be done.

3

THE MASTER CELL

> You will infiltrate across CFL [the Cease Fire Line] to op [operate] behind en [enemy] disposn [disposition] in TANGDHAR–TITWAL and cause max [maximum] attrition of en potential . . .
>
> I want you to note that you will not undertake set piece attks [attacks] and tie down [our] own tps [troops] unnec [unnecessarily] thereby suffering cas [casualties]. The pattern must be to conc [concentrate] at a preselected tgt [target] at a fixed time, carry out raids [to] inflict max [maximum] cas [casualties], cause max damage and disperse in different dirs [directions].[1]
>
> Operational Instruction 1 of 1965, dated August 29, 1965, from Major-General Akhtar Hussain Malik, 12 Division Headquarters, to Brigadier Fazle Rahim, MC, Officer Commanding the Pakistani column 'Khilji Force.'

Hayat Mir spent the night at the shrine of Baba Reshi, after crossing the 4000-metre Chor Panjal pass with the Pakistani column moving towards battle. In the morning, he discarded his uniform: his job was to live on to fight after the soldiers had died.

Over the next few weeks, Mir established himself in the mountains above Badgam. A resident of Bandi Abbaspur, from the Poonch area of Pakistan-administered Kashmir, Mir had spent two months before the offensive of August 1965, in special intensive training in the tradecraft of the covert operative. He had been taught how well-guarded enemy positions could be taken by stealth, and how ambushes were laid. He had been taught to cook in the forest without setting off a telltale column of smoke, to watch for signs of surveillance along his sources of water, to set up a system to signal danger to his associates and to live off wild vegetation, if necessary.[2] Now, he was to put that education to use.

Through the first weeks of the 1965 war, Mir established himself as a formidable presence among the peasantry in the rural areas of Badgam. Like similar units elsewhere in Jammu and Kashmir, Mir's cell carried out raids on the relatively affluent, in order to raise cash for distribution among the rural poor. Communal solidarities were manipulated to give such actions legitimacy, a model perhaps for the terrorists who would do much the same from the late 1980s onwards, although at infinitely greater levels of intensity. On August 18, 1965, for example, a covert unit raided the shop of an affluent Arigam grocer, Raj Nath, and looted cloth, provisions and cash for distribution among the poor. Interestingly enough, Nath's watchman, a Muslim, sought to fend off the raid, and was shot for his pains.[3] Mir used faith as the keystone of his efforts to build legitimacy among the peasantry. In one instance, he descended on the village of Wutligam, after hearing tales of the supposed promiscuity of a woman there. He gathered the entire village and condemned the woman for her un-Islamic conduct. In the sight of all, Mir then shot the woman through the head.[4]

As the war progressed, Mir proceeded to make even more of an impression. On August 28, 1965, Mir's group attempted to assassinate a pro-India politician, Ghulam Qadir, in the village of Arizal. Qadir fled, but his shop, like that of Nath, was looted and then set on fire. A day earlier, Mir had organized perhaps the most dramatic covert operation of the 1965 war, an assault on the Badgam police station. After shooting the sentry, the group seized sixteen rifles, and then blew up two bridges along the Badgam–Raithan road to prevent the inevitable reinforcements from interdicting them.[5]

Troops eventually began to move into the Badgam area and, as the Pakistani assault forces began to retreat back across the CFL, the time came for Mir to leave. Armed with false papers identifying him as a Srinagar businessman – and a diary detailing the supposed transactions of the businessman, to good effect – Mir moved to Srinagar.[6] The time had come for him to make contact with a shadowy covert organization known as the Master Cell, and to begin fighting the war he had in fact been sent to wage.

A decade earlier, as we have seen in the Chapter 2, Major-General Akbar Khan and others in Pakistan's strategic establishment had passionately argued for the need for a sustained covert war against India. The Master Cell was the flower of their efforts, and of the lessons learned during the Informal War.

Impasse

Less than two years before the outbreak of the war of 1965, India had released Sheikh Mohammad Abdullah from jail, responding to the storm of protests provoked by the disappearance of the *moe-e-muqaddas,* the holy relic claimed to be a hair of the Prophet. To the minds of Indian policy-makers, the release of Sheikh Abdullah offered the opportunity to break the long-running, and diplomatically damaging, political impasse over Jammu and Kashmir. While the events leading up to the Hazratbal crisis were one element in shaping policy, however, strategic concerns also had an important role in influencing its making.

After India's crushing defeat in its 1962 war with China, efforts to find a permanent solution to the India–Pakistan conflict had gathered importance in the agendas of the two major Western powers, the United Kingdom and the United States of America. Fought with varying degrees of intensity between September and November 21, 1962, the war had been, in the words of an official history, "a debacle" for India.[7] It also rang alarm bells in the West. Both the United States and the United Kingdom saw a united response by India and Pakistan as imperative if communist China was to be contained. The key to enabling such a response, quite obviously, was a resolution of the conflict in Jammu and Kashmir. At the end of November 1962, the United Kingdom's Secretary of State for Commonwealth Relations, Duncan Sandys, and the United States' Assistant Secretary of State for Far Eastern Affairs, Averell Harriman, were dispatched to South Asia to determine what military assistance India needed, and to reassure Pakistan that these supplies would not be used to its detriment.[8]

Sandys and Harriman were able to use India's desperate position to push it into a dialogue with Pakistan. Lord Louis Mountbatten travelled with Sandys on this mission, hoping to persuade Nehru to agree to a demilitarized and independent Jammu and Kashmir. India's Cabinet, however, shot down the idea. The diplomats did, however, succeed in pushing Nehru and the Pakistani military dictator General Ayub Khan, who had taken power in 1958, to the negotiating table. What happened during their discussions laid the foundations for disputation on Jammu and Kashmir for several decades. India suggested that the CFL demarcated at the end of the war of 1947–1948 become a permanent border. Pakistan, in turn, proposed that a boundary be drawn far to the east of the CFL, leaving India with only the Hindu-majority parts of the province of Jammu, which fell in Indian-held territory.

After two rounds of talks, an event took place which, for all practical purposes, destroyed the prospects of an accord. Zulfikar Ali Bhutto, who had taken over as Pakistan's Foreign Minister, travelled to Beijing in March 1963. There, he ceded to China several thousand square kilometres of the territories of the pre-Independence state of Jammu and Kashmir. Despite Indian anger over the affair, the talks dragged on until May 1963, when the two sides issued a communiqué regretting that they had been unable to arrive at an agreement. By October 1963, Pakistan had taken the issue back to the United Nations Security Council. In the spring of 1964, that body was to discuss the question of Jammu and Kashmir for the 110th time in 15 years. At that meeting, Bhutto made it clear that Pakistan was in no mood to accept the status quo, and asserted that it would be prepared to discuss the issue a thousand times.[9]

While the discussions continued, however, the managers of Pakistan's military and intelligence services had other plans up their sleeve. India's humiliating defeat at the hands of China, they believed, had opened the door of opportunity for another attempt to achieve Pakistan's objectives in Jammu and Kashmir by force. Another war was just a matter of time.

The origins of the Master Cell

Rehmat Ullah Khan had struggled since childhood with economic hardship and personal tragedy. His covert life was his one real chance to have found greatness. Had Pakistan won the war of 1965, his name may well have figured in history books instead of intelligence dossiers buried deep inside Indian intelligence archives.

Khan was born in the city of Jammu in 1925, the son of a contractor who died when the boy was just six months old. He was raised by his mother and, in 1937, joined his brother's cement concern in Lahore. On the eve of Partition, in 1945, he set up his own small business in Calcutta. While his business flourished, Khan seemed to have something of a self-destructive streak. In December 1948, in the wake of the violent Hindu–Muslim confrontation that tore apart the city, the young businessman was arrested for robbing a postman at gunpoint. He was tried, convicted and sentenced to eight years imprisonment.[10]

Incredibly, despite spending six years in jail – Khan was released in April 1954 – the young man was able to rebuild his life in short order. He was able to join the business of one of his other brothers, who worked as a Public Works contractor in Srinagar, and within a year had saved up enough to set up an independent concern operating from the Old Pataudi House in Darya Ganj, a commercial centre in old Delhi. In less than a decade, the enterprise had flourished. Khan employed over 100 people, and "was regarded as a prosperous and respectable citizen of his locality".[11]

Yet, the jail time tore apart other important aspects of Khan's life. Shortly before Partition, in April 1947, he had been betrothed to a young woman, Kaneez Fatima. Their wedding took longer than anticipated to arrange, and Khan soon went to jail. Soon after Partition, her family moved to Pakistan. Amazingly, Kaneez Fatima waited for her would-be husband for seven years. Khan struggled hard to arrange for an Indian visa for Fatima, and finally succeeded in January 1955. But the years apart had evidently taken their toll, for the couple quarrelled to such an extent that a separation took place within months of the marriage.

Khan remarried in 1962. It was shortly after this second marriage that he made the fateful contact who would lead him to a covert life. Mohammad Amin Beg, ostensibly responsible for the maintenance of the Pakistan High Commission's vehicle fleet but suspected by Indian authorities of being an intelligence operative, was married to one of Khan's cousins, Mahmooda Begum. The two became good friends; Nath's report describes their relationship as "quite intimate".[12] Soon after, Beg introduced him to another Embassy official, Raja Ashraf. Ashraf, in turn, asked for a favour. He requested that Khan contact Ghulam Mohiuddin Kara, a key member of the Plebiscite Front and a close associate of Sheikh Abdullah, and ask him about the organization's future course of action. Ashraf also asked Khan to carry instructions to Mubarik Shah, a member of the Sheikh's Legal Defence Committee, to visit the High Commission. Ashraf insisted on paying Rs. 50 – then a not inconsiderable amount of money – to Khan for his troubles.

Surendra Nath had no uncertainty about the purpose of the transaction:

> It is significant that in this first mission, the Pakistani masters gave this new agent tasks which were extremely simple and, on the face of it, innocuous. . . . Rehmat Ullah Khan could very well have carried out this mission without accepting any payment and, in fact, he refused it in the first instance. An Intelligence Agency, however, feels on safer ground when the agent employed by it accepts payment, even though it may only be a token, for two reasons. Firstly, it demolishes, to some extent, the self-respect of the agent in his relationship with his employers and to that extent he becomes more pliable. Secondly, it creates and whets an appetite for easy money which can later be exploited for getting the agent to undertake more difficult missions.[13]

Perhaps without quite understanding what had happened, Rehmat Ullah Khan had become a Pakistani spy. Even as the world attempted to find a peaceable resolution to the conflict in Jammu and Kashmir, Pakistan was laying the foundations for a new round of covert warfare to keep pressure on India.

Hopes of peace, clouds of war

Hope was in the air in the spring of 1964. Pakistan's covert campaigns of the past decade-and-a-half had yielded few results. Even Sheikh Mohammad Abdullah's prolonged incarceration had not led to the kind of mass rebellion upon which Pakistan's intelligence services had predicated their actions. Prime Minister Jawaharlal Nehru sensed the possibility of using the changed circumstances to break the impasse with Pakistan.

Six days before the Kashmir Conspiracy Case was withdrawn, and Sheikh Mohammad Abdullah was finally released from jail, Nehru invited him to New Delhi. "Yesterday's criminal conspirator", wrote the Indian bureaucrat Y.D. Gundevia, wryly, "stayed in the Prime Minister's house, as the Prime Minister's guest". During his conversations with Abdullah, Gundevia found no "sign of bitterness against any one in India over the treatment meted out to him and the long years of imprisonment". The Sher-i-Kashmir seemed concerned only about the "communal situation and he kept reverting to the need to bring about some understanding with Pakistan".[14] Nehru, for his part, had evidently put aside the deep mistrust the Indian Ministry of Home Affairs had of Abdullah, although he continued to be worried about the political stand his friend and comrade had adopted. "His attitude", wrote Nehru's biographer S. Gopal, "was a blend of guilt at having allowed him to have been kept so long in detention, and of concern at the consequences of his activities".[15]

Whatever Nehru's concerns, he was willing to see where Abdullah's efforts would head. In May 1964, Abdullah left for his first – and only – visit to Pakistan, hoping to persuade its military ruler, Field Marshall Ayub Khan, to open negotiations with Nehru. If Pakistan harboured some suspicions about just why India had allowed someone it until the other day had charged with being its agent to arrive as an envoy of the Prime Minister, it did not allow the sentiment to show. A crowd estimated at 500,000 greeted Abdullah in Rawalpindi, and one commentator hailed him as the "leopard of Kashmir, who had finally changed his spots".[16]

By all accounts, Abdullah and Ayub found no meeting ground. In his memoirs, Ayub charged Abdullah and Mirza Afzal Beg with bringing him an "absurd proposal of confederation between India, Pakistan and Kashmir". "I told him plainly", Ayub Khan wrote, "that we would have nothing to do with it. It was curious that whereas we were seeking the salvation of Kashmiris, they had been forced to mention an idea which, if pursued, would lead to our enslavement."[17] Ayub's ire was provoked by his belief that a "confederal arrangement would undo the Partition and place the Hindu majority in a dominant and decisive position".[18] In a letter to Ayub written in response to the General's memoirs, Abdullah said his advice to the Field Marshall had been "not to reject any proposal outright, but to discuss its pros and cons in a friendly manner across the conference table and convince the other side that a particular solution would not lead to ultimate peace".[19]

Ayub did agree, however, to meet Nehru. The meeting never took place. On May 27, 1964, India's first Prime Minister passed away. Abdullah and Pakistan's Foreign Minister Zulfikar Ali Bhutto – who would go on to be President of his country – flew back to New Delhi together. On the flight back, according to the account of the scholar Stanley Wolpert, Abdullah dramatically changed tack. He suggested that Bhutto press India to hold a plebiscite in all of Jammu and Kashmir, but then settle for a partition of the state along the Chenab River.[20] The Chenab Plan would essentially have divided the Hindu- and Buddhist-majority regions of Jammu and Kashmir from its Muslim-majority areas. This would, in essence, have replicated the logic of the Partition of India within Jammu and Kashmir. First proposed by the United Nations mediator on Jammu and Kashmir, and then advocated by Ayub Khan, the partition idea had been flatly rejected by Nehru as an "Alice in Wonderland of vague proposals".[21]

In retrospect, it is somewhat incredible that Abdullah would have pushed for ideas so repugnant to Nehru on the day of his death – ideas, moreover, which ran in the face of secularism and the repudiation of the two-nation theory that both men seemed to share. The fact is, though, that many within the National Conference backed the idea. Even Ghulam Mohammad Bakshi had supported an ethnic-religious partition in 1953, arguing it was "the only practicable, advantageous and honourable solution of the dispute".[22] Since then, however, few within Jammu and Kashmir had pushed the partition plan. Bhutto, by Wolpert's account, was elated by the position Abdullah adopted only days after proposing the idea of confederation.

Over the coming decades, the idea of an ethnic-religious partition in Jammu and Kashmir was to steadily gather momentum until, in October 2004, it found the express and public endorsement of yet another military ruler in Pakistan, General Pervez Musharraf. For the moment, however, Bhutto's glee at Abdullah's willingness to break ranks with Nehru's vision was underpinned by hope: the hope that at least some forces within Jammu and Kashmir would participate in the war Pakistan was about to initiate. Pakistan's intelligence services were convinced that conditions were ripe for a mass uprising in the state.[23] One credible insider account by Lieutenant-General Gul Hasan Khan suggests several in Pakistan's Kashmir-policy establishment were persuaded by claims that Abdullah, and several others in Jammu and Kashmir, would support a Pakistani military adventure:

> I had raised the question of sounding the leaders on both sides of the Line [of Control] in one of our Cell meetings. The reply I received from the Chairman, [Foreign Secretary] Aziz Ahmad, was that they would not be averse to our intended enterprise. And then followed a long list of names of Kashmiri leaders who were well-disposed towards Pakistan. And in this connection, the visit of Sheikh Abdullah in the recent past to Pakistan was mentioned. I was not aware what this visit had been about, much less what assurances were given by him to our Government, if any. My question was not answered even in part, but as everyone else seemed content with the Chairman's statement, I decided not to press the issue further. I thought perhaps some steps had been taken in this respect but that these were of no concern to me.[24]

Based in part on this perception, Lieutenant-General Khan has noted in his memoirs, the Government of Pakistan "decided to extend some form of moral support to the people of Indian-held Kashmir". On precisely what this "moral support" was, General Khan, unlike legions of official, media and academic propagandists, has left little doubt. "Consequently", he wrote, "the Army was ordered to train volunteers in carrying out sabotage activities across the Cease-fire Line".[25] There were to be two elements to the plan. In the long term, the Pakistan Army would "train guerrillas and induct them across the CFL, with a view to disrupting conditions in the Valley and eventually arming the locals and helping them rise against the Indian Army of occupation".[26] Before this, however, another task had to be undertaken. Pakistan's Army was told that the country's President wished it to prepare individuals bring about an escalation of the covert war: an "intensification of the firecracker type of activity that was already current", General Khan called it.[27]

The cells develop

When he was finally arrested at the age of 32, Mian Ghulam Sarwar was, in Nath's words,

> found to possess a highly romantic disposition, and was carrying on with three different girls. His taste in girls was fairly cosmopolitan, in

> as much as one of these girls was Muslim, the other Hindu and third Buddhist. His tastes were expensive, and his living extravagant. With all these traits in his character, it was perhaps inevitable that he should ultimately find himself in the twilight world of espionage.[28]

Unlike most others involved in the growing covert circle around Rehmat Ullah Khan, Mian Ghulam Sarwar had little apparent interest in politics. The son of a wealthy Srinagar family – his father Mian Ghulam Mohammad had served the Jammu and Kashmir Government in various senior capacities, and retired as Chairman of the State Public Service Commission – Sarwar had studied to be a veterinarian at Patna, in Bihar, but failed to complete his degree because of an illness. He began working in the Government Tourist Bureau in 1956, but the meagre salary failed to meet his somewhat extravagant lifestyle. It is unclear from the available material just why Sarwar decided to be a spy, but it appears that he made a conscious decision to acquire skills he thought would be useful to a covert life, notably photography.

In May 1964 – at around the time of Sheikh Abdullah's visit to Pakistan – Sarwar met a Visa Officer at the Pakistan High Commission, armed with photographs of the Kashmiri leader that he suggested be sent to Pakistan for publication. A conversation ensued. "It did not", Nath records, "take the Visa Officer long to understand that this approach implied a willingness to work for Pakistan".[29] Sarwar was passed on to Raja Mohammad Arshad, the First Secretary of the Pakistan High Commission in New Delhi. At this stage, Arshad made no espionage-related offer. He suggested that the negatives of Sarwar's pictures be passed on to an individual named "Afzal" who, Arshad said, would contact the new prospect in Srinagar.

"Afzal" was, in fact, Rehmat Ullah Khan. Khan first met Sarwar in the last week of June 1964, at what was eventually to become the covert cell's safe-house. Khan handed over the prints Sarwar had left behind at the Pakistan High Commission. The two discussed how best they could make contact with Sheikh Abdullah, and evidently decided on making a direct approach. The next day, Khan met Abdullah at the National Conference's headquarters. He had to wait some time, as Abdullah was meeting with Mridula Sarabhai, the Sheikh's most energetic supporter and sponsor outside of Jammu and Kashmir.[30] When Khan was finally ushered into Sheikh Abdullah's room, he introduced himself as a representative of the Pakistan High Commission's First Secretary. Khan asked about Abdullah's course of action, and what line the Sheikh would adopt on a forthcoming visit to New Delhi.

According to Nath's account, Abdullah gave a far from clear answer. "For the present", the *Report on Pakistan-Organized Sabotage* records the Sheikh as saying, "tell the Pakistan High Commission that they should not worry because we have a common God, a common Prophet, a common Religion, and a common Holy Book".[31] He evidently refused to elaborate on just what this somewhat cryptic remark might mean, but Khan's handlers may have drawn some comfort from the formulation. To those inclined to believe that conditions existed for a rebellion in Jammu and Kashmir, Abdullah's remark may have seemed to suggest he would back an uprising. A few days later, on July 7, 1964, Sarwar received a telegram from "Afzal" asking him to report to the Pakistan High Commission in New Delhi. At the meeting that followed, he was told to remain in touch with a radical youth organization known as the Students and Youth League, so as to build his influence with it and assess the potential of its cadre for covert work.

The SYL recruits

In March 1964, enthused by the mass discontent seemingly demonstrated by the Hazratbal crisis, radical elements at the edges of the National Conference set up the Students and Youth League to push the case for separation from India more aggressively than their parent organization. Although the SYL was a legitimate overground organization, several members of its inner core believed political activity in itself would be inadequate to free Jammu and Kashmir from Indian rule.

Branches of the SYL were set up in most of the Kashmir valley's main towns, as well as in the district of Poonch. The organization's cadre often held student protests against Government policies, and organized sit-ins in front of the offices of the United Nations Military Observer Group to draw attention to India's refusal to hold a plebiscite. Then, on April 8, 1964, just a month after the SYL was formed, Sheikh Mohammad Abdullah, Mirza Afzal Beg and others accused in the Kashmir Conspiracy Case were released from jail. This opened up enormous political space in Jammu and Kashmir, creating a climate in which the SYL had more leverage than a small student organization might otherwise have possessed.

Little hard evidence ever emerged that the SYL had a formal role in the covert campaign, but three of its members played a key role in subsequent events. Ashraf Batku, like Sarwar, came from an influential family; his father, Mohammad Sultan, served as Deputy Secretary to the Government of Jammu and Kashmir. Batku helped found the SYL, along with Zafar-ul-Islam, a close college associate and, again, the son of a senior Jammu and Kashmir Government bureaucrat. The third member of the SYL-affiliated covert unit was Bashir Ahmad Kitchloo, a son of an affluent family with substantial business interests in the district of Doda. Kitchloo had long supported Sheikh Abdullah, and had been arrested in 1958 for organizing protests against the leader's arrest.[32]

Batku had not been arrested when Nath wrote the *Report*, but interrogations of his associates suggest he made contact with Raja Ashraf, the Pakistani diplomat in New Delhi, before May 1964. He had visited Pakistan in 1962 and 1963 on personal business, and may have been recruited there. In June 1965, on the eve of war, Kitchloo again made a trip to New Delhi, this time to ask Ashraf for funds for the SYL's agitational programme. He was asked to return to New Delhi with Zafar-ul-Islam and Batku for further discussions on July 26, 1965.[33]

Four days earlier, Rehmat Ullah Khan met Ashraf near Delhi Gate, in the old city, a crowded business area. He was assigned his most important – and last – covert mission. He was to meet Abdullah's associate, Ghulam Mohiuddin Kara, as well as the Srinagar cleric Mirwaiz Mohammad Farooq, with messages of support – and hard cash. Kara was offered the Pakistan High Commission's sympathies for an incident of police brutality which had led to the loss of the politician's teeth, as well as a gift of Rs. 5000.[34] Maulvi Farooq received another Rs. 5000, and was asked to arrange a massive procession on August 9, in defiance of the police if necessary. Sarwar held a similar meeting with Kara at the Ahdoos Hotel in downtown Srinagar, and again asked for identical demonstrations to be held on the same date.

This was a date of obvious significance: Operation Gibraltar, the Pakistani offensive of 1965, commenced on August 8, and large-scale demonstrations would have given perfect cover to the irregulars who spearheaded the campaign in Jammu and Kashmir. Meanwhile, the SYL members made plans to send Kitchloo to Pakistan to learn how to use a wireless transmitter. The several pieces of the covert enterprise were now in place.

The Master Cell and its subsidiaries

Mian Ghulam Sarwar and his three key colleagues – Ashraf Batku, Bashir Ahmed Kitchloo, and Zafar-ul-Islam – set up a central organization to supervise their covert campaign against Indian rule in Jammu and Kashmir. To distinguish its hierarchical position from the string of other cells the unit would control, it was called the Master Cell.

The members of the Master Cell met on the first floor of Sarwar's old-fashioned wood house, just off the Budshah Bridge in Srinagar – almost directly opposite the contemporary location of the Police Headquarters in Srinagar. Arms and explosives were kept in a specially constructed shelter in the basement. Sarwar, the Jammu and Kashmir Police later recorded, changed his lifestyle radically. He "adopted a completely self-contained mode of life. He employed no servants, cooked no meals, and did not encourage even his relations living in Srinagar to visit him."[35]

Over time, the Master Cell spawned several subsidiary cells, the existence of each of which was known only to its own small circle of members. Five of these seem to have been particularly active:

- The Students' Cell: Charged with organizing strikes in colleges and demonstrations.
- The Poster Cells: Divided into sub-sections I and II, recruited staff at Government offices to print and issue posters on behalf of the Revolutionary Council, using typewriters and reproduction equipment there. One Poster Cell was run from the Power Project Generation Division's offices; the other from the Cultural Academy.

- The Narwara Cell: Named for a neighbourhood in Srinagar, the Narwara Cell was led by Mohammad Hussain Wazir, an employee of the Public Works Department. The nine-member cell included two other Department employees, a government architect and three teachers. Wazir, who received training in using hand grenades, instructed other cadre in the use of weapons.
- The Buchwara Cell: This six-member cell, like the Narwara Cell, was specific to a neighbourhood in Srinagar, in this case a particularly sensitive one, since it was home to the Chief Minister. During the 1965 war, the Buchwara Cell guided Pakistani irregular forces to the Government Cement Factory in Wuyan and an Army fuel dump at Khunmoh. The Buchwara Cell also ferried weapons intended for the use of the Master Cell.
- The Infiltrator Liaison Cell: Unlike other cells, the Infiltrator Liaison Cell had no operational role. It was meant to facilitate the working of stay-back agents like Hayat Mir, all of them residents of Pakistan-administered Kashmir, who remained on in Jammu and Kashmir in the event of the defeat of Pakistani irregular forces.[36]

The social composition of the Master Cell offers some insight into the class character of the organization. One member, Abdul Hamid Khan, was a junior technician at the Government Medical College in Srinagar; another, Mohammad Yusuf Butt, was an Assistant Engineer in the Electrical Department; while Ali Mohammad Malik and Mehboob Husain were both students of the Regional Engineering College in Srinagar. Broadly, this pattern is mirrored through the overall composition of the cells. Of a total of 44 residents of Indian-administered Jammu and Kashmir identified as cell members, the largest single category, 20, were government employees or contractors. The second largest category were students, of whom there were 19. Only five cell members had independent businesses or professions. All but three were from Srinagar or its immediate vicinity; none was a peasant.[37]

One thoughtful analysis of political contestation in Jammu and Kashmir has suggested that the National Conference's successful efforts to widen education may have created a class whose aspirations it could not meet. The scholar Sumit Ganguly has pointed to a dramatic growth in education in Jammu and Kashmir from the 1960s onwards, a process that took place in a growing economy and which at once found expression in a dramatic escalation in newspaper circulation.[38] In Ganguly's view:

> Young Kashmiris acquired a modicum of education and became aware of improved economic prospects. They were no longer content to seek employment in the traditional sectors of the economy, namely the handicraft industry or the tourist trade. The National Conference did little to expand employment growth in new sectors of the economy. Inevitably a reservoir of discontent among the quasi-educated and largely prospectless youth of the Kashmir valley.[39]

It would be misleading, of course, to draw a causal relationship between the actions of the members of the covert Cells and the economic climate of the time. Yet, the limited data available on the composition of the 1964–1965 covert Cells certainly appears to affirm Ganguly's description. The membership of the Master Cell seemed to represent a new, literate urban class that had acquired some degree of affluence, but was yet to find political representation. Soon, the young members of the covert Cells would assert their political presence, in the most dramatic way possible.

Operation Gibraltar

Early in 1965, it would have been evident to the covert cadre of the Master Cell and its subsidiaries that the time for action was nearing: the thunder of guns had resumed both along the CFL in Jammu and Kashmir and at the southern-most fringes of the India–Pakistan border, along the Indian Ocean.

In the spring, Pakistani regular forces initiated a series of limited border incursions in the Rann of Kutch, a massive sprawl of desert salt marshes on the southern end of the border with India. India lost the army post of Biar Bet in these early clashes, leading to several other Pakistani assaults against forward positions. Firing and shelling along the CFL in Jammu and Kashmir also escalated dramatically. From January to May 1965, India recorded 1347 cease-fire violations by Pakistan, compared to 522 during the same period of the previous year. India responded, in mid-May, by opening a new front. Its chosen theatre of operations was the Kargil sector in Jammu and Kashmir, where Indian troops vacated a small Pakistani encroachment and then reoccupied the strategically important Kala Pahar area.[40]

Although these early skirmishes provoked full-scale troop mobilizations by both India and Pakistan, ceasefire negotiations were initiated behind the scenes. Active offensive actions ceased in May, although heightened patrolling by both sides continued. India also placed on alert its formations in Punjab – the launching pad for a full-scale armored offensive deep into Pakistan. Nevertheless, a formal ceasefire agreement was signed on June 30, 1965, mediated by the British High Commissioner in New Delhi in the name of Prime Minister Harold Wilson. The agreement resulted in troop withdrawals on both sides. Pakistan, however, thought it had learned a valuable military lesson from the Kutch operations. Pakistani military officials now believed that India's armed forces, enfeebled by their defeat at China's hands in 1962, were in no position to resist aggression. The notion dovetailed neatly with Ayub Khan's communal assumption that Hindus, in general, had no stomach for a fight.[41]

Operation Gibraltar took its name from one of the most valourized battles in the history of Islam. In AD 711, the Moorish general Tariq ibn-Ziyad launched his invasion of Spain from the Mediterranean island. What is now known as the Rock of Gibraltar was called the Jabal Tariq, or the Mount of Tariq, in honor of the general and his campaign. Legend has it that Tariq burnt his boats upon arrival in Spain, ensuring that his troops had no means of retreat. Fittingly, the Pakistani military columns that were about to invade Jammu and Kashmir also drew their names from those who had come to be represented as heroes in a battle between Islam and unbelief. One column was named Salahuddin, after the hero who had taken Jerusalem from the crusaders; another Ghaznavi, for Mahmud of Ghazni, who invaded India seventeen times, annexing the Punjab and destroying the temple of Somnath in Gujarat.

Nath's *Report on Pakistan-Organized Subversion* offers one of the few detailed accounts of the precise operational tactics and structure of the Pakistani covert groups who spearheaded Operation Gibraltar. In general, Pakistani strategists seem to have made considerable effort to ensure deniability. In essence, the bulk of each company – 120 men or so – was comprised of Razakars and Mujahideen, especially trained volunteers drawn from the Pakistani-administered areas of Jammu and Kashmir. Officers and a component of men from two paramilitary formations, the Azad Kashmir Rifles and the Northern Light Infantry, as well as a small number of Special Services Group commandos, made up the rest of the company. Groups of four to six companies were combined into units under the command of a Major-rank officer.[42]

Under the overall command of Major-General Akhtar Husain Malik, the General Officer Commanding of the Pakistan's Army's XII Division, the troops who were to be deployed in Operation Gibraltar trained at four major centres: Sinkiari, Mangbajri, Dungi and Sakesar, all in Pakistan-administered Jammu and Kashmir. Additional training was also carried out by personnel of the Azad Kashmir Rifles at Nikial, Khuiratta, Darman, Tarkundi, Bohri Mahal, Pir Kalanjar, Hajira, Kotli and Bher. Interestingly, many of these locations would be used to train cadre for the jihad of 1989. By Indian intelligence accounts, Chinese instructors were involved in some parts of the training, although it is unclear whether this collaboration was part of general military cooperation or specific to Operation Gibraltar.[43] Much of the training was in irregular combat techniques and sabotage.

On August 5, 1965, a force of some 30,000 Pakistani infiltrators was unleashed across the CFL. Their objectives were to establish bases, carry out acts of sabotage and create conditions that would be conducive to the intervention of regular Pakistani troops. The first assault column, made up of about 1,500 infiltrators, was expected to encourage a local rebellion. Ironically enough, its presence was betrayed, and that by one of the supposedly insurrectionary residents of the Kashmir valley. Mohammad Din, a Gujjar shepherd, informed Indian police authorities that infiltrators, subsequently identified as the Salahuddin column, had offered him bribes in return for guiding them through the Pir Panjal range.[44] My interviews with intelligence personnel serving at the time suggest that Mohammad Din's information was passed by the Jammu and Kashmir Criminal Investigation Department head, Pir Ghulam Hassan Shah, to the state's home minister, D.P. Dhar. Dhar, in turn, alerted military authorities, who proved skeptical that Pakistani irregulars had penetrated their defences until this information was verified by contact. More important than his role as an informant, though, is that Mohammad Din's actions turned out to be illustrative of the sentiments of most Jammu and Kashmir residents. The rebellion upon which General Malik's dream was premised on never materialized, forcing him to commit additional troops. Between 5000 and 6000 additional forces entered Jammu and Kashmir in the third week of August. A third wave of equal proportions followed in early September.[45] It proved of little use: bar a small faction of the Plebiscite Front, no group proved willing to support the irregulars, and those who survived the fighting with Indian troops were tracked down by the police with the aid of local political activists.

As Operational Gibraltar began to bog down, some in Pakistan's military believed the time had come to wind down the operation and accept failure gracefully. Bhutto, however, insisted that this course of action would be a disaster, one that could "threaten the existence of Pakistan".[46] While Bhutto's claim was clearly theatrical, he succeeded in pushing through plans for an aggressive escalation of operations directed at cutting off Jammu and Kashmir from the rest of India. Ayub Khan seems to have gone along with this assessment despite the risk of massive Indian retaliation.[47] On August 29, 1965, Pakistan's army chief, General Mohammad Musa received secret orders to initiate the attack, orders that included the observation that as "a general rule, Hindu morale would not stand for more than a couple of hard blows delivered at the right place and the right time".[48]

Operation Grand Slam, as this second offensive was called, began on September 1. As it turned out, it was initiated at both the wrong place and the wrong time. Although Indian troops were taken by surprise, Pakistani advances in Chhamb and Jaurian were successfully checked. By September 5, Indian troops were able to launch a retaliatory offensive to the south in Punjab, attacking along the Icchogil Canal, the bridge across the Ravi River at Dera Baba Nanak, and the Khem Karan-Kasur sector. Strategists in Pakistan had evidently believed that the fighting would remain confined to Jammu and Kashmir, and thus failed anticipate an assault in Punjab. Despite its lack of preparation, however, Pakistan was in turn able to check the Indian advance. Its counterattacks forced some Indian divisions to fall back – the Dera Baba Nanak Bridge and the Khem Karan pocket were lost – but when hostilities ended, India's XI Corps had captured approximately 362 square kilometres of Pakistani territory, and had yielded only 51 square kilometres.[49] Even the Pakistani territorial gains at Khem Karan turned out to be pyrrhic. On September 11, India opened its dam floodgates, trapping some 100 Pakistani tanks.

The successes of India's XI Corps were mirrored by even more decisive triumphs further south. The new I Corps, tasked with combat in the Sialkot region, succeeded in capturing almost 500 square kilometres of Pakistani territory, in the process destroying 144 tanks and killing 693 troops. By September 8, Indian troops were also able to open a new front in Rajasthan, to tie down Pakistani units in Sindh and keep reinforcements from heading north. A war of thrust and parry followed. While much of the fighting was of little overall significance, Pakistan was unable to hold significant amounts of desert territory, imperative to giving it strategic depth in Sindh.[50]

What had gone wrong? The scholar Victoria Schofield has offered four cogent explanations. For one, General Malik's core assumption, that the presence of irregular troops in Jammu and Kashmir would spark off a rebellion, had been proved wrong. Second, Pakistan created problems within its own chain of command by sacking General Malik immediately after initiating Operation Grand Slam. The officer was replaced by Major-General Agha Mohammad Yahya Khan – the military ruler who would preside over the next India–Pakistan war – on the ground that General Malik's will had been broken by the failure of Operation Gibraltar. Third, Pakistan misread India's defensive posture in the Chhamb–Jaurian region, failing to understand that troops who were withdrawing were in fact regrouping in defensive positions. Fourth, by violating the accepted India–Pakistan international border in the Suchetgarh area while initiating Operation Grand Slam, Pakistan provided the opportunity for India to use its superior forces in the vital Punjab theatre.[51] Pakistani military accounts differ from this interpretation; General Gul Khan, for example, has attributed the reversal to delays in initiating Grand Slam.[52]

Ayub Khan now attempted to play one last, desperate card. From September 4, days after Operation Grand Slam commenced, China had issued a series of statements in support of Pakistan. On September 16, with Indian troops in control of the pace and structure of the 1965 War, China issued a dramatic ultimatum, warning against what it claimed were Indian raids launched from the state of Sikkim. The ultimatum provoked Prime Minister Harold Wilson to promise India the support of both the United Kingdom and the United States if China intervened. What had begun as a desert skirmish was threatening to turn into a global war. Nonetheless, on the night of September 19, Ayub Khan made a visit to Beijing, to confer with Premier Chou En-Lai. He received a promise of Chinese help – but also a warning that the war could be prolonged, and lead to the loss of major Pakistani cities. Pakistan's President, aware that his forces were in no state to fight a protracted war, left Beijing without taking up the offer of assistance.[53]

Days later, India and Pakistan entered into a ceasefire. Within Jammu and Kashmir, however, another war was still going on.

The covert campaign

"*Khabardar*!" [Beware], begins a poster issued by the Inqilabi [Revolutionary] Council on August 28, 1965. "The history of the world shows that only those who have paid the price of their freedom have attained salvation."[54]

"Thirty four years ago", the Revolutionary Council asserted,

> we started the war that continues until today. We have sacrificed a lot for our beliefs, and have colored every corner of this land red with our blood. Today, our freedom struggle has entered your neighborhood. In order to emerge from this successfully, we have to be prepared to sacrifice everything. If today we allow laziness and lack of courage to come near us, we will never be free. Nor will we have any right to live.

"Stay cautious of traitors", the poster warned, "for they will soon be dispatched to hell". It ended with a verse from the Quran: "have courage and you will have victory".

Two weeks after Pakistan launched its 1965 campaign – Operation Gibraltar commenced on August 4 – posters of this kind, invariably calling on the people of Jammu and Kashmir to overthrow the government, began appearing on the streets of Srinagar. They were authored by the Revolutionary Council, a front organization of the Master Cell that had proclaimed itself the representative of an insurrection in Jammu and Kashmir. The posters were, in general, put up at night. Local authorities had no idea just who was defying the wartime curfew to wage this propaganda campaign.

Within days, the Master Cell demonstrated that its competence extended beyond the literary skills of its members. On August 29, in the midst of war, students at several Srinagar colleges went on strike. The students, the Jammu and Kashmir Police learned, planned to hold a protest in front of the United Nations Military Observer Group [UMOGIP] offices in Srinagar and to deliver a memorandum claiming that the people of Kashmir supported the Pakistan-backed infiltrators. In the event, police dispersed the demonstration well before it reached the UNMOGIP offices, and a plan by four members of the Students' Cell to ignite Molotov cocktails during the demonstration had to be dropped. However, a grenade – later found to bear Pakistan Ordinance Factory markings – went off at the crowded Regal Chowk shopping area later that afternoon, and shops were set on fire along the Hotel Road after dark.[55]

Through September, the terror campaign in the Kashmir valley gathered momentum. On September 6, students of the Government Medical College – an institution from which the Master Cell had drawn several recruits – went on strike. At 2:30PM, someone within the crowd threw a grenade at police personnel posted for the purpose of preventing violence. Two officers and four spectators were injured. Less than a week later, on September 11, another grenade was thrown into the middle of the crowd at Lal Chowk, Srinagar's commercial hub. That night, efforts were made to set the Fateh Kadal bridge in downtown Srinagar on fire, along with the Indian National Congress' provincial office in Maisuma, an old-city neighbourhood.[56]

More attacks followed as the war progressed. A grenade was thrown at a police picket in Maisuma, the same neighbourhood where the arson attack on the

Congress office was executed, on September 12. This grenade also bore Pakistan Ordinance Factory markings. The same night, the Syed Mansoor Bridge was set on fire, and a portion destroyed before the blaze could be extinguished. There was then a nine-day lull. But on September 21, yet another hand grenade was thrown at the police guard on the Nowpora bridge. An identical attack took place at the Dalgate crossing on the morning of September 30, on a route traversed by the Chief Minister at least twice a day. There were several other cases of arson during this period. A row of shops in Dalgate was set on fire, as was the Vasanta High School. The most serious arson attack was on the Old Secretariat Building, which then housed a number of Government offices as well as both Houses of the State Legislature. The Legislature buildings were saved, but many offices were gutted.

Hayat Mir, who had made contact with Sarwar soon after his departure from Badgam, was claimed by Indian counter-intelligence to have been involved in several of these attacks. Nath credited three major incidents, the Maisuma grenade explosion, the Regal Chowk fire and the destruction of the Old Secretariat Building, to Mir. “After committing these acts”, Nath wrote, “he would tell the young men who had joined him accurate details of each incident so as to impress them, enthuse them, and inspire a greater degree of admiration and loyalty towards him”. Abdul Majid and Ahmad Yunus, two Pakistani operatives who had made contact with the Master Cell, were tasked to instruct new recruits in the use of hand grenades.[57] A fourth agent, Gul Zamaan, made contact with the Master Cell later, leaving his rural hideout in the Harwan area only with the onset of winter. Zamaan had dumped a stengun, three rifles, grenades and ammunition for use in the city.[58]

Unsurprisingly, the Cells’ agenda focused as much on providing a long-term infrastructure for Pakistani covert warfare as on actual acts of terrorism and insurgency. One of the key members of the Cells, Fazl-ul-Haq Qureshi, admitted in one interview that they “worked as a nodal agency for collecting monetary, moral and social support for ‘freedom fighters’ who had come from Pakistan”.[59] In the interview, Qureshi described himself as a founder and leader of one of the Cells, a claim that finds no substantiation in Nath’s account which accords him a role, so to speak, only in the footnotes. None the less, as we shall see in the subsequent chapter, Qureshi was one of many peripheral Cell members who would go on to play a key role in the politics of Jammu and Kashmir.

The politics of the covert war

Meanwhile, the question arises, what position did the major political forces in Jammu and Kashmir take on the activities of the cells? What were their linkages, their motives, and their objectives?

From the point of view of those who ran Pakistan’s covert war, Sheikh Abdullah’s incarceration offered many opportunities. One poster, issued to infiltrators for propaganda purposes, showed Abdullah behind bars, along with images of an Indian soldier, armed with a rifle, crushing a Kashmiri peasant under his boots.[60] The Revolutionary Council's printed "Proclamation of [the] War of Liberation", too, referred expressly to the fact that "our great leaders Sheikh Mohammad Abdullah and Mirza Afzal Beg have been languishing in Indian prisons, but their determination to throw off the yoke of Indian imperialism remains unflagging".[61] This was certainly a remarkable departure from traditional Pakistani propaganda, which had reviled Abdullah for choosing India at the time of Independence.

Operatives working for the Master Cell hoped the National Conference connection would be of practical value, too. Tasked with contacting the pro-Plebiscite politician Mohiuddin Kara, the Master Cell operative Rehmat Ullah Khan had turned for help with his first covert mission to his wife's sister's husband, Abdul Jabbar Butt. Butt had long been involved in political activism. He joined the Muslim Conference in 1935, when it was still headed by Maulvi Mohammad Farooq's father, Maulvi Yusuf Shah. He later served two-and-a-half months in jail for his participation in the 1937 freedom movement led by Sheikh Abdullah. Although he remained ideologically affiliated to Abdullah, Butt emigrated to Pakistan after the Partition, carrying with him Rs. 19,000 which he was able to raise by selling off his family assets. He was, perhaps unsurprisingly, detained by Pakistani Customs and Police on his arrival at Lahore Airport.[62]

Butt's unhappy experience at Lahore Airport seemed to have set the tone for his time in Pakistan, for he returned to India in 1949 and hitched his wagon to Sheikh Abdullah's political fortunes. After Sheikh Abdullah was deposed in 1953, Butt became a key mediator between the jailed politician and Mridula Sarabhai. His work as a courier earned him a seven-year term in the Central Jail at Jammu in 1956; undeterred, he spent several weeks as a guest at Sarabhai's residence after his release.[63] As the Kashmir Conspiracy Case dragged on, he worked with the Legal Defence Committee as a caterer, earning the confidence of Mohiuddin Kara and Begum Abdullah. Unsurprisingly, Butt was able to secure a rapid appointment with Kara through the offices of another Defence Committee member, Mubarik Shah.

Kara had been out of jail just a few days when he met Khan, having served time for his role in the holy relic agitation. Kara, by Nath's account, told the Pakistani agent that his pro-Plebiscite organization desperately needed funds to sustain party workers, who were starving. A first payment of Rs. 10,000 was delivered to Kara soon afterwards; the politician gave Butt a receipt for the cash and, perhaps optimistically, advance receipts for four more payments he expected to receive for the same amount.[64] Although all but one of the advance receipts had to be destroyed, Kara did indeed receive Rs. 49,000 from the Pakistan High Commission eventually – two payments of Rs. 10,000 and a final payment of Rs. 29,000, routed by Ashraf through the Master Cell.[65]

According to Nath's account, Abdullah himself received a single payment of Rs. 50,000 from the Pakistan High Commission. At a meeting with Rehmat Ullah Khan and Sheikh Abdullah's nephew, Sheikh Nasir, in New Delhi, Ashraf handed over five bundles of Rs. 10,000 to be couriered to Srinagar. Rehmat

Ullah Khan told his interrogators that he had taken the cash, wrapped in newspaper, to the Delhi Railway Station, and handed it over to Sheikh Majid, who was waiting to travel to Jammu on the Kashmir Mail. There is no evidence in Nath's investigation, though, that the cash was actually handed over to Abdullah. If it was, all it succeeded in extracting was the ambiguous promise of religious solidarity described earlier in this chapter.

One key subsequent event, however, could have some connection with the alleged pay-off. Soon after his release from jail, and just months before the 1965 war, Abdullah left for an extended tour of Europe, Asia and Africa. There was protracted controversy over remarks he supposedly made in the course of that tour, which his many critics saw as assaults on India's position in Jammu and Kashmir. Some of these the Sheikh denied; others he did not. One article written by Abdullah for a magazine in the United States, for example, called for India and Pakistan to concede to the Kashmiris "the substance of their demand for self-determination".[66] His indiscretions provoked a political furor, provoking applause from Pakistan and outrage from the Hindu-nationalist Jana Sangh in India's Parliament.[67]

Matters deteriorated further when Abdullah chose to meet the Chinese premier, Chou-en-Lai, in Algiers, at the Second Asian–African Conference. China had invaded India in 1962, and memories of India's humiliating defeat were still fresh. Although Abdullah subsequently claimed that he had reported the contents of his conversation, mainly about the status of the northern territories of Gilgit, to the Indian Ambassador, the meeting inflamed both the establishment and public opinion in India.

One commentator otherwise sympathetic to Abdullah has provided this narration of events:

> Chou-En-lai [*sic*] had been going about all over Algiers as the stark enemy of India – because India was the enemy of China, according to Chou. If the Sheikh had had a single, even amateurish, secretary to advise him, the latter would never have allowed him to call on Chou En-lai. The Sheikh, if he reads this book, will undoubtedly say that it is the bureaucrat in me now talking; and he is quite right. Long afterwards, when he was under "house arrest" in Delhi, I asked him why he had called on Chou En-lai in Algiers. Sheikh Adbullah perfectly innocently said to me, "I did not ask to see him. Chou En-lai sent me an invitation, so I went and saw him. What was wrong with that? After all, Kashmir has a common border with China, hasn't it?" That simple it was, and I am sure he still looks at it with the same simplicity. But how many people outside Kashmir are prepared to call him "innocent" or "simple."[68]

Not many, subsequent events suggest. India's Foreign Minister, Swaran Singh, announced that Abdullah's conduct in "seeking China's support" with regard to Jammu and Kashmir was "highly objectionable".[69] In Jeddah, Saudi Arabia, for the Haj pilgrimage, Abdullah was told that the validity of his passport would end on April 30, 1965. Despite a whisper-campaign that Abdullah would choose not to return, he did so, and was promptly arrested. He was taken to the hill resort of Ootacamund, 2000 kilometres south of Delhi; Mirza Afzal Beg and Begum Abdullah were held in India's capital city. Abdullah's decision to return to India has widely been cited as evidence of his innocence, a wholly plausible proposition. To this suggestion, Nath had a response. "There is a strongly held view", he wrote in the *Report*, that:

> in early 1965, when Sheikh Abdullah had gone abroad, ostensibly for the purpose of performing the Haj pilgrimage, he was sounded by Pakistani emissaries in this regard [Operation Gibraltar] in a general way, although the details of the plan may not have been disclosed to him at that time. It appears that his reaction with regard to further attempts by Pakistan to keep alive the Kashmir issue and focus world attention on it, even by the use of force, was elicited and carefully gauged. His reaction is reported to have been favourable.[70]

Did Abdullah, then, know that India and Pakistan would soon be at war? If what Indian intelligence had learned was indeed true, it raises the possibility that Abdullah returned to India, and a near-inevitable term in jail, hopeful that he would soon be freed by the force of arms. It is a harsh conclusion, and one that most commentators sympathetic to the Sher-i-Kashmir would find absurd. Yet, the *Report* suggests, it is a possibility that played on the minds of Indian intelligence, building on the suspicions that had developed in the first phase of the covert war, from 1947 onwards.

The lion and the goat

On the face of it, Sheikh Mohammad Abdullah and Maulvi Mohammad Farooq, the Mirwaiz of Srinagar, represented two diametrically opposed tendencies in Jammu and Kashmir's politics. As the covert war progressed towards its climax, though, an odd set of alliances began to form: alliances against which Indian intelligence suspicions about Abdullah need to be read.

Neither the radical nationalist nor the cleric had reason to love each other. Their street armies had often clashed in skirmishes that came to be described as fights between the *Sher*, a reference to Abdullah's favoured honorific of "The Lion of Kashmir", and the *Bakra* or goat – a droll reference to the long beards worn by the orthodox Muslims who made up Maulvi Farooq's following.[71] Sheikh Abdullah's critics contended that the paramilitary campaign – led, ironically enough, by his eventual tormentor Bakshi Ghulam Mohammad – was similar to the depredations "of the storm troopers in Nazi Germany".[72]

Whatever the truth, both *Sher* and *Bakra* made common cause as the war of 1965 began to approach. In the fallout from the Hazratbal crisis, the two camps had begun fighting for political space, each accusing the other of fracturing the mass religious solidarity generated by the disappearance of the holy relic. At this point, however, a colossal miscalculation by Bakshi's successor as Chief Minister, G.M. Sadiq, reshaped the politics of the state. Hoping to more deeply integrate politics in Jammu and Kashmir with that of India as a whole, Sadiq dissolved the National Conference. In January 1965, the Indian National Congress, which had ruled India since Independence, amended its party constitution, and enabled the setting up of a state unit in Jammu and Kashmir which in turn subsumed the National Conference.

Political integration was part of a larger effort to bind together the destinies of New Delhi and Srinagar. In late 1964, an order issued by the President of India allowed the central government to take charge of Jammu and Kashmir's administration in the event of the collapse of the constitutional machinery. Prior to this order, the imposition of emergency powers required the concurrence of the state legislature; now, New Delhi had the final say over the legislature's own existence. Then, in April 1965, the state legislature approved the renaming of Jammu and Kashmir's *Sadr-i-Riyasat* as its Governor and its *Wazir-e-Azam* as Chief Minister, in line with all-India practice. The change was semantic, but significant nonetheless. The term *Wazir-i-Azam*, for example, was generally translated as Prime Minister, suggesting that there was some degree of parity between the head of government in Jammu and Kashmir and the head of government in New Delhi. Now, the head of government in Jammu and Kashmir had the same status as that of the Chief Minister of any Indian state. A wide variety of central legislation, on issues ranging from social security to the rights of trade unions, was made applicable to Jammu and Kashmir. Sadiq also passed legislation making it necessary for high officials and candidates contesting elections to "uphold the sovereignty and territorial integrity of India".[73]

New Delhi's pursuit of integration – perhaps driven by the increasing prospect of war with Pakistan – enabled Abdullah and the Mirwaiz to join hands, opposing what they saw as a predatory erosion of the state's autonomy. Both leaders started addressing joint meetings. Claiming that New Delhi's policies threatened the existence of Kashmiri Muslims as a category, they called for a boycott of all Congress supporters. Kashmiri Muslims sympathetic to Abdullah and the Mirwaiz responded by boycotting marriages, funerals and religious ceremonies hosted by the families of Muslim Indian National Congress members. At Abdullah's headquarters at the Mujahid Manzil in Srinagar, posters of freedom struggle icons like Nehru, Mohandas Karamchand Gandhi and Abul Kalam Azad were replaced with images of Pakistan's founder, Mohammad Ali Jinnah. In the passport he was issued for his travels abroad, Abdullah insisted that his identity be entered as a "Kashmiri Muslim", conflating nation with religion. Navnita Chadha-Behera has perceptively noted of this time:

> The clock had turned full circle. In the 1940s, the Sheikh had joined hands with Indian nationalism in order to challenge Muslim nationalism, and now he joined forces which stood for the Muslim identity in order to challenge the Indian identity.[74]

When war broke out, Maulvi Mohammad Farooq gave unconcealed religious and political legitimacy to the Pakistani campaign. At Friday prayers on September 3, 1965, the Mirwaiz called for god to "give victory and glory to Islam" and "rid us of the *zulam* [oppression] of *zalims* [traitors]". "Make those friends who have come here with hopes successful", he said, a fairly obvious reference to the Pakistani forces. A week later, on September 10, by which time it was clear the Pakistani offensive was being rolled back, the Mirwaiz again invoked god, asking him "to take revenge on the enemy, make them eat dust, create an earthquake to destroy the enemy". "If you cannot come", he said, "send Hazrat-i-Ali [the prophet Ali] who may draw out his Zulficar (sword) and finish all the Kafirs [unbelievers]". As the tide of battle turned, the Mirwaiz's speeches grew more desperate. On October 12, for example, he asked that god "give help to Muslims now as you had mysteriously done for a handful of Sahabs [gentlemen] at Jang-i-Badar [the battle of Badr[75]]. Make the Muslim Lashkar (Army) victorious."[76]

Despite such public postures, however, the Srinagar clerical establishment stopped short of providing actual material aid to cell members. The Pakistani stay-back cell operative Hayat Mir, for example, made contact with Maulvi Farooq in September 1965, told him of his covert role and asked for assistance with the Cells' anti-India activities. By the account of Indian counter-intelligence, Maulvi Farooq listened with great interest to Mir's account. He gave him nothing tangible, though, other than Rs. 20 as a token of support, an insultingly small amount of money. Mir correctly "came to the conclusion that the political leaders would confine themselves to making speeches only and would not be of any active help in his field of work".[77] This experience was of a piece with that of the infiltrators. Having "contacted supposedly sympathetic Mullahs, they found that most were reluctant to help".[78]

Religious propaganda, though, continued apace after the end of the 1965 war. On October 18, 1965, protestors threw stones on police from inside one of Srinagar's most important shrines, Hazratbal, home to the moe-e-muqaddas relic, said to be a hair of the Prophet Mohammad. The police entered, provoking charges of sacrilege. Then, on October 27, Cell members dug up the grave of a Saint interred at the Ziarat Rozabal, a shrine in downtown Srinagar. Leaves of the Quran in an adjoining mosque were torn, a transparent attempt to provoke Muslim religious fury. Allegations were made that the police had beaten members of congregations at Jama Masjid mosque and the Khanqah-e-Maulla shrine, both centres of enormous religious significance. The idea, Nath believed, was:

> to create a situation similar to the one created at the time of [the] theft of the Holy Relic from Hazratbal in December 1963. They expected to exploit the religious emotions of the masses aroused in this manner. . . . The grand design behind these incidents was only dimly discernable at the time and it came into focus after the Master Cell and its ramifications were uncovered.[79]

Despite the fact that the Master Cell and its subsidiaries drew heavily on cadre with affiliations to the secular traditions of the National Conference, then, they were at once mired in communal politics and a right-wing vision of Islam. Islamist themes and ideas were a basis not only for mobilization and recruitment, but for actual practice. Islam was a central feature of the Cells' ground-level strategy for mobilization and recruitment. "The band of activists", Fazl-ul-Qureshi recalled of his own unit, "would also go to mosques and preach for the secession of Kashmir to Pakistan".[80] In effect, the available evidence suggests the distinction traditionally drawn between the nationalist and the Islamist is problematic. Both ideologies drew on each other and, indeed, suffused the practice of the other to the point where distinctions remained useless. The jihad for the defence of the Muslims in Jammu and Kashmir was also the national struggle for liberation; the struggle for national liberation, in turn, a jihad for the defence of the faith.

Over the coming decades, these ideas would come to be understood more clearly by participants, and to manifest themselves in new forms of practice. For the moment, however, the curtain was starting to fall on the second phase of the covert war.

The end of the underground cells

On September 22, 1965, India and Pakistan entered into a ceasefire, which on the face of it brought an end to the war. Both sides had registered some territorial gains, but in Jammu and Kashmir itself, the Pakistani campaign had been a failure. India now held the strategically crucial Haji Pir Pass. More important, it was clear that hopes of a mass pro-Pakistan uprising in the Kashmir Valley, one of the key premises of the 1965 campaign, were misplaced. The Master Cell, however, was intact, along with its subsidiaries and the Pakistani stay-back agents – poised to do what it had done right through the war itself.

Maulvi Farooq's political posturing was, interestingly enough, to give the Jammu and Kashmir Police its first breakthrough in the hunt for the Master Cell. A source told authorities that some of the posters that had appeared in Srinagar were being prepared at the Mirwaiz Manzil, the cleric's hereditary home. Given what the cleric had to say during the war, Indian counter-intelligence officials would have found the claim wholly plausible. In the event, the tip-off led authorities to Sofi Ghulam Ahmad, the editor of the *Huriyat* [Freedom], the house organ of the Mirwaiz's political organization, the Awami Action Council. From Ahmad, police investigators were able to establish that the posters emanated from multiple sources – information which was, through a painstaking process of investigation, to lead to the Poster Cells, the Students' Cell and on to the Master Cell.[81]

It is possible that the tip-off put an end to an armed struggle that may have had more serious consequences for India than the 1965 war itself. Interrogations led investigators to believe that the Master Cell had plans to assassinate several pro-India politicians, including Chief Minister G.M. Sadiq, his key political associate

Mir Qasim and the Home Minister D.P. Dhar. One member of the Students' Cell, Ghulam Hassan Mala, had been given a hand grenade for an attempt on Qasim's life scheduled for October 24, 1965 – after fighting with Pakistan had more or less wound down. Mala was instructed to hide amongst a group of student demonstrators, who would gather to protest an anti-Pakistan speech Qasim had made before the United Nations in New York during a discussion of the war. Police, however, prevented the demonstration – and the planned assassination – from being executed.

Interestingly, the Cells made no secret of the fact that their war continued, no matter what the fate of Pakistan's military enterprise. Officials received a number of letters issued by the Revolutionary Council through October 1965, warning of physical harm. One, mailed to Home Minister D.P. Dhar, referred to him as an "Indian Dog", and warned that if he did not change his "attitude, we shall punish you as we must".[82] Another letter, issued by the "Execution Squad" of the Revolutionary Council and mailed on October 20, 1965, just three days before the Qasim assassination attempt, described Sadiq as a "filthy and bloody Indian stooge". "The Revolutionary Council has taken a grave view of your activities and we have been forced to include your name in the first execution list", it read.[83]

Such threats were wholly credible. Three days after the abortive attempt on Qasim's life, a grenade was found concealed just 150 metres from Sadiq's residence. Investigators concluded that the grenade was hidden in a manner so as to enable an assassin to pick it up just as the Chief Minister's car would pass by. Authorities in Jammu and Kashmir came to believe that six short-barreled 0.38 revolvers given to the Cells during the war were also intended for use in assassinations. The revolvers were brought by two Cell members, Bashir Ahmad Kitchloo and Mohammad Ali Malik along with several hand grenades, and stored in the premises of a timber contractor in the Chhandji Forests.[84]

Subsequent investigations made clear that the Master Cell had the necessary hardware for a prolonged terrorist campaign. One raid on a weapons cache left behind by Pakistani irregulars for the Cell led to the recovery of 40 boxes of plastic explosive and three boxes of ammunition. Nath observed:

> Even without the training and the explosives, the underground cells had succeeded in creating a fairly strained situation in the valley by their acts of terrorism. Now, with the training received by two members of the Master Cell, which they would have imparted to other members, and with the availability of high explosives on such a large scale, they could have brought the normal life in the Valley to a standstill.[85]

What course might events have taken had the Master Cell not been shut down? Some of the most interesting testimony on the issue came from Sattar Khanday, who along with his fellow covert operatives Shaban Khanday, Samad Wani and Ghulam Nabi had played a key role in enabling Hayat Mir's operations in

Badgam. Khanday's behind-the-scenes support was so crucial to Mir's cadre that they used to wryly describe him as the Deputy Commissioner of Badgam, the designation of the senior-most civil servant in an Indian administrative district. Shaban Khanday was eventually arrested, and led investigators to the rifles looted in Badgam, but Sattar Khanday made his way back to Pakistan, travelling with a Pakistani column over the Pir Panjal mountains and then across the CFL through Rajouri.[86]

Back in Pakistan, Sattar Khanday was treated as a hero. He was feted by senior military personnel, and given a personal meeting with a senior Pakistani bureaucrat responsible for Kashmir policy. The then President of Pakistan-administered Kashmir, Abdul Hamid, gave him a Rs. 500 cash prize. The fall was, however, nearing. Using what the *Report* somewhat coyly describes as "certain sources", Khanday and Razzak Wani were lured across the CFL on January 28, 1966, believing their old comrades-in-arms would be waiting for them. Instead, five days later, they found police personnel waiting for them at the agreed rendezvous.

Khanday told his interrogators that the military stalemate of 1965 had done nothing to end the war over Kashmir:[87]

> He reported that in [the] course of a visit to the training camp at Sinkari, he found 15,000 Mujahids receiving training in Guerilla warfare. Among the instructors he counted 22 Chinese. The Mujahids under training were drawn mainly from Pakistan Occupied Kashmir and from among the people who had crossed over from Rajouri-Poonch area [in Indian-administered Jammu and Kashmir] into Pakistan-Occupied Kashmir. He also revealed that in any future infiltration, the Pakistanis will send in wave after wave of infiltrators. He gave it as his impression that the younger generation of army officers in Pakistan regarded [Prime Minister] Mr. Z.A. Bhutto as representing their aspirations better than President [General] Ayub Khan. These disclosures made by Sattar Khanday were particularly significant in the context of the Tashkent Agreement and the expectations of a peaceful policy by Pakistanis in the future.[88]

Hayat Mir may have hoped to be in Srinagar when this second, promised wave of mujahideen would arrive. Counter-intelligence personnel had little success finding him; a master of his tradecraft, he never slept more than two nights in one location, and used a plethora of pseudonyms. Eventually, Criminal Investigation Department personnel used what the *Report* describes as "a highly placed source in the Plebiscite Front" to trap him.[89] Mir missed a first meeting with the police's informant, part of a well-established practice of randomly skipping appointments. A second meeting was arranged, this time at Srinagar's famous Ahdoos Hotel, on November 17, 1965. The plain-clothes police personnel at the premises knew Mir always carried a hand grenade with him; their first action was to pin his arms behind his back. "You are lucky", Mir told the men who now held him prisoner. "Had I even a moment's warning, I would have blown you up. This is the first time in my life I have failed."[90]

Failure and the future

Mir was not the only one to have failed. The death of the Master Cell illustrated the profoundly ill-conceived ideas the Pakistani intelligence establishment, and their covert allies within Jammu and Kashmir, had both of their immediate theatre of operations and of India at large. Many of those involved in the Master Cell went on to make their peace with the Indian state: Batku, for example, went on to serve as a senior official in the Jammu and Kashmir Tourism Department, while Zafar-ul-Islam worked as an engineer for the government.

What did the second phase of the Pakistani covert war in Jammu and Kashmir hold out for its architects? In purely military terms, Pakistan had learned from the reverses it had suffered in the 1947–1948 war. Almost all of the criticism levelled by Major-General Akbar Khan – who had commanded critical elements of the 1947 offensive – was met. Pakistani troops and irregulars operated together from the outset of operations, preventing the many crises of communication and command which had led to a crippling loss of momentum in 1947. Training in guerilla tactics was provided systematically. Intervention was made in the Jammu region at a relatively early stage, in an effort to block Indian military lines of logistics and communications into the Kashmir valley. This time around, India did not have a monopoly on the use of air power – and had to deal, furthermore, with a well-entrenched campaign of sabotage.

And yet, the second covert campaign, and the India–Pakistan war which was one of its elements, failed. As during the first phase of the war, there had been critical misjudgements of politics by Pakistan's intelligence establishment. The rebellion upon which Operation Gibraltar had been premised had not taken place. During the war between wars, the covert campaign that ran through the 1950s, this may have been attributed to Pakistan's failure to decisively intervene on the side of anti-India forces in Jammu and Kashmir. Yet, when it did so in 1965, the operation was betrayed by a Muslim, and a Kashmiri to boot. Although there was discontent in Jammu and Kashmir – witness the *moe-e-muqaddas* affair – those arguing against Indian rule in the state had neither the legitimacy nor the resources to lead an effective revolt against India. Covert warfare was not in itself adequate to capitalize on conditions as they existed on ground, and transform them into a mass uprising.

Aside from politics, the Indian Army and intelligence establishment had proved more resilient than Pakistani military experts had anticipated. More important, the notion that India itself was a fragile state, torn by tensions between ethnicities which would explode in the face of war, stood exposed. As I have noted in the last chapter, Major-General Khan had happily expressed the view that while Indians were in "the habit of submission and servility", Pakistan's peoples possessed "a self-confidence and ready willingness to march forward into India".[91] Racist and religious chauvinist ideas of this kind, given birth to by colonial propagandists, had been nourished by Pakistan's military and suffused propaganda material distributed by the covert cells. "We expect all the sane and freedom-loving elements in India", the Revolutionary Council had said, "and particularly the brave Sikhs, the South Indians and the Rajputs who have always given us moral support to lend us active assistance".[92]

India, it turned out, was not on the verge of collapse: and as long as it did not fall apart, the chances of Pakistan successfully taking Jammu and Kashmir by force were slim. Over the coming decades, both Islamist ideologues and intelligence strategists would develop ideas on how ethnic and religious faultlines in India could be made to work in the way they had expected, a programme that would unfold through the 1980s. For the moment, though, the strategists planning Pakistan's war in Jammu and Kashmir had no choice but to go back to the drawing board. Over the coming years, India and Pakistan would be engaged in negotiations in Tashkent – negotiations that, we shall see, were mired in acrimony from the outset, and almost predestined to fail.

On the night of September 22, just before signing the ceasefire agreement, Bhutto had told the United Nations that Pakistan was prepared to wage war with India "for a thousand years".[93] He meant it. As Pakistan talked peace in Tashkent, its intelligence services began working to set up the infrastructure for yet another covert campaign – and yet another war.

4

AL-FATAH

> From the present indications, there appears to be little doubt that Pakistani [*sic*] will continue to press her claim to Kashmir by all possible means, including those of violence and subversion. In these circumstances, it is profitable to study and analyse the policy and tactics employed by Pakistan in the past and fashion our strategy accordingly. While one may or may not agree with the various conclusions drawn in the report, the need for an intelligent analysis of past events, a logical estimate of a potential enemy's future plans and the formulation of our own policies in the light of that, is obvious and inescapable.
>
> Surendra Nath, Inspector-General of Police, Jammu and Kashmir Criminal Investigation Department, 1966[1]

Fortune, goes the old adage, is a fickle goddess. On November 17, 2004, three of Srinagar's most important residents, all veterans of the Master Cell, had good reason to contemplate the exceptionally curious courses their lives had taken.

That morning, India's Prime Minister, Manmohan Singh, had arrived in Srinagar on his first official visit to Jammu and Kashmir. To signal that his government was serious about the pursuit for peace in the state, Singh had ordered the withdrawal of thousands of Indian troops from southern Kashmir.[2] Addressing a rally of some 10,000 people in Srinagar, he also announced a massive financial package to aid economic reconstruction, and offered unconditional dialogue with "all those who have concern for Kashmiris".[3] "Kashmir", he said, "needs prosperity with peace. But peace without dignity is meaningless".[4]

Singh had begun his speech two hours late, the consequence of a shootout with two terrorists behind the Suleiman Shopping Complex, just 200 meters from the stadium where the audience was gathering to listen to the Indian Prime Minister's speech. Armed with assault rifles, one equipped with an attachment that let it launch grenades, the terrorists had intended to disrupt Singh's rally. Their presence was, however, detected by a police official, and a spectacular exchange of fire followed. Three soldiers sustained minor injuries in the encounter; both terrorists were killed.[5]

Fazl-ul-Haq Qureshi, an energetic young member of the Master Cell who had gone on to become an elder statesman of political secessionism in Jammu and Kashmir, might perhaps have regarded the men who were about to die with some degree of empathy. Two of his comrades-in-arms, however, might have had very different sentiments. One of them was Javed Ahmad Mukhdoomi, the Inspector-General of Police with command of his force's operations in the Kashmir valley, the man with overall responsibility for Prime Minister Manmohan Singh's safety during his visit. Appendix VI of Surendra Nath's classified *Report on Pakistani-Organised Subversion Sabotage and Infiltration* records that he was a contact of the stay-back agents assigned to the Master Cell.[6] Then a student of the Sri Pratap College in Srinagar, Mukhdoomi is not mentioned in the main body report itself, which suggests that his activities were minor. But the ironies are only too obvious: an associate of terrorists determined to bring down the Indian state was now commanding a force committed to stopping them.

One of the men Mukhdoomi was responsible for protecting was in much the same position. Jammu and Kashmir Law Minister Muzaffar Husain Beig had come to power as a representative of the People's Democratic Party in 2002. Beig, listed in police records as a member of the Narwara unit of the Master Cell, had gone on to establish an enormously successful law practice in New Delhi before turning to politics.[7] In his second political innings, Beig was studiously pro-Indian; on one occasion he bitterly condemned politicians who sent "money and gifts to militants on the occasion of Eid so as to save their own skin".[8] As the de facto second-in-command in his party, and the third-in-command in the Cabinet, Beig attracted the wrath of terrorist groups. Within two years, he was under relentless attack, like dozens of other politicians who had chosen to participate in mainstream democratic politics.[9]

Others of the Master Cell, too, chose to make their peace with the establishment. Bashir Ahmad Kitchloo, for example, had been one of the organization's most active operatives, but then built a career for himself as a successful National Conference politician.[10] If some had abandoned the struggle to throw India out of Kashmir, though, another generation was readying itself to take up the baton.

For intelligence strategists in Pakistan, this pool of recruits was to offer new opportunities at a time when it seemed the war for Kashmir had been lost. The challenge before them was considerable. The battles of 1947–1948 and the Informal War had demonstrated that Jammu and Kashmir could not be won through means modelled on tribal ways of war. Building on the lessons of this period, Pakistan's covert services had forged new strategies. However, the experience of the Master Cell and the war of 1965 had made it clear that the conditions existed neither for a general uprising in Jammu and Kashmir nor for the defeat of India in the battlefield. Indeed, a full-blown war with India held out the risk of the destruction of Pakistan itself. What then could be done? For the next phase of the covert war, Pakistan's covert services would turn to struggles underway to its west, in Algeria and Palestine, for inspiration.

Red Kashmir

An obscure student activist would be the central instrument of this new phase of the long jihad in Jammu and Kashmir. During the war of 1965, Ghulam Rasool Zahgir had made no secret of which side he was on. Along with a couple of other students at the Kashmir Univeristy in Srinagar, he had made clear his sympathies for the mujahideen who had arrived from Pakistan. Behind the scenes, Zahgir worked with Poster Cell-1, one of the major subsidiaries of the Master Cell. It took little time for the Jammu and Kashmir Police to find the would-be insurrectionary, and Zahgir was detained on October 21, 1965. However, his activities were deemed to not have "been of a dangerous nature" and just a few months later, in January 1966, Zahgir was released on parole.[11]

Unlike many of those involved in the Master Cell and its subsidiaries, Zahgir did not have a political background. Nor was he, by any stretch of imagination, from amongst the ranks of Kashmir's oppressed. Having graduated from the Nawakadal High School in Srinagar in 1958, Zahgir had joined the Amar Singh College, but dropped out after just a single year of studies. For the next several years, he hopped from job to job, first working as a travelling salesman and then for a Kashmiri businessman in the Indian commercial hub of Mumbai. He left this job soon afterwards, and returned to Kashmir to work as a clerk at a local government office in Sonawari, but this job did not last long either. On the eve of the beginning of his political career, in November 1964, Zahgir was appointed as a clerical assistant at Kashmir University. It was an unchallenging job, and Zahgir turned again to the covert life which gave him a sense of purpose and direction that everyday life did not provide.

Zahgir was to prove himself considerably more dangerous than the authorities had imagined. Unlike Mukhdoomi or Beig, Zahgir had no intention of switching sides. Neither the time in jail nor the signing of the Tashkent Declaration, which paved the way for his release, did anything to still Zahgir's political beliefs. Upon his release, he made contact with intelligence operatives working out of the Pakistan High Commission in New Delhi, using connections that dated back to the operations of the Master Cell. Among the first tasks he was given was to develop contacts with underground student groups who were committed to fighting the Indian state. Zahgir's new political circle centered around the Students' Revolutionary Council, set up by Srinagar resident Syed Sarwar in 1966. Until Zahgir arrived on the scene, the group had confined itself to merely talking about revolution; he was to give it a map for action. Zahgir began discussing plans for arson, sabotage and acts intended to inflame religious fury, like the desecration of shrines. He was soon acknowledged as the leader of the Students Revolutionary Council.

Fazl-ul-Haq Qureshi, who Zahgir had met during his Master Cell days and again in jail, emerged as another key leader of the group. Qureshi had long been associated with the covert war to throw India out of Jammu and Kashmir. In 1964, he had joined a fringe anti-India organization, the Muslim Youth Federation, but soon left to throw in his lot with the operatives of the Master Cell. In 1965, he was arrested for distributing seditious posters in the name of the Muttahida Mahaz-e-Azadi (MMA, United Movement for Freedom). Like Zahgir, Qureshi had been released from jail in 1966 as part of the effort to bring about political normalization, and had promptly resumed his covert campaign. Along with a neighbour, Nazir Ahmad Wani, Qureshi began discussing ways in which cadre could be recruited for continuing the war against India. Zahgir enlisted both into his group.

By December that year, Zahgir and his group initiated a modest project launch, mailing posters bearing a map of India with Jammu and Kashmir marked as a separate entity in red ink. Sent out to government officials, politicians and influential private individuals, the "Red Kashmir" posters created a minor sensation, particularly since elections to the Jammu and Kashmir Assembly were scheduled to be held in 1967. Soon, however, Zahgir was told by his handlers at the Pakistan High Commission in New Delhi that putting out posters would not yield the kind of conditions needed to destabilize the elections. After some discussion, the members of the Red Kashmir group decided to carry out an attack on a sentry guarding a bridge in Srinagar – the first act of violence in what they hoped would develop into a war.

Tashkent and its wages

Pakistan had advertised the war of 1965 as a humiliating defeat of India to its people, an impression that the official media continued to encourage even after the ceasefire went into place.[12] The illusion – and its collapse – were to shape events for several years to come.

Although the war of 1965 had, in purely military terms, ended in a stalemate, its true meaning soon began to become evident to Pakistan's people. India now flatly asserted that the status of Jammu and Kashmir was non-negotiable. It was, quite clearly, unwilling to concede in peace what Pakistan had not been able to wrest through war. If, in 1947–1948, Pakistan had succeeded in winning a third of the territory of Jammu and Kashmir it was to come away from the 1965 conflict with nothing. Worse, from Pakistan's point of view, it had become clear that the alliances it had built over the past decade-and-a-half were of little practical value. China was unwilling to intercede militarily; the United States, for its part, had responded to the outbreak of hostilities by imposing an arms embargo on both India and Pakistan.

In January 1966, Indian and Pakistani delegations met in Tashkent to discuss the terms of a post-war settlement. The Prime Minister of the Union of Soviet Socialist Republics, Alexei Kosygin, acted as an informal mediator. The document that emerged, signed by Prime Minister Lal Bahadur Shastri and Pakistan's President General Mohammad Ayub Khan, is known as the Tashkent Declaration. Clause (ii) of the agreement committed India and Pakistan to withdraw their troops "not later than 25 February, 1966 to the positions they held prior to

5 August, 1965".[13] In purely territorial terms, India was the loser. It had to hand back a total of 1920 square kilometres of land to Pakistan. Of this, 700 square kilometres were areas of Jammu and Kashmir, including strategically important areas like Haji Pir, a high-altitude pass across the Pir Panjal mountains connecting Uri with Poonch. Pakistan, for its part, returned the 540 square kilometres it had occupied, 490 square kilometres in the Chhamb area west of Jammu, and another 50 square kilometres in the Khem Karan area of Punjab.[14]

Pakistan, in return for this advantageous territorial swap, made a fundamental political concession, altering its stated position on the problem of Jammu and Kashmir itself. The word "Kashmir" figured precisely once in the nine clauses of the agreement, surrounded by a series of significant caveats. Clause (i) of the Tashkent Declaration recorded the agreement of Shastri and Ayub Khan that "the interests of peace in their region and particularly in the Indo-Pakistan sub-continent, and, indeed, the interests of the people of India and Pakistan were not served by the continuance of tension between the two countries". "It was against this background that Jammu and Kashmir was discussed", the next sentence notes, "and each of the sides set forth its respective position". Clause (ix) committed both countries to hold high-level meetings to address "matters of direct concern to both countries", but there was no reference to the United Nations resolutions on Jammu and Kashmir nor a plebiscite. Furthermore, both countries agreed not to intervene in "the internal affairs of each other".

Ayub Khan, it would appear from the tenor of the Tashkent Declaration, had decided that the liberation of the people of Jammu and Kashmir was not a project that justified risking the interests of Pakistan. War with India, 1965 had made clear, could not be won; if it ran for any length of time, it would pose considerable hazards to Pakistan itself. To at least some in Pakistan, this assessment seemed suspiciously close to treachery. In early 1966, student riots broke out in Lahore and Sialkot; two students were killed when police opened fire on demonstrators.[15] The reaction, one commentator has observed, was particularly intense in those areas of the province of Punjab with substantial populations of Mirpuri migrants from the Pakistan-administered portion of Jammu and Kashmir. "For them", wrote Morrice James, "Ayub had betrayed the nation and had inexcusably lost face before the Indians".[16]

From the point of view of Pakistan's military establishment, the situation presented a complex problem. It had to continue the covert war in Jammu and Kashmir, or risk both public furore and the loss of the sole instrument of leverage Pakistan had over Indian policy in the state. At once, the conflict had to be calibrated to intensity short of levels which could provoke war with India. Could these apparently irreconcilable objectives be achieved?

The Nawakadal murder

Soon after the war of 1965, Pakistan's covert response to this complex problem began to unfold. Constable Charan Das, of the 45 Battalion of the Border Security

Force, was on duty at the Nawakadal Bridge in Srinagar on the night of February 3, 1967. He was armed to defend himself, carrying a .303 army-issue rifle, loaded with five rounds in its chamber. Not one of those rounds was fired that night, suggesting that Constable Das did not see Zahgir and Sarwar creep up behind him at around 10:00 PM. Armed with daggers, the two men stabbed the Border Security Force guard in the chest. Another member of their group, Abdul Hamid Shah, kept watch. Constable Das died after crawling some distance towards his post. Officials at the Maharajganj Police Station registered a First Information Report, 21 of 1967, recording Das' murder, but as to who had carried out the crime, they had no clue.[17]

Police investigation of the Nawakadal murder moved no further even after Red Kashmir claimed responsibility for the murder in the next issue of its newsletter. In fact, the man the Jammu and Kashmir Police was looking for was already in their custody. Soon after the Nawakadal murder, Zahgir was arrested under the Defence of India Rules, a draconian colonial law that allowed for preventive detention in times of crisis. Zahgir's arrest was effected because of intelligence reports that he had resumed his anti-India activities. Since investigators had no idea that Zahgir was involved in the Red Kashmir posters, however, he does not appear to have been interrogated on its activities or the Nawakadal murder. Nonetheless, the remaining members of the Red Kashmir group appear to have stopped their activities for a few months, perhaps apprehensive that Zahgir's arrest would lead Indian counter-intelligence to them.

By September, however, the second-rung leaders of Red Kashmir had recovered enough confidence to resume operations. In September Nazir Ahmad Wani and Syed Sarwar crossed the CFL into Pakistan, travelling through the mountains of Bandipora with the aid of a small-time smuggler, Mir Ahmad Gujjar. Pakistani officials greeted the group with great enthusiasm. Both men were trained in the use of small arms, as well as in the tradecraft of espionage and subversion. Much of the training related to the collection of military intelligence, the means to avoid surveillance and the protection of safehouses and communications. Wani and Sarwar knew their instructors as Major Habibullah, Major Tufail, and Major Kaiser Qureshi. A meeting was also organized with a senior officer, who called himself Brigadier Asghar. It is likely, of course, that all these names were aliases.

Wani and Sarwar returned across the CFL in November 1967, this time across the Sialkot–Ramgarh border in the Jammu region. A Pakistan Army officer, Major Nazir Malik, made arrangements for the crossing. Abdul Rehman, the courier who had acted as a liaison between Sarwar and his handlers in Pakistan, accompanied them. The major instruction given to the two Red Kashmir movement operatives was that they needed to recruit more cadre who could be trained in spycraft. Soon after his return, Wani recruited six more members to the group. Three, Mohammad Ashraf Manhas, Iftikhar Ahmad Paul and Sahibzada Mohammad Amin, were sent to Pakistan for training. Manhas was to prove particularly valuable to the Red Kashmir campaign. An employee of the Jammu and Kashmir Government's Secretariat, he was well placed to gather sensitive information.

Investigators of the Nawakadal murder, meanwhile, continued to grapple with the case without success. In April 1968, Zahgir was released from prison again, a fact that suggests counter-intelligence officials in Jammu and Kashmir had not managed to learn of his connection with either the Red Kashmir enterprise or the murder during his detention. Almost immediately after his release, Zahgir resumed his contact with his handlers. In July 1968, Wani travelled across the Ramgarh–Sialkot border into Pakistan for the second time, now with Zahgir himself. The two met a Pakistani Intelligence Bureau official, identified as Zafar Iqbal Rathore, who was to become their principal handler. The two men also met Major Tufail, who had greeted Wani and Sarwar on the earlier expedition, as well as an officer who identified himself as Colonel Bashir. This visit marked the beginning of a new stage in the covert group's life. Zahgir and Wani were instructed to send small groups of men for military training, and were themselves to return again for a longer period of specialized training.

Back in Srinagar, by the middle of the year, Zahgir was able to show the group a set of pencil-bombs that had arrived from Pakistan. Hoping that a religious incident would spark off a crisis like the one that followed the disappearance of the *moe-e-muqaddas* from Hazratbal, the group planned an attack on a shrine of Hazrat Sultan-ul-Arifeen, dedicated to the venerated Sufi mystic Sheikh Hamza Maqdoom. Sheikh Hamza's name is widely associated in the Kashmir valley with miracles. At the home of the saint Sheikh Khwaja Ishaq, one legend has it, Maqdoom was served roasted birds to mark the end of the dawn-to-dusk fast observed by the faithful during the month of Ramzan. Maqdoom ate the birds, and then collected their bones together in his hands as he raised them for prayer. The bones joined together, the birds came to life and flew out of the window.[18]

On the night that Zahgir's group arrived at the shrine of Hazrat Sultan-ul-Arifeen, a crowd had gathered there, maintaining a vigil to commemorate a holy day. The attack had to be aborted. Not too many months into the future, another encounter with the shrine would lead to a crippling reversal in al-Fatah's fortunes.

Lessons learned

Political change in India would, as time passed, prove one of the most important problems confronting the jihad in general, and al-Fatah in particular.

Hours after signing the Tashkent Declaration, Prime Minister Lal Bahadur Shastri died of a massive heart attack. Divided amongst themselves, the senior leadership of the Indian National Congress chose a successor who they thought would threaten no one – Jawaharlal Nehru's daughter, Indira Gandhi. The Congress' bosses described her as a *goongi gudiya* – a dumb doll. Indira Gandhi would soon demonstrate she was nothing of the kind. Her aggressive leadership would have fateful consequences for the covert war in Jammu and Kashmir. Soon after Indira Gandhi took power, a new sense of resolve became evident in the Indian armed forces. On September 11, 1967, while Zahgir was still under training in Pakistan, Chinese troops opened fire with machine-guns on Indian soldiers engaged in works along the Nathu-La pass in the eastern mountain state of Sikkim. Indian forces responded by using mortars against the Chinese positions, signalling that they were not willing to give ground. Chinese border troops responded in kind. At this stage, the General Officer-Commanding of the 17 Mountain Division, Sagat Singh, ordered his batteries of 5.5-inch medium guns to bombard the Chinese positions. Sagat Singh also ordered additional units to move forward, indicating he was willing and prepared to engage in an escalation of hostilities. After five days of fighting, the Chinese accepted a ceasefire. They had suffered an estimated 400 dead and wounded, compared with Indian casualties of 65 killed and 145 injured.[19]

To even those in Pakistan who believed that the war of 1965 could have been won, the Nathu-La fighting must have sent a clear message. Indira Gandhi's response had been enabled by measures initiated by her father, and pursued by Shastri. In the wake of the 1962 debacle, India had announced a massive rearmament programme. The plan envisioned increasing the strength of its army to a million men within five years. Large scale purchases of modern weapons had also been sanctioned.[20] Indeed, the increased preparedness of Indian forces may have been one reason why China backed down from its 1965 war threat. Contemporary photographs show that by October 1965 Indian troops in Nathu-La were armed with automatic weapons, not the antiquated .303 rifles they had faced the Chinese with three years earlier.[21] Much of this modernization process was complete by the time al-Fatah was becoming operationally viable, tilting the balance of military force decisively in India's direction.

In 1968, Indira Gandhi gave further indication of her intentions. Responding to concerns about the performance of the Indian intelligence services during the wars of 1962 and 1965, she ordered the creation of a new organization dedicated to external espionage. While India's Defence Minister, V.K. Krishna Menon, had taken much of the blame for the 1962 defeat, the Intelligence Bureau and its chief, B.N. Mullik, also faced criticism.[22] Headed by R.N. Kao, an ethnic Kashmiri, the new RAW functioned as part of the Cabinet Secretariat, with the Secretary (R), as India's spymaster is known, reporting directly to the Prime Minister. The Intelligence Bureau, inherited from Imperial Britain, retained responsibility for domestic espionage and counter-intelligence, and continued to function as part of the Ministry of Home Affairs. RAW soon had a formidable technical intelligence capability. The United States provided state-of-the-art surveillance equipment, for example, for RAW's fledgling aerial surveillance wing, the Aviation Research Centre, in return for India carrying out operations targeting China.[23] Within a short while, RAW was well on the way to developing a covert threat to Pakistan – to return the compliment, so to speak, that its intelligence services had inflicted on India since the 1947–1948 war.

Pakistan's military leadership, meanwhile, had problems other than Kashmir engaging its energies. Ever since General Ayub Khan took power in 1958, relations between the two wings of the country, sundered by 1600 kilometres of Indian territory, had been deteriorating. The coup that brought Ayub Khan to power had in part been motivated by concerns that elections would put in place a regime in which East Pakistan, where over 55 per cent of the country's population lived, had a dominant say. This was unacceptable to Punjab, home to the country's military and feudal elite, and the largest and most affluent province of West Pakistan. Ayub Khan's regime attempted to suppress the voices of protest in East Pakistan, provoking a backlash. In the elections of 1965, held to legitimize Ayub Khan's rule, 47 per cent of the East Bengal electorate voted against him.

Shortly after Zahgir's return to Srinagar, relations between East and West Pakistan deteriorated even further. In December 1967, Pakistani counter-intelligence unearthed a plot to assassinate Ayub Khan and establish an independent state in East Pakistan called Bangladesh. The Agartala Conspiracy, named after the town where the plot was planned, was hatched by a group of East Pakistani military officials alleged – quite plausibly, as we shall see – to have been in contact with India's covert services. The officers were overheard by Pakistani counter-intelligence at a club in Chittagong, while they were discussing their assassination plan.

Ayub Khan's regime was eventually to put over 50 civil servants, military officers and politicians on trial. The General, however, overplayed his hand. He attempted to implicate Sheikh Mujibur Rahman, the head of the Awami League and the most popular politician in East Pakistan. The backlash was immediate:

> Before this case few in East Pakistan dared to discuss secession in public, but as the papers printed more and more details of the proceedings, debate about breaking away became a normal part of public discourse. Mujibur Rahman, meanwhile, secured his place as a political martyr, and his support base became ever more solid.[24]

Ayub Khan could not but have noticed just how far the wheels of history had turned. Pakistan had long attempted to capitalize on the anger caused by the detention of Sheikh Mohammad Abdullah by going to war; India would retaliate after the arrest of Sheikh Mujibur Rahman. Demonstrations and protests became a regular feature of life in Dhaka over the coming months; the Army had to be called in to restore order on a regular basis. Pakistan's military had a difficult choice to make. It could escalate the covert war in the north to contain India, but any serious escalation could provide India with a pretext to launch hostilities in the east. The future of Zahgir and his associates depended on which side of this problem Pakistan would finally fall on.

A manifesto for war

From Zahgir's point of view, the situation could not have been worse. With his patrons preoccupied with the brewing rebellion in the east, support that had been promised to the group had yet to materialize. Worse, after the botched attempt on the Hazrat Sultan-ul-Arifeen shrine, calamity after calamity had befallen his associates.

Shortly after his return from Pakistan, Zahgir had brought two new recruits on board, Mohammad Aslam Wani and Zahoor Ahmad Shahdad. The two planned to rob rifles from an armoury used to store rifles for the National Cadet Corps, a government-run organization that provides elementary military training to student volunteers. In the event, the Islamia College Dacoity, as it came to be known, went badly. Although the terrorists who attacked it were dressed in pilfered National Cadet Corps uniforms, the security guards at the armoury refused to let them in. One security guard was stabbed in the ensuing melee, and his cries of distress attracted the police. One of the attackers, a college student named Ghulam Mohammad Baba, was arrested by the first police personnel to reach the scene, Constables Mohammad Yusuf and Abdul Aziz.

Baba's interrogation soon led the Jammu and Kashmir Police to Mohammad Aslam Wani and Zahoor Shahdad. It did not take investigators a great deal of time to work out their connections with the leaders of the Red Kashmir group and its role in the Nawakadal murder. Syed Sarwar, who was in New Delhi to meet officials of the Pakistan High Commission, was arrested. Before his interrogation was complete, however, several key members of the Red Kashmir group had gone underground. After successfully making contact with his handlers at the Pakistan High Commission, Nazir Ahmad was told to return to Srinagar, and maintain a low profile so that the remnants of the Red Kashmir group could make contact with him. Zahgir, Fazl-ul-Haq Qureshi and Musadaq Husain, for their part, fled across the border into Pakistan.

Notes maintained by Zahgir, recording the content of the instructions he received from the Pakistan Army's Field Intelligence Unit from November 1968 to January 1969, give us some insight into the evolving character of the covert war in Jammu and Kashmir. Like his predecessors, Zahgir saw the struggle in Jammu and Kashmir as Islamist in character. It was, he wrote, a battle to "upkeep [*sic*] our prestige and Islamic honour, as Algerian Muslims have done".[25] Unlike the Pakistani military strategists of an earlier period who believed India's "Hindu" forces could be easily swept away, Zahgir was taught that the battle was instead one between profoundly unequal adversaries. A plan was needed that would enable the enemy to be slowly weakened, just as "a mosquito does while fighting with an elephant".[26]

Zahgir's instructors identified for him the three separate planes on which the covert war was to be waged. First, there had to be a political party, "based on non-violence for the time being", which could place pressure on mainstream forces to oppose Indian rule in Jammu and Kashmir.[27] But, Zahgir was told, "no enemy has been overthrown by political means alone".[28] If India was to be defeated, there had to be a sustained campaign of economic sabotage, targeting airports, railway lines and industrial estates. Within Jammu and Kashmir, the principal targets of the covert campaign were to be military facilities, but movie theatres along with tobacconists and liquor shops were also to be shut down, since the "major portion of the profits goes towards [*sic*] India".[29] Movie theatres and liquor stores were indeed shut down by Islamist terror groups in the 1990s. Interestingly, however, Zahgir's notes do not seek religious legitimacy for such action.

Finally, Zahgir's notes outlined the means through which India could be militarily undermined. Before force could be used effectively, he observed, it was first necessary to study India's weaknesses.

> We have to be fully equipped with the complete information of the locations of military depoes [*sic*, depots], ammunition depoes, ordinance factories, petrol depoes and armed forces and their number. We have to know what are the different tactical nos. [numbers] being used in different coys [companies], and code signs and code nos. of different divs. [divisions], brgs [brigades] and bns [battalions]. We have also to know fully what are [the] different roads, bridges and national highways most important from the military point of view, so that at any odd [*sic*] time we can damage these and make the movement of the enemy impossible. After having such information, we can make the enemy hurt in all respects, whenever he may desire to attack on us.[30]

Zahgir's exphasis on military intelligence points to the growing integration of the covert war in Jammu and Kashmir with Pakistan's larger strategic imperatives. In 1965, Pakistan's efforts to take Jammu and Kashmir had been frustrated not just by the failure of the covert campaign that preceded it, but by India's willingness to widen the war outside of the state. Now, Pakistani military strategists had understood, it was impossible to initiate offensive covert operations in Jammu and Kashmir without preparing for full-scale hostilities. Major-General Akbar Khan, whose seminal plans for covert warfare with India have been discussed at length earlier in this book, had seen irregular forces operating in Jammu and Kashmir as near-autonomous of the regular Pakistan Army, serving mainly to harass and tie down enemy troops. As understood by Khan and his successors, their role was to precipitate a rebellion, not to actually spearhead a military campaign. The training imparted to earlier generations of covert operatives had been minimal and, unlike Zahgir, they had no mandate to execute classic military intelligence work. Now, the warriors of the jihad in Jammu and Kashmir would be an integral part of Pakistan's overall military structure. In effect, what distinctions had earlier existed between the war for Kashmir and the war against India were melting away.

How was such a war to come about? According to Zahgir's notes, his instructors saw the covert campaign as a four-stage process. In the initial stage, covert forces had to go through a painstaking process of recruitment, training, collection of information and the setting up of an organizational structure. "The history of guerilla tactics", Zahgir wrote, "teaches us that anybody whosoever [*sic*] has failed so far in achieving their objectives and aims through this type of war has failed due to the fact that he has left the initial stage incomplete".[31] To substantiate this proposition, Zahgir's instructors drew his attention to successes – Ernesto "Che" Guevara's campaign in Cuba, the Algerian struggle against French colonialism and the defeat of the United States in Vietnam. Their thinking may have been influenced by the growing interest of Pakistan's officer corps in doctrines of non-conventional war, driven in part by their growing interaction with the United States military, for which finding means to defeat insurgencies had become a major concern.[32] Learning from the defeat of Pakistan in 1947–1948 and 1965, Zahgir was also taught that inspirational leadership or tactical brilliance could not compensate for organizational and strategic weaknesses.

Apart from the content of the tradecraft skills that had to be imparted to cadre – the setting up of safehouses and hideouts, for example, or the collection of military intelligence – recruitment occupied a considerable portion of Zahgir's education. If his tactical inspirations came from left-wing anti-imperialist struggles, Zahgir's new organization had no room for communists. His covert organization's cadre would be made up of "educated youth only", who moreover "believe in the ideology of jahad [jihad] and liberation of the country by continuous armed struggle".[33] In this vision, therefore, the means of praxis of left insurgencies were to be dyed with the deep-green colour of Islam. Once this jihadist organizational core had consolidated itself, a new round of recruitment had to target those who could be persuaded to the course of jihad. Eventually, a wider mass of people, "whether literate or illiterate, boys or girls, youths or old or teenagers", had to be brought on board: "at least 5 per cent of the total population", Zahgir was told.[34]

It is interesting to consider what impact these ideas may have had on those who planned the war that began in 1989. Zahgir's scheme of recruitment was closely followed, with a core of highly trained operatives building a base of educated urban youth, and then opening the doors for cadre from the countryside. I shall discuss this issue in subsequent chapters, but once again we see the ways in which ideas generated in the course of the long-running covert war in Jammu and Kashmir were to influence events decades later. Zahgir also anticipated another key element of the war that was to come, the systematic subversion of the state apparatus. No grand project of ideological infiltration of the bureaucracy, he suggested, needed to be initiated to this end. Corrupt government officials could simply "create resistance by not working honestly, efficiently and zealously", while businessmen could easily be incited not to pay their taxes: ideas not dissimilar, in their outlines, to those Central Intelligence Agency blueprints for sabotaging the regime of Fidel Castro in Cuba were advocating.[35]

Unlike his predecessors, then, Zahgir actually had a map not just for the defeat of the coercive forces of the Indian state, but its entire apparatus of power and control. In his vision, tax strikes, protests by the unemployed and demands by bureaucrats for higher pay all had a role in the larger political struggle against India. Almost any act that disrupted the functioning of everyday life – a piece of iron inserted in a machine to make it inoperable, or sugar stuffed into the fuel tank of an engine – also had political meaning. Eventually, he was told, the Indian state would be compelled to take action not only against covert groups, as it had done in the past, but the "nation as a whole, thereby creating anger and [a sense of] violation among the herd [*sic*] of people".[36]

Zahgir now had to choose a name for the organization he had been told would spearhead the revolution: he picked al-Fatah, which in Arabic as in Urdu carries meanings of liberation, salvation and conquest. It was a name that had acquired fame ever since an until then obscure new Palestinian group that used it had, in March 1968, beaten back a full-scale Israeli military attack on the small Jordanian town of al-Karameh. Yet, Zahgir's notes, written just a few months after the battle of al-Karameh, do not suggest he was even aware of this famous victory; there is not a single mention, indeed, of Palestine. Instead, the name may have been chosen because of its religious resonance. In Islamic history, al-Fatah referred to the day when, in AD 630, the Prophet Mohammad had entered the Kaaba and smashed the idols that were then displayed within it.[37] Pitted as they were against mainly Hindu India, the event would have had an obvious significance to the men who joined al-Fatah.

Abdullah's retreat

Pakistan and India were not the only players whose appreciation of the problem of Jammu and Kashmir had been transformed by the war of 1965. As Zahgir was busy receiving instructions for the working of al-Fatah, Sheikh Abdullah was engaged in reconsidering his political position.

Arrested on the eve of the 1965 war for his controversial flirtation with the Chinese, Abdullah had watched Chief Minister G.M. Sadiq crush the opposition with little resistance. His key lieutenants, Maulana Masoodi and G.M. Kara, had been jailed for attempting to organize protests against his arrest, as had his new-found ally, Maulvi Mohammad Farooq. Mirza Afzal Beg's Plebiscite Front remained alive, but had no formal contact with Abdullah, and the organization's ability to agitate was limited. The 1965 war had pushed Sadiq's regime to impose drastic restrictions on media and political freedoms. New Delhi threw its weight behind the Chief Minister, in time-tested fashion. During the 1967 elections to the state legislature, the nomination papers of 118 opposition candidates were rejected. As a result, 22 candidates were elected unopposed. All but one of the unopposed candidates was from the politically sensitive seats of the Kashmir valley; all without exception were from the ruling Indian National Congress.[38]

Without even the spectre of Pakistani intervention to haunt Indian policy-making, Abdullah's position was now weaker than at any time in the past. Released from jail in 1968, Abdullah turned for support to parties outside Jammu and Kashmir. In October 1968, he called a conference, the All-Kashmir State Peoples' Convention, to discuss new ideas for its constitutional relationship with India. The Indian Government, as well as major political parties, reacted warily. Across the border, Bhutto had unleashed violent anti-India rhetoric, unprecedented in its biliousness. Public opinion in New Delhi was, predictably, hostile to both Abdullah and the Pakistani establishment. "In this foul atmosphere", Y.D. Gundevia observed, "Abdullah a little too consistently preaching friendship with Pakistan as also maintaining his friendly contacts with the Pakistan High Commissioner in New Delhi was viewed with immense suspicion."[39]

Sadiq, to his credit, allowed the Convention to be held. In the event, four major parties – the Indian National Congress' Indira Gandhi-led faction and a rival grouping of her opponents from within that fold, the Hindu nationalist Jana Sangh, and the Swatantra Party, the platform of the economic right-wing – stayed away. None the less, the convention ended up attracting some 250 people, and, more importantly, the support of an opposition politician with enormous all-India legitimacy, Jayprakash Narayan. Inaugurating the convention, Narayan made clear that after the 1965 war, a solution to the problem was only possible "within the framework of the Indian Union".[40] He repeated this assertion at a public meeting on October 11, to the annoyance of not only Abdullah, who wanted all options to be kept open, but also the audience.

Narayan's speech was built on ideas he had long advocated, so it should have surprised neither Abdullah nor others at the conference. At the height of Abdullah's Islamist adventures in 1964, Narayan had written a newspaper article suggesting that

> the right and constructive approach is not to deny to Kashmir the *right* to self-determination, or to assert that the right has already been exercised, but to show rationally how impractical and imprudent it would be to exercise that right now

He continued:

> The following facts might be put forward; first, there is the hard fact of aggression by Pakistan which they show no intention to vacate; second, a referendum might have serious consequences for the minorities both in India and Pakistan; third, a referendum might lead to further disintegration of the State of Jammu & Kashmir; fourth, the effects of the foregoing would have serious consequences for India's defence. There may be other factors; but these four are weighty and reasonable and neither Sheikh Abdullah nor any other Kashmiri leader, because they are as much concerned with India's good as with that of Kashmir, would brush them aside.[41]

Sheikh Abdullah, given his political conduct at the time, would more likely than not have paid little attention to these ideas in 1964. In 1968, he had no choice but to take them seriously. The Convention ended after eight days of deliberation, and adopted a resolution that any solution to the problem of Kashmir must keep in mind the interests of all regions of the state. Only such a resolution, it argued, could bring peace both to Jammu and Kashmir and South Asia as a whole. A steering committee was set up to tabulate the different points of view about the state's future that had been articulated, after which another convention was to be held to prepare a roadmap for a "peaceful, democratic, just, realistic and lasting solution".[42] In essence, Abdullah had recognized that the Kashmir valley alone could not decide the state's future, and that the pro-Indian sentiments that prevailed in the Jammu and Ladakh region would also have to be taken into account.

How had this come about? The weakening of the ethnic-Kashmiri domination of the politics of Jammu and Kashmir, one of the byproducts of the 1965 war, had unleashed new political forces through the state. In 1966, a socialist-affiliated forum in Jammu began demanding greater political autonomy for the different regions of the state. At the other end of the political spectrum, Karan Singh, the state's last *Sadr-i-Riyasat* and Maharaja Hari Singh's son, called for a sundering of Jammu and Kashmir along its ethnic-religious fault lines. Karan Singh called for Jammu to be cut away from Kashmir, and merged into the Hindu-dominated mountain state of Himachal Pradesh. In Ladakh, meanwhile, local political forces began arguing for direct federal administration, in response to the domination of administration and business by ethnic Kashmiris.

Sadiq's government did appoint a Commission of Inquiry, headed by P.B. Ganjendragadkar, to investigate the issue and recommend measures for an equitable sharing of resources between the regions. At once, however, Sadiq encouraged the rise of chauvinist forces to undermine the movement for greater regional autonomy. To prevent the rise of a secular opposition which could challenge the Indian National Congress in Jammu, he promoted the Jana Sangh, describing it as "a lesser evil".[43] In Ladakh, he sought to create a new leadership of Buddhist Lamas to undermine the position of the region's most important leader, Kushak Bakula. By 1969, Sadiq's efforts at cultivating chauvinists had led to violence. Incidents like the alleged desecration of a Buddhist flag by a Muslim and the stoning of mosques by Buddhist demonstrators led to rioting and arson. This, in turn, sparked suspicions among Kashmiri chauvinists that a grand plan was underway to transform the demographic balance of the state.

In 1970, after the meeting of the second All-Kashmir State Peoples' Convention that June, Abdullah dropped his emphasis on holding a plebiscite. Pakistan received relatively little attention; of 62 draft solutions considered for a resolution of the Kashmir problem, only one proposed accession to that country. In his inaugural address, Abdullah insisted that "there could be no solution that did not strengthen the secular and democratic forces that stood for equal rights to members of all religious communities, both in Kashmir and the rest of India".[44] He argued that this could be achieved through a programme of decentralization, which would give powers to each region, right down to the level of village local bodies. Accepting that "fear and suspicion of one region regarding the other" separated the visions of the state's citizens, he pointed to the need to first put "our own house in order" before deliberating Jammu and Kashmir's future relationship with India.[45]

The war of 1965, and his many years in jail, had clearly led Abdullah to take a pragmatic position, accepting the realities of the situation as they stood. He continued, however, to display the Pakistan card on occasion. Even as he dropped the plebiscite demand, and spoke of the need for regional autonomy, Abdullah asked that the substance of his demand for self-determination be conceded, if not its method. He suggested that a conference be held between representatives of the state, India and Pakistan, to work out the state's future. The Sheikh had bowed. In coming years, however, it would turn out his submission was not abject enough for New Delhi.

al-Fatah

To Pakistan's intelligence services, the direction the All-Kashmir State Peoples' Convention had taken would have held out little hope of large-scale popular support for escalating the jihad. Nonetheless, they could ill-afford to allow events to drift, now that political circumstances were less than favourable for Pakistan's cause.

Zahgir, Fazl-ul-Haq Qureshi and Musadaq Husain returned to India from Pakistan in January 1969, travelling through the Punjab border. Using the services of a cross-border trafficker named Roshan, they reached the town of Dina Nagar, in the frontier province of Gurdaspur, and then travelled to New Delhi. Husain dropped out of the group there, while Qureshi and Zahgir proceeded on to Srinagar. Nazir Ahmad Wani, the intellectual and organizational powerhouse of al-Fatah who had been forced to hide out in Srinagar as the Red Kashmir organization was being destroyed around him, was waiting for them. Old Red Kashmir associates were reactivated, and contacts were made with potential new recruits. By May 1969, Zahgir, Qureshi and Wani were ready to travel back to Pakistan, with news that the core of their group was in place.

For al-Fatah, this visit to Pakistan was of particular significance. In the course of meetings with Pakistani officials – Zafar Iqbal Rathore, "Brigadier Asgar" and "Major Tufail," among others – the organization now received clear instructions to initiate an intense phase of covert activity. Wani received military instruction for the first time, learning to operate machine guns, rifles, hand grenades and explosives. All three were also taught how to fabricate improvised explosive devices from easily available materials, such as potassium chlorate and arsenic sulphide.[46] In July 1969, the group finally returned to India, travelling through the Sialkot–Ramgarh border. A border trafficker, Chirag-ud-Din, was requisitioned by the Pakistan Army to facilitate the crossing.

Zahgir set about recruitment in earnest after his return. Each potential candidate was studied thoroughly, and reports were prepared on his potential strengths and weaknesses as a covert operative. Wani, for example, wrote a note on a recruit named Rasheed, recording the facts that he had three brothers, was a school dropout, and that while he had no "ideological tendencies", he wished to "rise against India as [it is] the enemy of [the] entire Muslim [community]".[47] In some early cases, recruits were also asked to sign an oath of allegiance to al-Fatah, and to make a financial contribution to the organization. Zahgir himself coached new recruits on ideological and organizational questions, and provided them with an elementary education in the use of codes and ciphers. Wani, for his part, held courses in guerilla combat tactics for eight recruits in the Hak-Khul forests above the village in Arizal, in Beerwah. This was, interestingly, the same village where the covert group of Hayat Mir, an affiliate of the Master Cell, had carried out a savage terrorist attack in May, 1965.

Al-Fatah's choice of the forests above Arizal was not coincidental. Zahgir and Qureshi had turned to an old prison-mate for help. Salim Jehangir Khan had worked as a Pakistani covert operative during the war of 1947–1948, acting as a guide to irregular troops. He subsequently smuggled in large numbers of hand-grenades into Jammu and Kashmir, which were intended for use during the terror campaigns of the Informal War. Most members of Khan's group had been arrested and served prison terms of various lengths; he, however, had escaped to Pakistan. Khan was finally arrested in Poonch in 1961, and brought to trial in the Kashmir Conspiracy Case. In 1964, when the Kashmir Conspiracy Case was withdrawn, the Jammu and Kashmir Government initiated charges against him in the 1948 affair. Given the considerable passage of time, however, gathering adequate evidence proved difficult, and the prosecution collapsed. Khan was released from jail in 1968, and opened up a poultry farm in Arizal, an area in which his old comradcs-in-arms had exercised not inconsiderable influence. This poultry farm would serve as cover for al-Fatah's training activities.

Considerable pains were taken to ensure that the organization ran along professional lines. The rules of covert organization were rigorously enforced. Zahgir was known to other members, for example, as "Rehman"; Wani as "Asad"; Fazl-ul-Haq Qureshi as "Faiz". Enquiries by al-Fatah members about the real identities of their comrades-in-arms were strongly discouraged. The idea was that even if one member broke down under interrogation and told investigators about the identities of other operatives, the information would be of little use to Indian counter-intelligence. Al-Fatah even named its safehouses and operational centres using codes, so that the names would make no sense to members who were not involved in specific operations. Most important of all, the "need to know principle" was strictly observed. Only those operatives involved in an operation knew of its existence, and only that much about it was as necessary to execute their particular role.

No actual operations, however, were initiated until Zahgir received clear instructions to act. In January 1970, he again travelled to Pakistan, this time with two new members of al-Fatah, Bashir Ahmed and Gulzar Ahmed "Khaki". Zahgir provided his handlers a detailed account of al-Fatah's organization-building activities, as well as some intelligence of military value the group had managed to acquire and copy on to microfilm. The group was now told it was time to initiate armed activities in Jammu and Kashmir, focusing on relatively soft targets like government offices, banks and treasuries. Rathore, the Pakistani intelligence officer, told Zahgir he would shortly be posted to the Pakistan High Commission in New Delhi, and would thus be available for regular consultation once the armed campaign got underway. After a fortnight, Zahgir and his associates travelled back across the CFL into Jammu and Kashmir.

Over the next few days, al-Fatah focused its energies on finding a suitable target to mark the beginning of its second phase of activity. It first considered robbing the Government Treasury at Anantnag, but then abandoned the idea, after considering the risks involved in overpowering the guards. Then, using information provided by a recruit from the Horticulture Department's offices in Anantnag, the group planned to rob its cashier. On the day planned for the operation, however, the cashier did not withdraw funds from the Treasury, and the ambush group waiting for him was forced to stand down. Finally, Zahgir selected 15 al-Fatah members to execute the crime that would announce its presence to the world. Nine of them would finally be chosen to execute the attack. It would target the office of the Education Department at Pulwama, an unlikely location for a terrorist act given its insignificant value. None the less, al-Fatah would succeed in igniting enormous concern in the corridors of power in Jammu and Kashmir.

The Pulwama Education Department dacoity

All through the morning of April 1, 1970, Mohammad Salim Gilkar kept a close watch on the employee who handled the cash transactions of the Education Department's office at Pulwama. The cashier, he knew, had withdrawn Rs. 91,000 from the treasury that morning to pay employees' dues. Gilkar hung around the office through the day, chatting with staff, so he would have some idea of how much cash would be left in the safe at the end of the day. Another al-Fatah operative, in the meanwhile, spent his day looking for four-wheel drive jeeps. He found three which could be easily stolen, one parked at the Women's Hospital, another near the Crown Hotel and a third near Budshah Bridge.

Late that night, Zahgir and nine other al-Fatah members travelled in the stolen Budshah Bridge jeep to Pulwama. Outside the Education Department office, they encountered three unarmed guards. Confronted with the group and their revolvers, two guards promptly surrendered. One, who put up more of struggle, was anesthetized with a rag soaked in surgical ether. The group then loaded the entire safe on to the back of the jeep, and drove down the deserted Awantipora-Koil road. After making an unsuccessful effort to force the safe open, the group hid it under a bridge near the town of Bijbehara, a few kilometres from Anantnag. The jeep was then dumped outside the village of Dantar. The next morning, Gul Mohammad, a sympathizer who owned a car repair shop in Anantnag, brought the tools needed to break open the safe. Inside, they found Rs. 71,847.60, an enormous sum by the standards of the time. Zahgir kept all but Rs. 90, which was handed over to the three members who had come from the District of Doda for the operation, to meet their expenses.

Not surprisingly, the Pulwama Dacoity provoked a sensation. Police investigations went nowhere. Official suspicion focused on local naxalites, members of a Maoist fringe group which at the time had a small network in southern Kashmir. The confusion helped ensure that Zahgir was able to spend the cash without arousing suspicion. He found a piece of land in the village of Barsoo, in the District of Anantnag, which seemed an ideal location for al-Fatah's headquarters. Perched on a hill, the land faced the Srinagar–Jammu highway on one side, the Jhelum River on another and a dense apple orchard on the third. A sawmill and rice-husking factory covered the fourth side. As such, anyone in the headquarters could keep an eye on the highway, and had several routes of escape if a threat arose. Since there was a regular flow of visitors to the mill, any activity at the headquarters would not seem unusual, either. Zahgir spent some Rs. 50,000 on the property to construct a building on it. Al-Fatah's headquarters was built with care, incorporating a special attic which opened out on all four sides, allowing it to be used as a watchtower.

Al-Fatah spent another Rs. 10,000 on developing long-term cover businesses. Pir Gul Mohammad, a member of the organization from Bijbehara, was told to put up a poultry farm, which could be used to offer legitimate jobs to al-Fatah members, and also serve as an emergency hideout at times of crisis. A Bhaderwah-based member, Abdul Hai, was given another Rs. 2000 to set up a cafeteria, a cover business which was to be used as a meeting place for al-Fatah operatives in the remote mountain region of Doda, and serve as a hideout if police pressure made it dangerous for operatives to remain in the Kashmir valley. Zahgir also rented a house in the Buchwara area of Srinagar, which served as a kind of political headquarters for the activities of the Young Men's League and the Students Federation. The rest of the funds were used to purchase equipment for al-Fatah, notably a camera, a tape recorder, a projector for viewing microfilm and a typewriter.

Surendra Nath, who later compiled the official reports on al-Fatah as he had done for the Master Cell, described the organization's handling of the Pulwama Dacoity with considerable respect:

> It is highly significant that the large amount which had been procured in this dacoity was utilized entirely for the purposes of underground organization and it was not distributed among the members of the gang as personal booty. This reveals a sense of duty on the part of the members of the organization to their cause. It also presents a reason why it was not possible to get any clue regarding this crime. Normally, criminals fall out on the distribution of the loot or due to other jealousies or rivalries. In the present case, Zahgir had kept a fairly strict control [*sic*] and the members were satisfied in having taken part in an operation that furthered the cause of their organization.[48]

Despite its new-found wealth, al-Fatah's technical resources and the training of its operatives in their tradecraft were meagre even by contemporary standards of espionage. Its written documents, for example, were encrypted with nothing stronger than the centuries-old substitution cipher, a system in which each alphabet is replaced with another chosen by the cryptographer. Abdul Hai, for example, became RDBLU KRF; the name of his father, Abdul Karim, was encrypted as RDBLU HRAFQ.[49] Cryptanalysts had deciphered these defences centuries earlier, using a method called frequency analysis. In essence, alphabets in each written language appear with certain frequencies. The letters "a" or "e", for example, are more used in the English language than "y" or "z".[50] By examining the frequency of alphabets in al-Fatah's documents, Indian counter-intelligence was able to decrypt its contents with great ease. None the less, the crude technical resources al-Fatah utilized proved adequate for one key task – avoiding leaks from within its own ranks.

By May 1970, Zahgir was ready to report on his activities to intelligence officials at the Pakistan High Commission in New Delhi. Travelling along with Abdul Hai, Zahgir carried a report on the Pulwama Dacoity Case, as well as some military intelligence product that had been transferred to microfilm. Little has been recorded about al-Fatah's successful military espionage operations, and no assessment of its potential impact has emerged in the public domain. Indian counter-intelligence found that the group succeeded in gathering 22 discrete sets of intelligence, including restricted Indian Army training-related documents which were on the microfilm Zahgir carried with him. Over time, al-Fatah also obtained documents containing the tactical numbers of units, the deployment of formations and troop movements. At least some appear to have been of considerable sensitivity. Nath's *Report on Underground Organisation Known As Al-Fatah*, contains among facsimiles of other classified material the first page of an Indian Army document issued by Army Headquarters in January 1968, bearing instructions that it "must not fall into enemy hands".[51]

Pakistani officials in New Delhi had little to offer Zahgir in return for his efforts, other than encouragement and comfort. His visit does, however, seem to have sparked off a serious effort by Pakistan's intelligence services to assess the credibility of al-Fatah. While the Pakistan Intelligence Bureau's Rathore had not yet been posted to the Pakistan High Commission, a suspected intelligence operative named Mufti Zia-ul-Haq, a resident of the village of Kreeri who had left for Pakistan after the Partition of India, visited the Kashmir valley and held meetings with Zahgir, Nazir Ahmad and Fazl-ul-Haq Qureshi, to discuss al-Fatah's expansion plans. While al-Fatah was focusing on recruiting new cadre in the Kashmir valley, its unit in Doda had taken up a vanguard role in covert activity. Run by Ghulam Hasan Bhat, an ethnic Kashmiri who lived in the remote Kishtwar area, al-Fatah's Doda unit generated much of the military intelligence the organization gathered. It also engineered a dramatic robbery of potassium cyanide from the laboratory of a college in Bhaderwah, for use as a means of suicide in the event that any of al-Fatah's operatives were captured.

Under Zahgir's charismatic leadership, al-Fatah was becoming a major force. Incredibly, Indian counter-intelligence still had little idea about what the organization was up to, and seemed unable to halt its activities. Al-Fatah could now turn to the next phase of the covert war Zahgir had outlined in his notes, giving political form and shape to al-Fatah's struggle.

The political war

In April 1968, soon after Zahgir was released from jail, he had received a visit from the Plebiscite Front leader, Mirza Afzal Beg. Beg had expressed his support for Zahgir's activities, although there is little to suggest that he knew precisely what these were at the time. The two men, united in their struggle to remove Jammu and Kashmir from the Indian Union, had maintained occasional contact. It was to prove a fruitful relationship.

Al-Fatah had two major concerns at the time. At around the same time that al-Fatah was planning the Pulwama Education Department robbery, the group had made the first, tentative steps towards building a political front organization of its own. Two senior al-Fatah members, Abdul Rashid Dar and Mohammad Yousaf Mir, were given charge of building the Young Men's League and the Students Federation. Both felt the need to affiliate the fledgling bodies with a mainstream political group. Second, al-Fatah had a pressing short-term problem. After the arrests in the Islamia College Dacoity, prosecutions had been initiated in that affair and the Nawakadal murder. Zahgir himself had been named an offender in the murder of Constable Charan Das, even though he continued to elude the police. A sympathetic lawyer was needed, though, to defend Syed Sarwar, Mohammad Aslam Wani, and Zahoor Shahdad, all of whom were behind bars.

Afzal Beg, an able lawyer who had energetically defended himself in the Kashmir Conspiracy Case, fit the bill on both counts. He had his own reasons to ally with al-Fatah, despite the obvious risks of association with Zahgir, who had been declared a suspect by the courts. Perhaps fearing arrest, Abdullah had for some time maintained a distance from the activities of the Plebiscite Front. In any case, the Sheikh's own political position was shifting towards the centre, leaving the Plebiscite Front with a diminishing pool of activists to work for its cause. A tactical alliance between the Plebiscite Front and al-Fatah's political bodies made tremendous sense to both sides. Abdul Rashid Dar was given charge of maintaining contact with Beg, both to discuss the defence of the al-Fatah members charged with the Nawakadal murder and to frame a common programme of agitation.

Soon after the Pulwama Education Department robbery, Zahgir, Dar and Beg held another meeting, this time in the Chashm-i-Shahi gardens on the banks of the Dal Lake in Srinagar. This time, Beg had matters other than the law to discuss. General Elections to India's Parliament were due in 1971, and elections to the Jammu and Kashmir Legislative Assembly in 1972. The 1967 elections had made it clear to Beg that the Plebiscite Front risked being marginalized, along with all it stood for, unless the organization acquired some degree of political power. He now hoped to contest the coming elections. Beg asked the Young Men's League and the Students Federation to issue manifestos supporting participation in the elections. This was necessary, he said, to persuade potential supporters that the Plebiscite Front had not sold out under pressure, and to give it legitimacy as the representative of the vanguard anti-India opinion in Jammu and Kashmir.

At a meeting on November 14, 1970, Beg finally told Zahgir that the Plebiscite Front was ready to contest the coming elections. His principal concern was that the Indian National Congress would, as it had in 1967, simply rig the elections. One major threat was the prospect of what Indians called "booth-capturing" – the takeover of voting centers by cadre of a major party, who would simply stamp all the available ballot papers in favour of their candidate, and then stuff the ballot boxes. Force, Beg argued, had to be met with force, and the activists of the Students Federation and Young Men's League were the only people available who could help the Plebiscite Front secure a respectable showing. Beg, in parting, asked Zahgir if he had any message to pass on to the Pakistan High Commission in New Delhi, where he was scheduled for a visit the next month. Zahgir said there was nothing to be conveyed, but did promise the political assistance Beg had requested.

It was the last time the two men were to meet. As Beg departed for Delhi, with politics on his mind, Zahgir prepared al-Fatah for its next military operation.

The Hazratbal bank robbery

Soon after the Pakistan envoy Mufti Zia-ul-Haq's departure, al-Fatah began considering its future course of action. For Zahgir, there were two major concerns. First, al-Fatah was strapped for cash. Second, the organization desperately needed new military hardware if it was to make a greater impact than that which could be achieved by a few robberies.

Zahgir's early plans for a second robbery were, therefore, considerably more ambitious than those al-Fatah had successfully executed in Pulwama. He planned to attack the sole branch of the State Bank of India in Srinagar, hoping that it would not only yield a large amount of cash, but also prove to be a blow to the prestige of the Government of India. The idea, however, was abandoned, after it became clear that personnel from the Kothibagh Police Station, located across the road, would intervene even if al-Fatah cadre succeeded in neutralizing the guards at thc bank. Some in al-Fatah argued that this problem, too, could be overcome, if the group managed to acquire automatic weapons. From Salim Jehangir, Zahgir knew that a large cache of sten guns already existed, dumped in a pit near Chrar-e-Sharif. However, Haji Jalal-ud-Din, the Pakistani operative holding the weapons, had strict instructions not to make them available to al-Fatah until he received orders from Pakistani intelligence. While Zahgir initiated efforts to get the sten-guns buried in Chrar-e-Sharif released, Nazir Ahmad was charged with putting together a relatively limited operation – robbing the Jammu and Kashmir Bank's Hazratbal branch, near the Kashmir University campus in Srinagar.

He did so with aplomb that would have made professional bank robbers envious. Pay day for government employees fell on January 2, 1971, since the previous day had been a holiday. Zahgir personally tailed the bank employee charged with drawing cash for the branch. Once the cash had reached the bank, it had been agreed, Zahgir would drive past the Lal Bazaar seated at the rear of a hired taxi; if there was a hitch, he would sit on the right-hand side next to the driver, with his arm hanging out of the window. Ahmad and three other al-Fatah operatives, all dressed in police uniforms acquired from a tailor who made them for personnel of the force, waited at the Lal Bazaar for the signal. Once Zahgir drove past them, seated at the rear to signal the "all-clear," the group made their way in another taxi to Hazratbal.

Dressed as a Deputy Superintendent of Police, Nazir Ahmad Wani confidently walked into the bank, claiming he was investigating an embezzlement that had just taken place at the Government Treasury. He demanded that the manager and cashier hand over all funds and record books at the bank, and then accompany them to their waiting vehicle. The apparently befuddled manager and cashier complied, without asking even for Ahmad's name or evidence of his authority. Short of Lal Bazaar, the group ordered the manager and cashier out of the car. Both now realized the situation was, to say the least, odd, and attempted to argue with the "policemen". It was a futile effort, for the al-Fatah operatives had guns.

After ridding themselves of their clueless hostages, the group drove to a pre-arranged rendezvous with Zahgir on the outskirts of Srinagar. There, they changed into civilian clothes, tossing their uniforms into a sack which was weighted and thrown into the Jhelum River. All the al-Fatah members now made their way back to Srinagar. Near the shrine of Hazrat Sultan-ul-Arifeen – the same building al-Fatah had planned to destroy not many months earlier – the group crossed the Jhelum, hoping to throw off police sniffer dogs that might be brought in to follow their scent. Zahgir now counted the money they had stolen, down to the smallest piece of change. It turned out to be a healthy Rs. 97,175.76, which was divided between Zahgir and another al-Fatah member, Mohammad Yusuf, for safe keeping.

Zahgir could have been forgiven for thinking that the operation had gone flawlessly. But once again, the encounter with the shrine brought bad luck to his group. Back in Srinagar, police authorities had got their first real break in the hunt for al-Fatah. One of the men in uniform, the bank cashier told investigators, seemed familiar, resembling a student he had known while at Srinagar's Sri Pratap College in 1967. When police officials presented the cashier with photographs of students who had been at the College around that time, he quickly identified the man he knew as Farooq Ahmad Bhat. Bhat, it turned out, was well known. As he was the son of a prominent politician who had served in the Jammu and Kashmir Legislative Assembly, Bhat's involvement in the Hazratbal robbery intrigued investigators. An outstanding student who had been elected to head his students' union – and had once been voted the best-dressed student – Bhat was pro-Indian in his public pronouncements and decidedly secular in his personal life.

Police officials promptly raided Bhat's home, but he had fled. They then searched the rooms he had occupied at the Medical College Hostel, even though these had been vacated some time earlier. The police were in luck. Among his books, they found a small diary with seven names scribbled on one page. One of these was of Bhat's cousin, Abdul Ghaffar. To the delight of police officials, Ghaffar was at home. During his interrogation, the contours and significance of the Hazratbal robbery began to unveil themselves for the first time. Abdul Ghaffar told the police that he had been recruited to al-Fatah some time earlier, by a Kupwara resident named Ghulam Nabi Mir. Shortly afterwards, he said, Bhat had told him during a casual encounter that he knew of his decision to join the covert organization. From this, Abdul Ghaffar had surmised that Bhat was a senior figure in al-Fatah. During a subsequent meeting with Zahgir, Ghaffar said, his speculation was confirmed.

Bhat's cousin, however, knew the names of only a handful of low-level al-Fatah operatives – and all of them, it turned out, knew their leaders only by their code-names. Of Bhat himself, there was no trace. What the police did learn was that all of al-Fatah's senior members were close to the owner of a Srinagar *paan*-shop, Mohammad Yusuf Mir. Through all of South Asia, *paan* shops, which sell betel leaf and tobacco substance, cigarettes and tea, are neighbourhood gathering places and centres for both political discussion and gossip. For all of al-Fatah's careful security measures, its top leadership had been careless enough to congregate together at one place. Mir was arrested and interrogated for several days. Finally, on January 16, 1971, he cracked and told police officials of the al-Fatah safehouse at Barsoo. Deputy Inspector-General of Police Pir Ghulam Hassan Shah, who was eventually to become one of the most highly regarded officials in the state, personally led the raid, along with another officer who acquired prominence at the outset of the phase of the jihad which would commence in 1989, Deputy Superintendent of Police A.M. Watali.

Nazir Ahmad and Farooq Bhat were both in the Barsoo safehouse when the police arrived. Both opened fire with their revolvers, but were taken alive. Ahmad, it turned out, was known to those living around the Barsoo safehouse as a servant, and actually did all the menial work in the home to maintain his cover. Inside, the police found explosives, large numbers of documents, microfilm and Rs. 8,600 in cash. Watali, however, did not rest on his laurels. Village residents were gathered, and agreed to keep the news of the raid a secret. With their help, Watali arranged for a watch to be kept on the house, hoping other al-Fatah members would drop by. He was right. Fazl-ul-Haq Qureshi was arrested there at 11:00 AM on the morning of January 17, and Abdul Hai was picked up later that very evening.

Fazl-ul-Haq Qureshi's interrogation was to finally lead the police to Zahgir. In the course of the day, he told the police about another safehouse that had been established in Srinagar's Shahidgunj area just before the Hazratbal robbery. Zahgir, he said, was living in the new safehouse, and had at least two firearms in his possession. Police officials in plain clothes promptly placed the premises under surveillance. Soon after Zahgir reached the safehouse at 10:00 PM, two police teams burst into his top-floor room and disarmed him. From Zahgir, police investigators learned for the first time of the details of the Pulwama Dacoity, and the existence of the political wing of al-Fatah, as well its links with Beg. They also found a treasure of documentation on the organizational structure of al-Fatah in the safehouse, as well as some classified Indian military papers. Soon afterwards, the automatic weapons cache stored by Haji Jalal-ud-Din at Chrar-e-Sharif was recovered – the largest haul of its kind until the ongoing phase of the jihad in Jammu and Kashmir.

What direction al-Fatah might have taken in the build-up to the 1971 war, had the Hazratbal bank cashier not recognized his one-time friend? We shall, of course, never know for certain. During his interrogation, Zahgir told the police that he had planned to initiate a series of kidnappings and assassinations when campaigning for the upcoming General Election began. Zahgir, the police recorded, said they had even "prepared certain rooms in their Headquarters at Barsoo for confining the kidnapped VIPs".[52] It is also possible that al-Fatah's political wing would have intensified its activities. According to Indian counter-intelligence, al-Fatah's Abdul Rashid Dar had met Beg at Jammu on January 7 and 8, 1971, days after the Hazratbal robbery. Beg, by this account, asked Dar for an escalation of anti-India activity. Zahgir received Dar's message on January 15; the same day, members of the Young Men's League and the Students Federation attempted to hijack the stage at Friday congregational prayers in Srinagar, Sopore and Anantnag to deliver anti-India speeches.

Like the Master Cell, al-Fatah had been shut down well before it could achieve its ends. Its members would watch the war of 1971 from behind prison bars, unable to intervene in a conflict that would transform the balance of power in South Asia and shatter, for over a decade, the dream of throwing India out of Jammu and Kashmir.

Shattered dreams

Pakistan's covert services had hoped that al-Fatah would serve as the nucleus of a revolutionary people's war – in other words, that it would succeed where the Master Cell and the war of 1965 had failed. The experiences of the winter of 1970–1971 laid these fond beliefs to rest: indeed, the destruction of al-Fatah would be mirrored by the dismantling of much of the infrastructure of anti-India mobilization in Jammu and Kashmir.

At the beginning of 1971, almost 350 cadre of the Plebiscite Front, were arrested under the Preventive Detention Act, a controversial piece of legislation that gave authorities enormous power to detain individuals they believed posed a threat to the security of the state. Zafar Iqbal Rathore, the Pakistan

Intelligence Bureau officer who had handled al-Fatah, did arrive under diplomatic cover in New Delhi, but far too late to help the child he had helped give birth to. A fortnight after the arrest of the al-Fatah conspirators, he was declared *persona non grata*, and ordered to leave India. Beg, for his part, was served with orders prohibiting him from entering Jammu and Kashmir for three months. Sheikh Abdullah himself was ordered out of Jammu and Kashmir, although he had no demonstrable connection with al-Fatah, along with his son-in-law, Gul Mohammad Shah.[53] With war looming, the authorities evidently did not want to take chances. Sadiq, who had taken power with a commitment to democratizing Jammu and Kashmir, had been pushed towards the same authoritarian postures as his predecessor, Bakshi, by Pakistan's continuing covert offensive.

At one level, al-Fatah was merely a continuation of the existing Pakistani policy on Jammu and Kashmir; of the search to end Indian rule over the state through both conventional and sub-conventional military means. Yet, in several respects, al-Fateh was very different from its predecessors. Unlike the covert warfare units formed in 1947–1948, or the Master Cell and its subsidiaries in the build-up to the war of 1965, it was not an enterprise intended to be a mere ancillary for regular Pakistani forces. It was, rather, designed to develop and sustain the capabilities to wage a war of liberation by itself. Pakistani nationals played no major role in its actual operations, while that country's intelligence services confined themselves to providing logistical and material support on its own side of the CFL. Even Mufti Zia-ul-Haq's fact-finding mission to Kashmir did not result in a flow of funds or equipment to al-Fatah, forcing the organization to resort to the criminal enterprises that eventually caused its undoing.

How does one account for Pakistan's evident unwillingness to unleash the weapon it had worked so hard to build? As we have seen, Zahgir's desperate efforts to acquire automatic weapons *already* in place in Kashmir were shot down by Pakistani intelligence. Members of al-Fateh have also made it clear that they had succeeded in recruiting the cadre needed to use those weapons. "I traveled to every village and town in the Valley", Fazl-ul-Haq Qureshi told an interviewer, "recruiting boys to send them across the Line of Control for arms training".[54] By his account:

> Three hundred boys in six batches were sent to Pakistan by 1971 for arms training by his outfit. The boys came back to the Valley after training in guerilla warfare. But they had not been given weapons, as Pakistan had decided that the time was not ripe to arm the jehadis [jihadis].[55]

Indian intelligence functionaries involved in the al-Fatah investigation dispute Qureshi's claims, suggesting that no more than 200 recruits were tapped, of which less than 25 received training in Pakistan. Whatever the figure, though, it is clear that the war of 1965 had compelled Pakistan to keep the numbers involved in the secret jihad, as well as its intensity, at minimal levels. India had demonstrated its willingness to respond to Pakistani military provocation in Jammu and Kashmir by crossing the international border in the Punjab, and had threatened that country's most important cities in the process. General Yahya Khan clearly understood that support for war aimed at the liberation of Kashmir could, very easily, turn into a war leading to the annihilation of Pakistan. By 1970, with India's military modernization and expansion processes well underway, Pakistan had no stomach for a fight. Also, as we shall see in the next chapter, Pakistan's military and political establishment was in any case focused on managing an insurrection in its eastern wing, which is now an independent country, Bangladesh. Even as Pakistani intelligence was considering how best to calibrate sub-conventional warfare, their Indian counterparts were busy waging a successful covert campaign that would end in the vivisection of Pakistan. On the eve of this coming war, Pakistan had no desire to be the author of a provocation that would legitimize Indian military intervention.

Provocation, however, would indeed be provided – and would, some years down the road, bring al-Fatah's story to a close. After the end of the war of 1971, Prime Minister Indira Gandhi capitalized on India's decisive victory to negotiate a political settlement with Sheikh Abdullah. Abdullah eventually agreed to some limited political autonomy for Jammu and Kashmir, renouncing his demand for a plebiscite once and for all. As part of the political settlement, Sadiq's successor in office, Syed Mir Qasim, launched an enterprise to win over elements of the group. Pir Ghulam Hassan Shah, the officer who had broken the al-Fatah network, was now charged with winning its cadre over. Those charged with minor crimes were released; the rest were provided with special facilities for their education and care at the Central Prison in Srinagar. Qasim later wrote:

> I was under pressure from the parents of the arrested youths to give a humanitarian consideration to their case. When I studied their case histories, I was upset at the kinds of crimes they had committed. [But] I was after all a father and, therefore, could not take refuge under the cold crime-and-punishment principle. I told the State Assembly on March 25, 1972, I can swear that I suffer the same pain as do the parents of these young people. It is not the fault of these youths; but it is the responsibility of those institutions and practices which were not constructive.[56]

Shah's psychological operation worked. In 1975, the bulk of al-Fatah's cadre went mainstream, forming the Inquilabi Mahaz, or Revolutionary Union, which supported the Indira Gandhi–Sheikh Abdullah agreement. Fazl-ul-Haq Qureshi, along with Nazir Ahmad Wani, Hamidullah Bhat, Mohammad Shaban Vakil and Farooq Ahmad Bhat, was among the small group who rejected the deal.[57] He would continue to fight for the cause he had served since the mid-1960s with unswerving zeal. By 1980, however, he was no longer considered a significant threat to the Indian state – and the years of struggle had taken their toll. "Long years of absence from employment had made a major dent on my financial status", he told an interviewer, "with children to raise, I wanted to stay on the job".[58] Qureshi resigned his membership of a secessionist political organization, the People's League, and in return the government dropped his prosecution and gave him back his job.

It was a decision the Indian government – and the police officers Pir Ghulam Hassan Shah and A.M. Watali – would have plenty of opportunity to regret.

5

YEARS OF RETREAT AND REVIVAL

Ek Haal, ek umang, guerilla jang, guerilla jang;
Azadi ka ek hi dhang, guerilla jang, guerilla jang
[Our one objective, our one desire, guerilla war, guerilla war;
There is only one way to freedom, guerilla war, guerilla war]

Anthem of the Jammu and Kashmir National Liberation Front[1]

For a few hours in the autumn of 1966, it may have appeared to villagers near Sopore that war had broken out again. Shots were fired; men were killed. Journalists came in to witness the firefight, which, though small in scale, provoked a sensation. Then, almost as suddenly as it had begun, the affair was forgotten. "Least did anyone realize", a journalist later recalled, "that this would become a major turning point in Kashmir's recent history".[2]

In November 1965, just after the end of the war which had turned out so badly for Pakistan, the newly formed National Liberation Front [NLF], headquartered in Muzaffarabad, decided to initiate its first armed operation in Jammu and Kashmir. One group, led by Mohammad Maqbool Butt was to recruit personnel and execute political tasks, including the setting up of covert cells. Another led by Major Amanullah, a former soldier in Pakistan's Azad Kashmir forces who hailed from Kupwara, was to train new cadre in the use of explosives and small arms. Two other former soldiers, Subedar Kala Khan and Subedar Habibullah Butt, joined the group under Major Amanullah's command. Using the services of a Kupwara-based guide, Ayub Khan, the two NLF groups crossed the cease-fire line on June 10, 1966.

The NLF's first experience of sub-conventional warfare was to prove less than happy. While the military members of the NLF unit busied themselves preparing rudimentary training facilities in the forests above Kupwara, along the Cease-Fire Line, Maqbool Butt travelled through major cities and towns attempting to find recruits for them. The enterprise rapidly led to disaster, most likely as a consequence of betrayal. Two months after Maqbool Butt had entered Jammu and Kashmir, he and a new recruit, Mir Ahmad, were interdicted by a police patrol. Amar Chand, a police official, was shot and killed in an exchange of fire.[3] Police authorities responded by initiating a massive search-and-cordon operation. The NLF group was soon tracked down and, after a firefight claimed the life of one of its members, they surrendered. In August 1968, a Srinagar court sentenced Maqbool Butt and Mir Ahmad to death; Kala Khan was awarded life imprisonment.[4] Major Amanullah, for his part, succeeded in escaping the cordon and returned to Pakistan.[5] Indian authorities made several subsequent arrests, but otherwise paid little attention to the group.

Indian disinterest was in one sense well founded: in and of itself, the NLF was at worst a minor nuisance. Nor was it seen as a resource by Pakistan's covert services. During the life of the Master Cell, Pakistan's covert services did not lack cadre for their secret jihad in Jammu and Kashmir. Sattar Khanday, one of the Master Cell's key operatives, had subsequently told his interrogators that large numbers of new cadre were being trained, and that in a future war Pakistan would be in a position to "unleash wave after wave of infiltrators".[6] Members of al-Fatah, in turn, had made clear they had the trained personnel needed to wage war on India. In the wake of its 1965 defeat, however, Pakistan needed to calibrate the jihad carefully, to ensure that it did not escalate into full-blown war. In al-Fatah, it had its chosen instrument, tightly controlled and disciplined. Freelance expeditions of the kind the NLF had mounted were not the kinds of enterprises Pakistan's covert services were interested in encouraging.

During much of the first two decades of its existence, the history of NLF resembled the course of a kite without a string. Its destiny was, notably, to be guided by a war fought thousands of kilometres away. Ending in the vivisection of Pakistan, the war of 1971 brought an end to Pakistani military support for the secret jihad in Jammu and Kashmir. A defeated Pakistan, wracked by internal political and economic crisis, was in no position to act in a manner India would consider militarily provocative. Within Jammu and Kashmir itself, anti-India forces reacted with despair to the reality that Pakistan was no longer a credible military ally. However, a small core of covert operatives, despite being militarily ineffectual, were able to keep the idea of liberation from Indian rule alive until more favourable times arrived. Their contribution to the events of the late 1980s and the 1990s, as we shall see in subsequent chapters, was enormous. The life of Maqbool Butt has a key role in this narrative, just as his death would have an extraordinary influence on the imagination of those who succeeded him on the frontlines of the war against India.

February 11, 1984, the date of Maqbool Butt's execution, is commemorated with strikes and protests by Kashmiri nationalists across the world; one prominent secessionist organization describes him as "the first authentic martyr of the

Kashmiri independence movement".[7] "Each year", Rafique Khan has noted in a perceptive essay,

> more mythology is added as the legend of Maqbool Butt expands. . . . [he] may one day join the ranks among the 'rishis' (sages) of Kashmir who are believed to have such powers that they could travel by air, mounted on a stone boulder, when their mounts tired.[8]

This chapter is, in large part, about the events which enabled the myth of Maqbool Butt to be manufactured.

The making of the NLF

For a figure central to the covert war in Jammu and Kashmir – and one revered by a substantial spectrum of political opinion in the Kashmir valley itself – we know relatively little about Maqbool Butt. His life, more than that of most leaders, was wrapped in ironies: jailed in Pakistan on suspicion of being an Indian covert agent, he was eventually executed as a traitor to India.

Like many of those active in the early phases of the anti-India covert movement in Jammu and Kashmir, Butt was closely allied with Sheikh Mohammad Abdullah and the National Conference. The son of a tailor and farmer in the village of Trehgam, Butt was part of the post-Independence generation of rural ethnic Kashmiris who benefited substantially from the National Conference's rise to power. Butt's first experience with politics was through peasant resistance to the local *jagirdar* or feudal landlord. When he was eight years old, Butt wrote in a 1973 letter, a crop failure led to a brutal crackdown on tenant farmers who were unable to deliver their scheduled levies of grain. The *jagirdar*, Butt recalled, "insisted on having his usual share of the crops even if it meant that the children of the peasants had to starve to death".[9] Eventually, "all the children of the village were told to lie down on the road in front of the [landlord's] car by their elders". "Hundreds of children lay down in front of the *jagirdar*'s car and pleaded for a concession, for the writing off of a share of the levy, or for the car to drive over them", Butt recalled. The resistance worked: a concession was granted.

In 1958, Sheikh Abdullah was arrested for the second time, on charges of having collaborated with a conspiracy to evict India from Jammu and Kashmir. Butt, 19 years old at the time, was a student at the St Joseph's College in Baramulla (a part, interestingly, of the same institution where Pakistani irregulars had raped and murdered nuns during the war of 1947–1948).[10] Like thousands of young people beholden to the National Conference, he joined protests against Abdullah's detention. Under pressure from the police, he then spent some months underground and finally crossed the cease-fire line into Pakistan.[11] It is unclear from the available literature just why Butt chose this course of action. Most of those who were active in the protests against Sheikh Abdullah's arrest, as we have seen in earlier parts of this book, remained on in India, and continued their political life with varying degrees of enthusiasm. Whatever the reason,

Butt spent some time in Muzaffarabad, before moving to Rawalpindi, where he worked as a journalist and attended university courses in Urdu.[12]

Butt had no intention of abandoning his political interests, though. In 1962, he formed the Kashmir Independence Committee to lobby the government of Pakistan-administered Kashmir to take a more aggressive position on events on the Indian-controlled side of the Cease-Fire Line. Then, in April 1965, the Kashmir Independence Committee was merged into the newly formed Plebiscite Front of Pakistan-administered Kashmir, a branch of the organization led on the other side of the cease-fire line by Sheikh Abdullah's lieutenant, Mirza Afzal Beg.[13] Many in the new Plebiscite Front, inspired by struggles in Algeria and elsewhere, demanded armed struggle. The Executive Committee of the Plebiscite Front's Pakistan-administered Kashmir branch formally considered the demand on July 12, 1965, and rejected it by an overwhelming majority.[14] It is unlikely, as this event shows, that Butt and others in the Plebiscite Front had any knowledge of Operation Gibraltar, which was well underway by that time. Almost certainly, they had no idea of the activities or even existence of the Master Cell.

It was only on August 13, 1965, just a month before the end of the war of 1965, when a decision was finally taken to form a guerilla organization. Patterned, much as al-Fatah was, along the lines of the Algerian Front de Libération Nationale [FLN: National Liberation Front], the NLF was most likely born of the realization amongst Kashmiri nationalists that they needed to compete with the official jihad being run by Pakistan's covert services or risk marginalization. Major Amanullah was charged with heading the armed wing of the organization while Amanullah Khan (no relation) took charge of the political wing. The financial wing was headed by Mir Abdul Qayoom, and Butt himself was made responsible for co-ordination between these bodies. The heads of these four wings together constituted the NLF's Central Committee. At the first NLF Central Committee meeting, it was decided that whosoever decided to join the organization would have to sign a membership agreement in their own blood, and take an oath undertaking to lay their life down for the organization. The four also agreed on an anthem for the NLF. Through October and November that year, the NLF's founders toured west Pakistan seeking support for the organization. While the NLF does not seem to have secured any official backing, it had considerable success recruiting members from among the bureaucracy of Pakistan-administered Kashmir: the province's Defense Secretary, Mir Hidayatullah, and a Sub-Divisional Magistrate, Ghulam Din Ashai, were among those who joined.[15]

But the NLF's ill-conceived expedition across the cease-fire line in 1966 showed just how ill-prepared it in fact was for covert warfare: the organization had no real pool of cadre within Jammu and Kashmir, nor a functional infrastructure. Butt's arrest, moreover, put the existing group of anti-India activists in Jammu and Kashmir at risk. A number of university teachers sympathetic to the NLF, for example, were arrested, a loss of valuable sympathizers incurred without profit.[16] The Plebiscite Front's Pakistan-administered Kashmir branch reacted to the disaster by backing out of the enterprise. It demanded that the NLF disband itself, asserting that it did not believe the time was right for armed struggle "at this stage".[17] Under other circumstances, Butt's arrest might just have meant the end of the organization. In December 1968, though, two years after he was arrested, Butt escaped from jail along with Mir Ahmad, and a third prisoner held on separate charges, Ghulam Yasin. Having tunnelled their way out of the Srinagar prison complex, the three made their way back across the cease-fire line into Pakistan-administered Kashmir.

Much to the astonishment of the would-be heroes, their return was greeted with anything but delight by Pakistan's military and intelligence establishment. The three men were promptly detained in Muzaffarabad, where they were brutally interrogated on suspicion of being Indian agents.[18] Although the men were released three months later, suspicions seemed to have long persisted among Pakistani counter-intelligence – and among some Indian commentators. Ajit Bhattacharjea, for example, has asserted that Butt was a "colourful double agent used both by India and Pakistan".[19] In his statement at a subsequent trial before a special court in Pakistan, an officer of the military counter-intelligence organization, the Field Intelligence Unit, Major Naseer Gul again voiced suspicion that Butt's escape from jail had not been above-board. Butt responded by telling the court that Gul's claims were founded not on fact, but on ideological bias. His language drew squarely on Maoist concepts that were fashionable amongst contemporary revolutionaries. In Butt's vision, he was engaged in a peoples struggle against power, not a jihad against a Hindu state:

> Generals hate the concept of [a] peoples' army because it challenges the monopoly of [the] Generals on military resources. It is a historical fact that military dictators never supported any organization engaged in liberation war[s].[20]

Out of jail, Butt continued his efforts to build up the NLF. Stories of his dramatic escape from prison had fired the imagination of many young people in Indian-administered Kashmir. In March 1969, a young student named Mohammad Altaf Khan – later to become famous by the somewhat vainglorious pseudonym Azam Inqilabi or "Great Revolutionary" – made his way across the Cease-Fire Line. Altaf Khan had been involved in anti-India activism since 1966, when he formed an organization known as the Muslim Liberation Front. The Muslim Liberation Front attempted to fabricate explosives, but never succeeded in putting together anything more destructive than a firecracker. Neither Altaf Khan nor his associates were ever arrested, which suggests that their activities were neither of a scale or character likely to invite the attentions of Indian's security apparatus. Altaf Khan was, however, taken more seriously by security authorities in Pakistan. Shortly after he crossed the cease-fire line, Altaf Khan was arrested by Pakistani border guards and was held in Muzaffarabad jail on charges of having crossed into the country illegally. Altaf Khan had to spend six months in jail, during which time he was intensively interrogated. His interrogators were, presumably, satisfied with Khan's motives, for he was never charged with espionage. A court subsequently acquitted him on the charge of illegal border crossing, asserting that "for a Kashmiri, crossing the LoC [Line of Control] is not a crime".[21]

Out of jail, Altaf Khan made contact with Muzaffarabad-based NLF and Plebiscite Front activists. Amanullah Khan, the NLF's military chief, arranged some basic military training for him, and eventually put him in touch with Butt. Altaf Khan was, by his own account, seen as "a future leader of the movement" by his mentors.[22] It seems a reasonable claim, for in May 1970, along with three other NLF cadre, Altaf Khan was tasked to lead an NLF group across the cease-fire line. Like the first NLF expedition, this second foray was another disaster. The four-man NLF unit had to walk for a day and a half before reaching the last Pakistani border outpost in the Kupwara sector, at Ath Muqam. There, they had to wait for night to fall. "In those days", Altaf Khan was later to recall, "there was no support from the [Pakistan border] Rangers in providing cover fire for intruding into Kashmir. In fact, the threat of arrest loomed large on both sides."[23] Under cover of darkness, the group moved into the district of Kupwara, on the Indian side of the cease-fire line. By Altaf Khan's account, it was a miserable trek. The group had run out of food and water, and adding to their hardship, heavy snow made their progress near-impossible. Indian border guards from the Border Security Force [BSF] interdicted the worn-out group. "Shivering with cold and weak with hunger", Altaf Khan recorded, "we could not even lift up our arms in the gesture of surrender. Eventually, I managed to mumble some words in English to the commander of the [BSF] group."[24]

Once again, then, the NLF's plans to go to war had fallen apart. Its problems were clear: the organization had neither the funds nor infrastructure or support needed to make a real impact within India. How does one account for the small scale of the NLF's activities, and the apparent disinterest of Pakistan's intelligence establishment in encouraging them? One commentator has suggested, based on an interview with Altaf Khan, that Pakistan was not, at that juncture, "showing any interest in encouraging youth from Kashmir to cross over for training in guerilla warfare against India".[25] This proposition is not, as the story of al-Fatah makes clear, wholly correct. In fact, Pakistan was running a carefully calibrated covert campaign within Jammu and Kashmir at the time. Unlike al-Fatah, however, the NLF had no established relationship with the Pakistani intelligence establishment, which in addition may have been wary of its nationalist ideology and suspicious of the antecedents of its personnel. Although no archival material has so far become available on the issue, it seems reasonable to believe that the managers of Pakistan's covert war may in addition have been apprehensive that NLF activity may also have had the unwelcome consequence of sparking off Indian intelligence activity that could lead to the exposure of al-Fatah.

Most important of all, though, while the NLF was drawing up plans for the liberation of Jammu and Kashmir, Pakistan's military and intelligence services had other things on their mind. A crisis over a thousand kilometres away was demanding their undivided attention, and trouble with India was the last thing they needed.

Bangladesh

Travelling through East Pakistan in 1970, the bureaucrat Siddiq Salik observed that "the women had hardly a patch of linen to preserve their modesty"; the men, he recorded, "were short and starved". "I concluded", Salik wrote, "that the poor of Bengal are poorer than the poorest of West Pakistan".[26]

From the outset, Pakistan's founders had struggled to manage relations between its two wings, sundered by over 1500 kilometres of Indian territory, but also language, culture and economic disparities. As early as 1948, widespread protests had broken out as a result of the Pakistani state's decision to establish Urdu, little-used in the East, as the sole official language. On one occasion that year, students had shouted down Mohammad Ali Jinnah himself when he attempted to defend the one-language policy. A January 1952 speech along similar lines by Prime Minister Khwaja Nazimuddin had even more disastrous consequences, provoking a general strike and rioting.

Language was just one part of a larger problem between East and West Pakistan. Religion, it was turning out, was not enough to bind the two halves of the new nation together. As Owen Bennett Jones has pointed out, Bengali activists were perceived "anti-Pakistani conspirators infiltrated by Hindus", rather than "citizens with legitimate grievances".[27] Efforts by Pakistan to suppress Bengali nationalism after martial law was imposed by General Ayub Khan in 1958 also came to nothing. In 1966, Sheikh Mujibur Rahman, the leader of the Awami League, demanded a drastic restructuring of federal relations with East Pakistan, notably the restriction of the central government's powers to defence and external affairs. Among other demands, the six-points resolution also called for East Pakistan to have its own militia, and for a separate currency to prevent capital flight to the Western wing. Much of this was a response to the growing economic dominance of West Pakistan. Although Pakistan's foreign exchange reserves were heavily dependent on the East's jute industry, much of the wealth flowed West. In 1970, almost half of East Pakistan's industrial assets were controlled by six non-Bengali industrialists.[28] Paradoxically, the relatively high rates of growth during Ayub Khan's regime, inequitably distributed as they were, only served to fuel East Pakistani resentment. One commentator has offered this perceptive analysis:

> The country leaped forward in economic terms during the Ayub era, but the political rights of the people were buried in the process. Bengalis felt it more because their presence in the civil-military bureaucracy was only symbolic. In the army, the most important institution in the country, there were only 300 Bengali officers out of 6,000.[29]

Pakistan's establishment described the six-points resolution as treasonous. The then Law Minister, S.M. Zafar, vowed that its advocates would be "identified, hunted, crushed and destroyed".[30] An opportunity to attempt just such an enterprise presented itself in 1967, after Pakistani security personnel overheard a group of men discussing plans for a coup. According to Pakistani investigations, the conspirators intended to assassinate Ayub Khan on a Dacca–Chittagong flight, and then declare an independent state in East Pakistan. Mujibur Rahman himself was charged with complicity in the plot, and put on trial along with some 50 Bengali civil servants, politicians and military personnel. Pakistan's military believed that the Indian intelligence services had a role in the affair, although, as Hassan Abbas notes, "no evidence was ever made public".[31] This was because Mujibur Rahman's arrest triggered massive public protests, forcing the military regime to drop the prosecution. It was a political decision, just as the decision to drop the Kashmir Conspiracy Case against Sheikh Abdullah had been some years earlier.

Yet, the release of Mujibur Rahman did not still dissent. By the time General Yahya Khan replaced Ayub Khan in March 1969, Dacca was gripped by almost daily protests. Yahya Khan eventually arrived at a political accommodation with Mujibur Rahman by promulgating a Legal Framework Order which opened the way for elections. In return, Mujibur Rahman expressed his willingness to negotiate elements of the six-point resolution, and made clear "his objective was autonomy, not secession".[32] The Awami League won 160 of 162 directly elected seats in East Pakistan, and none in the West; Zulfikar Ali Bhutto's Pakistan People's Party took 81 out of 138 seats in the West and none in the East. In January 1971, a triumphant Mujibur Rahman demanded that the six-point resolution be enshrined in the new constitution that was to be framed by the Assembly. From Yahya Khan's point of view, this display of intransigence was a stunning volte-face. It was not, however, entirely unexpected. Pakistani intelligence had, in the late-1970s, bugged a meeting between Mujibur Rahman and one of his senior aides, where he asserted that his aim was to "establish Bangladesh; I shall tear the LFO to pieces as soon as the elections are over".[33] By the accounts of insiders, however, Yahya Khan had paid little attention to the tape presented to him by his intelligence services, asserting only that he would "fix Mujib if he betrays me".[34] The General's confidence may have been based on Pakistani intelligence assessments which had predicted that the Awami League would pick up only 60 per cent of the popular vote in East Pakistan – well short of the near-sweep it eventually recorded.[35]

If Mujibur Rahman's stance was hardening, Bhutto for his part made clear that the People's Party would not sit in the opposition despite its defeat. Neither side, it began to become apparent, was willing to give ground. Within the military, voices calling for an aggressive handling of the situation were gaining ground. Plans for using force in East Pakistan were drawn up. Two days before the National Assembly was scheduled to meet, Yahya Khan called off the opening session. On the streets of Dacca, the General's actions provoked a revolution.

Pakistani troops were overwhelmed by protestors, and students paraded through the streets of the city waving the flags of their new nation, Bangladesh. Mujibur Rahman soon proclaimed that he was taking charge of administration of East Pakistan, thus issuing a de-facto declaration of independence.

No one is quite certain just how many tens of thousands of people died in the carnage that followed. On March 26, 1971, Yahya Khan ordered the initiation of Operation Searchlight, a brutal military pogrom directed at people Pakistan claimed were its citizens. Pakistan's military had believed that a short display of force would serve its purpose; instead, the "Bengali population stood full square behind their arrested leader, Mujibur Rahman".[36] In response to this defiance, the Pakistan Army unleashed its full wrath, "raping, murdering and even massacring whole villages, women and children included". Elements of the Islamist Jamaat-e-Islami party, eager to demonstrate their support for the Army, joined in the killing of those they claimed were "enemies of Islam".[37] By the estimate of the officer who directed the campaign, General Tikka Khan, West Pakistani troops killed at least 30,000 people.[38] Among them were hundreds of soldiers of the East Bangladesh Rifles who had refused to disarm when ordered to do so at the outset of Operation Searchlight. Some 17,000 soldiers crossed the border into India over coming months, along with millions of civilian refugees – student activists, politicians, and ordinary people who had come under attack for no reason other than their ethnicity.

By May 1971, the Pakistan Army believed it had crushed resistance in East Pakistan. It was to prove a disastrous miscalculation.

The Ganga hijacking

At 1:05 PM on January 30, 1971, just three months before the East Pakistan massacres began, an Indian Airlines flight carrying 26 passengers and 4 crew members made an unscheduled landing at Lahore Airport. On a flight from Srinagar to Jammu, the Fokker Friendship F-27 had been hijacked by two men who claimed to be armed with a hand-grenade and a pistol. No blood was spilt in the course of the hijacking, but it was to have a central role in events that would claim a good many lives.

Towards the end of 1969, a young Srinagar resident named Hashim Qureshi had travelled to the Pakistani city of Peshawar in connection with arrangements for the marriage of his sister.[39] It is unclear just how he ran into Maqbool Butt, but the NLF leader, after his spectacular jailbreak, needed no introduction. Although Hashim Qureshi had little past knowledge of politics, and none whatsoever of Pakistan-based Kashmiri groups, he promptly volunteered to work for the organization. Before his return to India, Hashim Qureshi was given an extensive education in NLF ideology, as well as some rudimentary training in handling weapons and guerilla tactics. Since Butt's release from jail, and the dismal failure of Altaf Khan's mission, the NLF had been contemplating what it might do to draw global attention to its fight against Indian rule in Jammu and Kashmir. Its attention turned – as had al-Fatah's, some years earlier – to events in west Asia. In September, 1970, the Popular Front for the Liberation of Palestine [PFLP] had hijacked five aircraft, three of which it later blew up at Dawson's Field in Jordan.[40] The hijackings had provoked international outrage, but also drew global attention to the Palestinian liberation movement. A similar action, the NLF hoped, would yield this same result for the anti-India movement in Jammu and Kashmir. Hashim Qureshi – who had by then set up a local wing of the NLF made up of himself and his cousin, Ashraf Qureshi – was ordered to execute the hijacking.

As with so many of its past enterprises, the NLF's hijacking plan almost collapsed at the outset. The NLF's Muzaffarabad-based headquarters used the services of an individual they believed to be a reliable courier to send a pistol and hand-grenade to the hijackers. Unfortunately for the organization, the courier turned out to be a double agent, who promptly turned the weapons over to his Indian intelligence handlers.[41] It would appear, however, that the NLF had taken care not to inform him of the eventual destination of the weapons, for investigations did not lead Indian counter-intelligence personnel to the would-be hijackers. Undaunted by the defection of the courier, the hijack squad instead fabricated plausible-looking models of the weapons from wood. Carrying these faux weapons, the hijackers boarded the aircraft without difficulty. Hashim Qureshi entered the aircraft's cockpit and, brandishing his mock grenade, took control of the flight. Altaf Qureshi remained in the cabin, where he informed the passengers that they had been hijacked. The hijackers had hoped to take the plane to Rawalpindi, but a shortage of fuel forced them to land at Lahore. There, speaking for the hijackers, NLF member Farooq Haider announced that the hijackers wished to be granted political asylum in Pakistan and assurances that their families in Jammu and Kashmir would not be harmed.[42] Most importantly, they demanded that 36 prisoners, all of whom they claimed were members of the NLF, be released by Indian authorities.[43]

Unfortunately for the NLF, the hijackers lost their leverage soon after landing in Lahore. Lahore's administrative head, Agha Ashraf, and an army officer present with him, Major Rahim Shah, persuaded Hashim Qureshi and Altaf Qureshi to hand over their hostages. From that moment on, India quite obviously had little interest in meeting the hijackers' demands. "We were very young", Ashraf Qureshi was to ruefully recall, "and did not realize that the passengers were more important than the actual plane".[44] If nothing else, Agha Ashraf's good-faith intervention illustrates that Pakistani authorities had no knowledge of the hijacking – and, more important, no desire to see hostages harmed on their watch. Under other circumstances, the hijacking might have ended relatively uneventfully. Events, however, were to rapidly spin out of hand. Even while negotiations with the hijackers were underway, Lieutenant-General Gul Hassan Khan acidly noted in his memoirs, "[Zulfikar Ali] Bhutto, with his sixth sense for the dramatic, somehow contrived to show up in Lahore".[45] He was, in fact, there on a shuttle mission provoked by the growing crisis in

East Pakistan. In full public view, Bhutto "embraced the two hijackers as true champions of the Pakistani cause".[46] It was a disaster: having given the hijackers legitimacy in public, the Pakistani state had closed the door on any hopes of bringing a quiet end to the affair.

What happened next was vividly described by K.H. Khurshid, the former President of Pakistan-administered Kashmir, before a special court in Pakistan. When he arrived at Lahore Airport on the evening of February 3, Khurshid said, he was ushered into a room by the Senior Superintendent of Police responsible for the airport's security, Sardar Abdul Vakil Khan. Inside, Maqbool Butt and Hashim Qureshi were engaged in an energetic discussion. Qureshi, it turned out, had been asked by the Pakistani authorities present there to burn the plane down. He suggested instead that some minor damage be inflicted to its body, hoping, somewhat optimistically, that this would give the Indian Government some further time to consider the release of the prisoners. Abdul Vakil Khan then implored Qureshi to do as he was told:

> *Khuda ke liye hamari jaan chhorh do jahaz ko urha do* (for God's sake, spare our lives and destroy the plane). As we left, Hashim was taken away by the police officers whom I did not know. Sardar Abdul Vakil Khan and Baqar Ali Shah [another officer] both went with Hashim and we came to join the crowds awaiting there. As we did so, somebody said that the plane had been set on fire.[47]

With the aid of a little petrol, and egged on by the appeals of Pakistani officialdom, Hashim and Altaf Qureshi gutted the Ganga. Not a single one of their objectives had been realized, and the manner of the ending of the crisis was to bring considerable international opprobrium to bear on Pakistan. Just hours before the destruction of the aircraft, the head of the International Civil Aviation Organization had told Pakistan that the aircraft had to be released and the hijackers prosecuted.[48] No clear account exists of the official decision-making process that led to the destruction of the aircraft. Just days earlier, as we have seen, Pakistani officials had ensured the safety of the passengers; they had little to gain by now engineering the destruction of the plane. As Gul Hasan Khan has pointed out, it was widely rumored "that Bhutto himself gave the signal to the hijackers to blow up the plane".[49] There is no evidence to support the proposition, but it would have made some political sense for Bhutto to act in this way. An aggressive position on Jammu and Kashmir would have helped deflect attention away from the simmering crisis in the East and thus enabled Bhutto to unite his disparate West Pakistan constituency on a nationalist platform. Bhutto had, after all, taken a consistently hawkish position on Jammu and Kashmir, and was instrumental in pushing Pakistan towards war in 1965. His decision to applaud the hijacking, and support the Qureshis' request for asylum, was of a piece with his public position.[50]

For some time, Hashim Qureshi and Altaf Qureshi were feted indeed as heroes. In an official note, the Pakistan Government sought to legitimize the hijacking, claiming it was "the direct result of repressive measures taken by the Government of India in occupied Kashmir".[51] The hijackers along with several important NLF leaders were then arrested, ironically enough, by a regime led by Bhutto himself. One Pakistani account has claimed that this was because of the realization that "Indian intelligence [had] stage-managed the hijacking".[52] Much the same conclusion was arrived at by a judicial inquiry set up to consider the issue, held by a judge of the Sindh and Balochistan High Court. While other NLF members were acquitted of the treason charges, on the ground that they had been ignorant of the larger conspiracy behind the hijacking, Hashim Qureshi was to serve 7 years of a 13 year term before being released on a successful appeal to the Lahore High Court.[53] He would continue to play a key role in the NLF's affairs after his release in 1982. Altaf Qureshi, for his part, had enough. He pursued his education and then turned to academic, rising to become Professor of Kashmir Studies at the Punjab University in Lahore – the very city where the Ganga had landed.

An Indian plot?

Conspiracy theories have for long swirled around the hijacking of the Ganga. Mujib, for example, claimed that the affair was engineered to provide a pretext to stall the handing over of power to the triumphant Awami League.[54] Other commentators have charged that the hijacking was an Indian intelligence conspiracy. One prominent historian of the period claims that both Chief Minister G.M. Sadiq and Sheikh Mohammad Abdullah asserted that Indian intelligence were the architects of the affair. By another account, the Indian news agency UNI had made a similar claim on the day of the hijacking itself.[55] What do these accounts show us, and how credible are they?

Charges that Hashim Qureshi was an Indian agent were first seriously articulated in B.M. Sinha's 1981 book, *The Samba Spying Case*. By Sinha's account, Hashim Qureshi was dispatched across the cease-fire line by the Border Security Force [BSF] to spy on Maqbool Butt's activities. Soon after he made contact with the NLF, however, he switched sides. Then, for reasons that Sinha does not state, he returned to Jammu and Kashmir in January 1971, where he was arrested by his former handlers. Qureshi, according to Sinha, told the BSF that he had been trained in Pakistan to hijack an Indian Airlines flight piloted by Prime Minister Indira Gandhi's son and successor-in-office, Rajiv Gandhi. K.F. Rustomji, the BSF's Director-General, then persuaded Qureshi to go ahead with the hijacking plot – but as an Indian intelligence agent.[56] According to Sinha's version of events, Qureshi was to hijack a flight from Srinagar, and then "create the impression that he was a member of the al-Fatah and was hijacking the plane for liberating Kashmir from India". To ensure that Pakistan became enmeshed in the affair,

> he would refuse to hand over the possession of the plane to the airport authorities unless Zulfikar Ali Bhutto, then the leader of the Opposition in Pakistan, came to see him. After the meeting with Mr. Bhutto, he would blow up the plane.[57]

If Hashim Qureshi did receive such instructions, he quite clearly departed from the plan after the Ganga landed at Lahore. Nonetheless, writing a decade after Sinha, the historian Alistair Lamb arrived at much the same conclusions. "The Pakistani authorities", he claimed, "were extremely suspicious of the motives behind the hijacking, though public opinion obliged them to act with considerable circumspection".[58] Among the reasons for these suspicions was that the hijackers were "armed with toy weapons", that "the passengers were either Indian service personnel in mufti or their families", and that the Fokker Friendship, "was the oldest of its type" in the Indian Airlines fleet and could be "expended".[59] Lamb asserted that while Hashim Qureshi had indeed been a genuine NLF operative, he was recruited by the BSF in July 1970. By Lamb's account, Hashim Qureshi was used to monitor Kashmiri visitors to the Pakistan High Commission in New Delhi, and then "became involved in the scheme to hijack an Indian aircraft being considered by Indian Intelligence in Srinagar where it was seen to be a 'disinformation' device of great promise".[60] Beyond claiming that this version of events was endorsed by Lahore police authorities, however, Lamb provides no evidence in support of his assertions.

On the face of it, many of Lamb's claims do not stand the test of reason. Fake weapons, for example, have been frequently used by hijackers. Indeed, to argue that if the Qureshis had the patronage of India's covert services, they would have had no trouble getting *real* weapons and transporting them past airport security. Since civilian aircraft are covered by insurance, moreover, there is no particular reason why India should have found one as opposed to another more expendable. Hashim Qureshi's memoirs, unsurprisingly, challenge Lamb's rendition of events in no uncertain terms. In his own defence, Qureshi suggests that the historian had ignored critical evidence which emerged during the hearings of the judicial inquiry into the hijacking. Qureshi's memoirs contain detailed accounts from this special court's hearings, notably the testimony of Major Rahim Shah. Major Shah stated that he was denied permission to disarm the hijackers even after the passengers had been released – and, critically, that the orders were not revised even after it turned out the weapons the Qureshis were carrying were fake.[61] If Shah's testimony was accurate, it would suggest that high-level officials in Lahore shaped the course of events, rather than Indian covert operatives sitting several hundred kilometres away.

Contested court-room testimony does not, of course, settle the dispute either way. How credible, then, are suggestions that the hijacking of the Ganga was an Indian-run enterprise? On points of fact, both Sinha's and Lamb's accounts are problematic. Both, for example, conflate al-Fatah and the NLF, asserting at various points in their narrative that the first group was responsible for the hijacking.[62] Such carelessness does little to render their other claims credible. Lamb's sympathy for the Pakistani case also seems to have coloured his appraisal of the evidence. For example, he asserts that Pakistan only learned of the existence of the NLF at the time of the hijacking.[63] For this claim to be plausible, one would have to believe, in the face of considerable evidence to the contrary, that Pakistan's intelligence services were extraordinarily incompetent. Several NLF members, after all, had been arrested and interrogated *prior* to the hijacking. Moreover, neither Sinha nor Lamb, makes any effort to explain the several secondary mysteries which emerge from their accounts. It is hard to understand why, for example, the BSF would have assigned Qureshi to monitor the Pakistan High Commission in New Delhi. Espionage directed at foreign missions lies in the domain of the Indian Intelligence Bureau which, as its surveillance of the Pakistan High Commission during the al-Fatah investigation makes clear, had both assets and a competent surveillance staff available to it.

Significantly, the motive both authors attribute to the hijacking – that is, India's use of the incident as a pretext for cutting off air routes from West to East Pakistan – is less than satisfying. Pakistani over-flight rights had been restricted since 1965. Had India chosen, it could have denied permission for military flights irrespective of the Ganga affair.[64] Furthermore, while the suspension of air communication may indeed have created some logistical delays, Western and Indian media reports from the time make clear that Pakistani C-130 transport aircraft were able to supply East Pakistan through Sri Lanka.[65] Credible accounts by Pakistani military personnel also assert that the embargo did not hinder the flying in of personnel as the East Pakistan situation deteriorated. For example, a two-battalion brigade used during Operation Searchlight was airlifted into Dacca in late, February 1971.[66] Maritime transport, the principal means of logistical support for Pakistani forces in the East, was untouched, moreover, by the Indian embargo. Most importantly, there is no evidence to show that India was even considering war at the time of the Ganga affair. According to one of her leading biographers, Prime Minister Indira Gandhi only turned her mind to the East Pakistan issue at the end of March 1971, after being re-elected to office earlier that month.[67] As such, it seems profoundly unlikely that Indian intelligence would have been seeking to create a climate for a war which, as yet, was still to be conceived.

Nonetheless, it would be foolish to make hard and fast assertions: full archival disclosure is the only means by which the truth *might* – and the word might here deserve emphasis – become known. Even if the hijacking of the Ganga was not authorized or planned by India, moreover, elements of the National Liberation Front may well have had some linkages with Indian intelligence. Officials interviewed for this book asserted that Qureshi had indeed worked for the BSF on his return from Pakistan, and asserted that Surendra Nath himself had alerted authorities in New Delhi of this fact in the hours after the hijacking. According to these informants, Qureshi was recruited for the BSF by Ghulam Hassan Mir, a long-standing asset of the organization's intelligence wing, the "G"-Branch. BSF authorities, in this narration of events, persistently stonewalled the Jammu and Kashmir Government's efforts to interrogate Qureshi on his hijacking plans. After the hijacking, an alarmed Surendra Nath is thought to have leaked the essence of this information to UNI, leading to the first public report of Qureshi's relationship with the BSF. If these assertions are correct, it would help explain why Nath's career went into decline for some years after 1971, when he was dispatched to head a then-obscure forensic facility in the city of Hyderabad. However, any relationship which Qureshi may have had with the BSF does not necessarily mean the hijacking was carried out on the instructions of India's covert services: he could, after all, have acted in pursuit of his ideological commitments. In any event, the question is of little fundamental relevance: the future of Jammu and Kashmir was now to be shaped not by covert combat, but by a full blown war.

Operation Instruction

Major-General Surjit Singh Uban was later to claim that his guru, the Bengali mystic Baba Onkarnath, had prophesied that his men would be at war in Bangladesh a year before the 1971 war began.[68] If correct, it would help explain just why the officer had such a central role in the covert campaign that underpinned the war: he had more opportunity than most to prepare for what was to come.

On May 1, 1971, almost three months after the day when the Ganga was set on fire, India's Chief of Army Staff issued a secret order initiating the war that would end with the dismemberment of Pakistan. India's objectives were remarkably similar to those attempted by Pakistan in 1965, envisaging the use of a covert army as a catalyst for insurrection and a spearhead for regular forces. Its scale and objectives, however, were altogether more ambitious. Operation Instruction formally committed Indian forces to "assist the Provisional Government of Bangladesh to rally the people of East Bengal in support of the liberation movement", and "to raise, equip and train East Bengal cadres for guerilla operations for employment in their own native land".[69] Eastern Command was to ensure that the guerilla forces were to work towards "tying down the Pak [Pakistan] Military forces in protective tasks in East Bengal", "sap and corrode the morale of the Pak forces in the Eastern theatre and simultaneously to impair their logistic capability for undertaking any offensive against Assam and West Bengal", and, finally, be used along with regular Indian troops "in the event of Pakistan initiating hostilities against us".

Much of the guerilla war, however, was waged by the volunteers of the Gano Bahini, a volunteer force whose fighters received no wages. Indian forces initially set up seven camps for recruiting and training volunteers, two each in West Bengal and Meghalaya, and one each in Bihar, Tripura and Assam. Awami League officials who had fled East Pakistan were responsible for screening and recommending recruits. Although each camp was capable of handling 1000 recruits at a time for four-week courses in weapons, field craft, communications and sabotage, the Indian-run facilities found themselves swamped. At one camp, some 3000 young men had to wait up to two months for induction, although the "hygienic condition was pitiable and food and water supply almost non-existent".[70] By September 1971, Indian training operations had expanded dramatically in scale, processing a staggering 20,000 guerillas each month. Eight Indian soldiers were committed to each 100 trainees at ten camps. On the eve of the war, at the end of November 1971, over 83,000 Gano Bahini fighters had been trained, 51,000 of whom were operating inside East Pakistan – a guerilla operation perhaps unrivalled in scale until that time.[71]

Outside of Indian support to the Bangladesh forces lay several secondary layers of covert operations. Principal among these activities was a force of 1,800 commandos who extensively operated in the Chittagong Hill Tracts under Uban's command.[72] India's official accounts of Uban's operations refer to his acting under the administrative control of a Special Secretary to the Cabinet, one of the two direct subordinates of the head of India's external intelligence service, the Research and Analysis Wing. Starting out from bases set up in the Indian frontier state of Mizoram in October 1971, Uban's forces engaged in a series of low-grade border skirmishes before moving into East Pakistan on December 3, 1971, where they were to wage an extraordinary campaign of sabotage and harassment. At the cost of just 56 dead and 190 wounded, Uban's commandos succeeded in destroying several key bridges, and in ensuring that Pakistan's 97 Independent Brigade and crack 2 Commando Battalion remained bogged down in the Chittagong Hill Tracts.[73]

Some 580 members of Uban's covert force were awarded cash, medals and prizes by the Government of India after the war. Their names did not figure in lists of war honours, however, for one simple reason: Major-General Uban's fighting men were not Indian. The fighters were members of the Special Frontier Force [SFF], also known as "Establishment 22," which had been set up in November 1962, towards the end of the India–China war. Made up mainly of Tibetan refugees who had fought with the Dalai Llama's Chushi Gangdruk irregular forces from the mid-1950s, the 12,000-strong SFF was trained in guerilla warfare by both Indian military personnel and instructors from the United States Central Intelligence Agency [CIA]. Many of the first recruits into the SFF had earlier received training from the CIA at bases near Camp Hale, near Denver, Colorado, having been smuggled through the very East Pakistan where they were to return to fight in years later.[74] SFF personnel did not, ironically enough, ever carry out full-scale operations against China, although the CIA did use them to place sensors in the Great Himalayas to detect Chinese nuclear and missile tests.[75] There were, however, several unauthorized raids and cross-border firing, including a shootout which had claimed the lives of two SFF personnel in 1971.[76]

India's intelligence services learned several lessons of value during the limited incursions that did take place. Small-scale infiltration by covert agents, similar in character to what Pakistan had attempted in Jammu and Kashmir, was easily interdicted by the Chinese.[77] To India's spymaster, B.N. Mullik, it became clear that such operations could not generate the scale of resistance needed to pose a plausible threat to the Chinese state. Conversely, the experience of near-autonomous irregular groups operating out of Nepal's Mustang region taught Mullik that loosely controlled forces were disinclined to engage in high-risk, aggressive operations and were, furthermore, vulnerable to factional power-struggles.[78] After Nehru's death, Mullik was removed from the leadership of the Intelligence Bureau, but retained control of covert operations against China, including Establishment 22 and the unit responsible for air operations in its support, the Aviation Research Centre. In September 1967, control of these assets, as well as external intelligence in general, was formally handed over to the newly formed Research and Analysis Wing. Its first director, R.N. Kao, had earlier headed the Aviation Research Centre, and was thus intimately familiar with Mullik's operations – and, presumably, the lessons learned during their course.[79]

From the point of view of Kao and his staff, the use of the SFF in Bangladesh had several obvious advantages. Unlike the personnel used by Pakistan's covert services in Jammu and Kashmir, India could flatly deny any responsibility for Tibetan nationals who might be taken as prisoners in the Bangladesh operations. Since Pakistani military intelligence personnel could safely be assumed to have at best limited access to Tibetan-language interpreters, interrogation of any captured SFF personnel would have been well-nigh impossible, particularly given the short duration of the war. Considerable pains, interestingly, seem to have been taken to disguise the origins of Uban's commandos during their 1971 operations. They were, for example, armed with Bulgarian-made Kalashnikov rifles and United States-manufactured carbines. Neither of these weapons was in service with Indian forces.[80] There was no risk, therefore, of the kinds of embarrassment caused to Pakistan by the discovery of its Ordinance Factory markings on weapons found in Jammu and Kashmir. Unlike the fractious Mustang guerillas – and, potentially, the Mukti Bahini – the SFF would, however, have the advantage of being under direct Indian control.

By November 1971, when the Indian-backed low-intensity war in East Pakistan escalated to levels Pakistan found intolerable, it was finally pushed to act. On December 3, Pakistan attempted to relieve the pressure on its eastern wing by carrying out strikes on major Indian airbases. The Indian Air Force, which had anticipated such an attack, had already taken defensive measures, and was able to mount a successful counter-offensive the next day. Pakistan's major port at Karachi came under heavy air and sea attack. By mid-December, the Indian Navy had succeeded in blockading both wings of Pakistan. Indian land forces, meanwhile, made rapid progress. Six Indian divisions were involved in the parallel land war, an offensive of extraordinary speed that has been described as a "blitzkrieg without tanks".[81] In essence, Indian troops bypassed major Pakistani defences, leaving these beleaguered garrisons to be besieged by reserve units while striking out ahead to seize as much territory as possible.[82] After rejecting an offer for conditional surrender in the East, Indian forces entered Dacca on December 15. Resisting some pressure from hawks to continue the war in the West and secure further gains against Pakistan, Indira Gandhi promptly ordered a ceasefire on the western front as well: "if I don't do so today", she said of the decision to end war, "I shall not be able to do so tomorrow".[83]

Pakistan had ceased to exist – in the form and for the purpose for which it was born – in just two weeks of full-scale war. Some of the blame for this, an inquiry into the debacle led by the Pakistani judge Hamood-ur-Rahman was later to argue, lay in the domain of military decision-making. "This was a war", he asserted, "in which everything went wrong for the Pakistan Armed Forces. They were not only out-manned, but out-gunned and out-Generaled [*sic*]. Our planning was unrealistic, strategy unsuited, decisions untimely and execution faulty".[84] Hamood-ur-Rahman's report proceeded to suggest that Pakistani forces in the east should have fought on to the end rather than surrendered, suggesting this course of action would have meant General Niazi would then have been remembered "as a great hero and martyr".[85] Such an assessment seems grossly unfair to Pakistani forces that fought in the face of overwhelming odds, and were, moreover, paying for a series of political disasters which had alienated the population of East Pakistan beyond the point of no-return. As Owen Bennett-Jones has argued, had General Niazi chosen to fight on instead of surrender, he would have led "his men to certain death so as to satisfy public opinion in West Pakistan".[86] That Hamood-ur-Rehman should have made such a suggestion, Jones proceeds to point out, only "reflects the extent of Pakistan's humiliation".[87]

No one may have felt that humiliation more strongly than Niazi himself. As he signed the document of surrender, Niazi wrote, "sorrow rose from my heart to my eyes, brimming them with unshed tears of despair and frustration".[88]

Defeat and a deal

Pakistan's defeat generated dismay among anti-India forces in Jammu and Kashmir. Altaf Khan, just released from jail, described his feelings thus:

> I, too, was devastated by the defeat of Pakistan at India's hands. My entire life ambition of making Kashmir a part of Pakistan with the latter's covert support seemed shattered in one go. What would I do?[89]

Ever since he was interdicted on the India–Pakistan border in May, 1970, life had not been happy for the would-be liberator. By his own account, Altaf Khan had received soft treatment by the BSF personnel who captured him, but the Army personnel who subsequently interrogated him showed little mercy. His family succeeded in having the prosecution dropped, using the offices of a relative who served as a Superintendent of Police in Jammu and Kashmir.[90] In one interview, Altaf Khan said he had joined al-Fatah on his release from jail, but the claim seems somewhat dubious; he does not figure in the Jammu and Kashmir Police's classified records on that organization.[91] Whatever the truth, by 1973 Altaf Khan abandoned both political and armed struggle for spiritualism, and began spending much of his time visiting mystics, and in meditation. At one stage, he left home for the forests from where he was only persuaded to return, it is said, by the use of the mystical powers of a Sufi mystic, or *Pir*.[92]

Altaf Khan's despair was understandable. Under pressure from angry officers in the humiliated Pakistan Army, General Yahya Khan had been forced out of office. The disgraced military ruler demitted office on December 20, 1971; Bhutto took charge as martial law administrator the next day. His first priority was to negotiate a post-war settlement with Indira Gandhi. After a series of preparatory meetings, Bhutto and Indira Gandhi met in the Indian resort-town of Shimla between June 28 and July 2, 1972.[93] Bhutto secured several objectives, notably, the release of 93,000 prisoners of war captured in East Pakistan, the prevention of war crimes trials against Pakistani officers, and the return of the 5,000 square kilometres of West Pakistan territory lost to India. In return, he agreed on a new nomenclature for the CFL in Jammu and Kashmir: the Line of Control, which marks the border between the Indian and Pakistani-administered parts of the state to this day. Indian commentators have argued that Bhutto privately agreed to convert the LoC into a border over time, a proposition buttressed by the fact that he subsequently detached the Northern Areas from Pakistan-administered Kashmir and placed them under direct federal control.[94]

Driven by pragmatism, politics in Jammu and Kashmir had also begun to turn in a direction that could not but enrage separatists. With Sheikh Abdullah and the Plebiscite Front's cadre in jail, in part as a consequence of al-Fatah's activities, Mir Qasim comfortably won the elections to the Jammu and Kashmir Legislative Assembly in March 1972. Several commentators have argued that had the Plebiscite Front had a crack at power, it would have won; certainly, this was Qasim's own assessment.[95] One consequence of the decision to keep Sheikh Abdullah out of the fray was to create political opportunity in the Kashmir valley for the far-right Jamaat-e-Islami. Although the Jamaat-e-Islami won just 5 of the 22 seats it contested, it picked up 23.53 per cent of the popular vote in those seats, all of it in the Kashmir valley. This was, of course, just 7.18 per cent of the overall vote in all of Kashmir's 75 seats, but it illustrated that the Jamaat-e-Islami had succeeded in occupying the oppositional space left vacant by the absence of the Plebiscite Front. Similarly, in Jammu, the Hindu-chauvinist Jana Sangh emerged with significant pockets of support, winning 23.11 per cent of the vote in the 32 seats it contested, even though it won just 3.[96] Communal polarization was starting to become a significant feature of political life in Jammu and Kashmir.

Nonetheless, the sole fact of importance for the moment was that the Congress was firmly ensconced in power. From this position of power, it began to make overtures to Sheikh Abdullah. In April 1972, Begum Abdullah was allowed to return to the state; by June, Sheikh Abdullah and Mirza Afzal Beg were permitted to follow her. Orders proscribing the Plebiscite Front were also lifted. New Delhi, these signals made clear, wished for a rapprochement with Sheikh Abdullah, hoping this would enable a final resolution of the Jammu and Kashmir conflict. Sheikh Abdullah promptly reciprocated with gestures of his own. While he continued to insist that India and Pakistan had no right to negotiate a settlement without the consultation of the people of Jammu and Kashmir, Sheikh Abdullah was at once sharply critical of Bhutto's advocacy of self-determination. Among other things, he pointed out that the residents of Pakistan-administered Kashmir did not enjoy this right.[97] In an interview given to a London-based newspaper, Abdullah also made clear his willingness to negotiate a deal involving greater autonomy for Jammu and Kashmir within the Union of India. "There is no quarrel with the Government of India over accession", he affirmed, adding that "it was we who brought Kashmir into India". What remained to be resolved was "the structure of internal autonomy".[98]

Time for leisurely discussion was, however, running out. In May 1973, a student in Anantnag was appalled by an image he saw while leafing through an old encyclopedia stored in the local college's library, Arthur Mee's *Book of Knowledge*. The picture depicted the Archangel Gabriel dictating the text of the Quran to Mohammad, a violation of Islamic edicts prohibiting the representation of the prophet through graven images. The encyclopedia had lain in Anantnag for decades, first in the collection of a school run by Christian missionaries and then at the Degree College's library, without provoking the smallest contention. When clerics in Anantnag learned of the picture, however, it was denounced as blasphemous. College students in Anantnag went on strike, and the protests soon spread to Srinagar, where the Jamaat-e-Islami had won two seats the previous year. Protestors demanded that the author of the encyclopedia be hanged, "a vain demand", Katherine Frank has wryly noted, "since Arthur Mee had died in England in 1943".[99] The Government of India banned sales of the encyclopedia, again a futile gesture, since it was no longer in print. However, protests continued, and the police eventually had to use fire to disperse violent crowds, leading to four fatalities.

How does one account for the extraordinary outrage provoked by the *Book of Knowledge*? No empirical data exists on precisely who the participants in the affair were, but the event needs to be read against the slow growth of the Jamaat-e-Islami from the 1950s onwards. As Yoginder Sikand has pointed out, the Jamaat-e-Islami had set up a wide network of schools to counteract what it believed was "an Indian onslaught in the cultural sphere" because of which "many young Kashmiris had begun to lose their Islamic moorings".[100] Yet, these schools – like the Jamaat's industrial-scale production of propagandistic literature – also represented a political project. Sikand cites one insider as suggesting that the schools were "set up in order to lead a silent revolution, to keep alive the memory of Kashmiri independence and of India's brutal occupation of the state". Moreover:

> It was widely believed in JIJK [Jamaat-e-Islami] circles that a carefully planned Indian conspiracy was at work to destroy the Islamic identity

> of the Kashmiris, through Hinduizing the school syllabus and spreading immorality and vice among the youth. It was alleged that the government of India had dispatched a team to Andalusia, headed by the Kashmiri Pandit [politician] D.P. Dhar, to investigate how Islam was driven out of Spain and to suggest measures as to how the Spanish experiment could be repeated in Kashmir, too. Faced with what it saw as these menacing threats, the JIJK felt the compelling need for a comprehensive educational system of its own to save the Kashmiri Muslim youth from Indian cultural imperialism.[101]

For policy-makers in both New Delhi and Srinagar, the message of the *Book of Knowledge* riots would have been unmistakable: while the war of 1971 may have proved the undoing of Pakistan, anti-India forces within Jammu and Kashmir were far from spent. As during the Hazratbal crisis, Islamists had demonstrated their ideological authority, as well as the existence of an urban constituency who believed that their faith was under attack in India. The point was hammered home through 1974, as party cadre loyal to Mirwaiz Mohammad Farooq clashed with their Plebiscite Front rivals, in response to charges by the Srinagar cleric that Sheikh Abdullah was preparing to abandon Kashmiri claims to nationhood.[102] Amidst this street drama, Indira Gandhi began a series of closed-doors meetings with Abdullah in early 1974. Her principal adviser on Jammu and Kashmir, G. Parthasarathi, held a parallel series of discussions with Beg. Sheikh Abdullah pushed hard for fresh elections to be held in Jammu and Kashmir, hoping that a poll victory would enhance his bargaining position. Indira Gandhi would have nothing of it, but offered him the Chief Minister's position in place of Qasim.[103] Sheikh Abdullah took the bait. Qasim resigned on February 23, 1975. The next day, Indira Gandhi made public the six-point formula that Beg and Parthasarathi had signed, in secrecy, four months earlier.

The Delhi Agreement, as it came to be known, affirmed that New Delhi would "continue to have power to make laws relating to the prevention of activities directed towards disclaiming, questioning or disrupting the sovereignty and territorial integrity of India or bringing about cession of a part of the territory of India or secession of a part of the territory of India from the Union".[104] While such laws already existed, the agreement represented a commitment by Sheikh Abdullah that he would no longer seek independence. The Delhi Agreement went on to assert that Jammu and Kashmir was "a constituent unit of the Union of India". This again was not a novel formula; Sheikh Abdullah had said as much on several occasions. It did, however, mark a formal renunciation of the Plebiscite Front's *raison d'etre*, and paved the way for its return to mainstream politics. Critically, the Delhi Agreement mandated that "provisions of the Constitution of India already applied to the State of Jammu and Kashmir without adaptation or modification are unalterable". In effect, this meant Sheikh Abdullah concurred with the restructuring of Jammu and Kashmir's relationship with India, much of which had been carried out while he was in jail. No agreement could be arrived at on the sixth issue before Beg and Parthasarathi, Sheikh Abdullah's demand that the Governor and Chief Minister of Jammu and Kashmir be called its *Sadr-e-Riyasat* and *Wazir-e-Azam*, or President and Prime Minister, as had been the situation prior to 1965. It was therefore "remitted to the Principals" and was never to be discussed again until the late 1990s.

Indira Gandhi arrived in Srinagar for her first post-war visit in October 1975. "Her progress on the Dal Lake by boat", Victoria Schofield has recorded, "propelled by turbaned oarsmen, was reminiscent of the visits of Mughal emperors".[105] Earlier that year, besieged by growing opposition to her policies and an erosion of popular support, Gandhi had imposed draconian Emergency powers through India. Fundamental rights had been suspended, and most key opposition politicians were in jail. For several reasons, Sheikh Abdullah chose to remain distant from this pan-India crisis. His own political position was hugely complicated. Although the National Conference had been reborn, it had no representation in the Assembly, which remained dominated by the Congress. Years of struggle and inner-party ruptures, moreover, had shorn Sheikh Abdullah of aides and confidantes; for his first two years in office, he and Beg alone held office. Increasingly, Sheikh Abdullah operated through his family, notably his son, Farooq Abdullah and son-in-law, Ghulam Mohammad Shah. With these limited political resources, Sheikh Abdullah had to prepare for state elections scheduled to be held in 1978: elections in which the legitimacy of his accord with Indira Gandhi would be tested, as would his claims to speak for Jammu and Kashmir.

Heading into the elections of 1977, Sheikh Abdullah had one principal adversary: the Islamist right. During the Emergency, the Jamaat-e-Islami had been proscribed, along with several other communal organizations of both the Hindu and Islamist right. Much of the organization's leadership was jailed, and its publications were suppressed. Indira Gandhi's crackdown on the Jamaat-e-Islami had Sheikh Abdullah's enthusiastic endorsement; in one speech, he had described the Islamist organization's schools as "the real source for spreading communal poison".[106] Some 125 Jamaat-run schools, with over 550 teachers and 25,000 students, were banned. So were another 1,000 evening schools run by the organization, which reached out to an estimated 50,000 boys and girls.[107] However, the state offered employment to the teachers who had lost jobs as a result of the shutting down of the Jamaat's schools. This meant, in practice, that "teachers were absorbed into government schools that offered them a new and wider platform to propagate their ideology".[108] Furthermore, Jammu and Kashmir's Islamists proved adept at evading the law. As Navnita Chadha Behera has pointed out, most of the Jamaat's schools simply re-christened themselves and proceeded to operate freely as supposedly independent private institutions. Some claimed to offer English as the language of instruction, attracting the children of middle-class parents fed up with poor standards at state-run schools.

Soon, the scale of the Jamaat's influence was to be tested. In March 1977, Indira Gandhi withdrew the Emergency and called for General Elections. She was defeated. The coming to power of the Janata Party, a coalition spanning socialists, centrists and Hindu chauvinists, provoked a crisis within the Jammu and Kashmir Assembly, and elections in the state had to be called early. Having emerged more or less unscathed from the Emergency, and wearing, moreover, the halo of political martyrdom, the Jamaat-e-Islami sought to capitalize on the new situation. It allied itself with the Janata Party both at the national level and in Jammu and Kashmir. Sheikh Abdullah responded to the threat with unconcealed appeals to communal sentiment. A vote for the Jamaat-e-Islami, Sheikh Abdullah claimed, was a vote for the Jana Sangh, a Hindu-chauvinist constituent of the Janata Party whose "hands were still red with the blood of Muslims".[109] Islam, National Conference leaders insisted, would be in danger if the Jamaat-Janata alliance took power. Beg went one step further and appropriated the pro-Pakistan position traditionally taken by Mirwaiz Mohammad Farooq, the Srinagar cleric. At rally after rally, he would produce a green handkerchief with Pakistani rock-salt – as opposed to Indian sea-salt – contained in it, signalling support for that country to his audience.[110] National Conference cadre administered oaths on the Quran to potential voters, through which they pledged their commitment to the party. Clerics were imported from Uttar Pradesh and Bihar to campaign in Muslim-majority areas of Jammu. Sheikh Abdullah, wary of the consequences of pushing New Delhi too hard, was careful to assert that "Kashmir was a part of India and Kashmiris were Indians", but added that "if we are not assured of a place of honour and dignity in India, we shall not hesitate to secede".[111]

Sheikh Abdullah's incendiary campaign paid off: the National Conference won 47 of 75 seats in the Jammu and Kashmir Assembly, a decisive majority. Moreover, the National Conference secured over 46 per cent of the popular vote, an exceptionally high proportion in Indian elections. By contrast, the Jamaat-e-Islami could secure just 1 of the 19 seats it contested, and received only 3.59 per cent of the state-wide vote.[112] This was a poorer performance than even the fledgling Janata Party, which picked up 13 seats and secured 23.7 per cent of the popular vote. Sheikh Abdullah's victory had, however, come at a price. His aggressive use of Islamist themes and images during the campaign had cost him support in Jammu, particularly among Hindus. Just one of the seven seats the National Conference picked up in Jammu, that of Ramban, had a Hindu majority.[113] In effect, the National Conference had abandoned its historic project of building itself into a spokesperson for the entire state, and had retreated instead to its heartland in the valley. More importantly, the party had opened the gates for the large-scale use of religion in mass politics, a weapon that others, in time, would also learn to use.

All this, though, was in the future: at the time, Sheikh Abdullah's rule seemed unshakeable. Despite the anger of Islamists like Mohammad Farooq and the despair of figures like Altaf Khan, Sheikh Abdullah's return to political centre-stage put an end to visible anti-India protest. While the Sher-i-Kashmir himself was to regret his capitulation to New Delhi and decision to ally with the Congress, on the substance of the Agreement, however, there could be no withdrawal.[114]

Until Sheikh Abdullah's death in 1982, the secessionists would continue to stage an unhappy retreat.

The birth of the JKLF

Indira Gandhi, wrote the journalist and historian M.J. Akbar, had succeeded in giving Jammu and Kashmir "a wonderful decade of freedom and peace".[115] While his assessment is, perhaps, excessive, the fact was that Sheikh Abdullah's return to mainstream politics had stripped anti-India formations of both legitimacy and the ready pool of potential recruits that had fed both the Master Cell and the al-Fatah. Neither a party apparatus nor activists remained to be tapped.

Pakistan's own position was, moreover, deeply compromised – and not just by military defeat. On paper, it remained committed to regaining Jammu and Kashmir. In a speech made to Pakistan's National Assembly just weeks after the Shimla Accord was signed, Bhutto had asserted that

> if the people of Jammu and Kashmir want their independence, if they want to be liberated from the Hindu yolk, if they want to be a free people in friendship, and friendship and comradeship in Pakistan, they will have to give the lead and we will be with them.[116]

Maqbool Butt, however, was deeply skeptical of official Pakistan. In May 1973, still in prison, he wrote a pained letter to his niece Azra Mir, asserting that the

> Pakistani ruling class has never ever supported Kashmiris in their struggle for freedom, as they should have done. This class has never been interested in the liberation of Kashmiris. Whatever they say is merely lip service, and must not be trusted."[117]

After all, he pointed out, "rulers who declared war against their own people cannot offer anything to anyone else but injustice".

No accounts exist of what Butt made of the situation he was confronted with on his release from jail. Bhutto had a record of taking aggressive positions on Jammu and Kashmir, but his domestic position now was far from stable. In February 1973, Pakistan's Inter-Services Intelligence (ISI) discovered a consignment of arms shipped by Iraq's Embassy to members of the Marri tribe, who were seeking independence for Balochistan, a sensitive eastern province bordering both Iran and Afghanistan.[118] Bhutto promptly dismissed the provincial government. Baloch nationalists responded by launching a full-blown insurgency. In mid-1973, meanwhile, Bhutto also dismissed the government of the North–West Frontier Province, accusing it of allying with the pro-Moscow regime of Sardar Daud, which had taken control of Afghanistan. The NWFP government, he claimed, had been working to realize Pakhtunkhwa, a homeland for Pashtun tribesmen sprawling across parts of northern Pakistan and Afghanistan. The Pakistan government attempted to counter Pashtun nationalism by cultivating

Islamist exiles who had fled the country. Among them was Gulbuddin Hekmatyar, who would go on to play a key role in the anti-Soviet jihad in Afghanistan.[119] Problems were also evident in Punjab. In an effort to make peace with the clergy, Bhutto had declared Islam the religion of state, promised to introduce the teaching of the Quran and Islamic culture in government-run schools and set up a council to bring laws into conformity with religious injunctions. Bhutto's efforts at appeasement in fact gave the clerics the scent of blood. In May 1974, Islamist groups led massive protests, demanding that the heterodox Ahmadi sect be declared non-Muslim. A similar protest had been beaten back in 1953, but this time around, Bhutto caved in.[120] His actions bought a temporary peace, but laid the foundations for further crisis: "as populism lost its momentum to Islam", Seyyed Vali Reza Nasr has argued, "the fate of Bhutto's government was sealed".[121]

Embroiled in sectarian and ethnic crisis, then, there was little chance that Pakistan would risk opening another front in Jammu and Kashmir, however passionate Bhutto's advocacy of the cause might have been. Given Butt's suspicions about the Pakistani establishment, moreover, it is profoundly unlikely he would have taken Bhutto's polemic at face value. Resistance to India *within* Jammu and Kashmir had, however, far from collapsed. Galvanized by the emerging rapprochement between New Delhi and the Plebiscite Front, Altaf Khan set up the Students Islamic Organization, which was among several organizations involved in street protests against the Delhi Agreement.[122] Partly as a consequence of those protests, a fresh round of student radicalization had taken place. In May 1973, a small group of students led by Syed Nazir Gilani had abandoned their studies at the Kashmir University and made their way across the LoC to the nearest Pakistani military post. Gilani's reception in Pakistan was less than enthusiastic – "my dream processed me through military post, Muzaffarabad fort, Muzaffarabad jail and the High Court of Azad Kashmir", he wrote, referring to his prompt detention and incarceration by Pakistani authorities – but his actions suggest that the post-1971 mood of defeat in Jammu and Kashmir was again being replaced by defiance.[123] Soon after Gilani's abortive enterprise, three NLF cadre, Iqbal Qureshi, Altaf Qureshi and Hamid Lala, succeeded in throwing a small explosive at Indian troops parading to mark India's Republic Day.[124]

There were signs, too, that opposition to a deal with New Delhi was not restricted to student radicals. Many suspected, correctly or otherwise, that Sheikh Abdullah believed he had made a bad bargain, and wished to resile from it. During a visit to Pakistan-administered Kashmir in 1974, Sheikh Abdullah's son Farooq Abdullah, who had trained as a doctor in the United Kingdom, electrified audiences:

> he lifted a gun in his hand along with Maqbool Butt (Shaheed) [Martyr], Ashraf Qureshi, Abdul Khaliq Ansari and Amanullah Khan, [and] administered the oath that he would fight for the freedom and sovereignty

> of Kashmir till his last breath. He announced: "If my father makes a deal with Indira Gandhi at the cost of Kashmir's freedom, I shall be the first to rebel against him." A hundred thousand people and news reports of the day stand witness to these pronouncements.[125]

Farooq Abdullah has always denied this version of events, dismissing accounts such as these as propaganda. Whatever the truth, though, the large-scale protests which had greeted the Delhi Agreement gave the NLF good reason to believe it still had a constituency in Jammu and Kashmir. In May 1976, Maqbool Butt crossed what is now the LoC along with two other NLF cadre, hoping to set up a network of active cells. Others in the NLF had warned him that the enterprise was far too hazardous, given the chances he would be recognized and apprehended; Butt paid little attention to these warnings. That the enterprise had no support from Pakistan's intelligence services is made clear by the fact that his first target was a bank. The robbery in Kupwara, like each of the NLF's past operations, turned sour. A bank employee was killed in the course of the robbery, while Butt himself was arrested. Butt was tried, and received the second death sentence of his life. Given that he had managed to escape from Srinagar, authorities decided to incarcerate him at the maximum-security Tihar Jail in New Delhi this time around.[126]

Butt's arrest effectively broke the back of the NLF on both sides of the LoC. With both Hashim Qureshi and Butt in jail – albeit in different countries – Amanullah Khan decided to leave Pakistan for the United Kingdom. After Jammu and Kashmir itself, it was perhaps the perfect location for a politician from the region to locate himself. Migrants from the Mirpur area of Pakistan-administered Kashmir were among the largest components of the South Asian community in the United Kingdom. Many had been displaced from their agricultural lands because of the large-scale submergence caused by the construction of the Mangla Dam in 1966, and moved to the United Kingdom to tap the already well-established network of Mirpuri entrepreneurs there.[127] For political groups, moreover, the relative affluence of the Mirpuri community – its remittances in 2003 were estimated at UK £500 million – was powerful bait. Amanullah Khan was just one of many Kashmiri figures to make the journey. Altaf Khan's comrade in the Students Islamic Organization, Ayub Thakur, for example, acquired considerable influence among anti-India formations in Jammu and Kashmir before his death in 2005.[128] He was, shortly before his death, investigated for funnelling funds to Jammu and Kashmir terrorist groups through Mercy International, a charity active amongst the Mirpuri community.[129]

In London, the NLF began to reinvent itself. Set up initially as the Kashmir Plebiscite Front's United Kingdom chapter, the organization renamed itself in response to demands that it have a more revolutionary identity. At a meeting in May 1977, the Plebiscite Front was reborn as the Jammu Kashmir Liberation Front [JKLF], which took under its wing the old NLF, now renamed the National Liberation Army. Amanullah Khan took charge as the General Secretary of the

JKLF the following February. Drawing on the resources of new overseas ethnic-Kashmiri and Mirpuri patrons, Amanullah Khan was able to bring about a rapid expansion of the organization. JKLF branches were soon set up in Pakistan, Denmark, Holland, Saudi Arabia, the United Arab Emirates, Germany, France and the United States. In October 1979, while the United Nations General Assembly was in session, four JKLF cadre created a minor sensation by throwing anti-India leaflets and shouting slogans from the visitors' gallery. Actions like these raised the JKLF's profile among the global Kashmiri diaspora, as did conventions it organised at Birmingham in 1981 and Luton in 1982.[130]

Few appear to have paid much attention to these developments at the time: the activities of a few dissidents based in London, after all, were not a source of great concern to either Islamabad or New Delhi. In retrospect, the years of retreat had ended, and the foundations for the revival of the covert war in Jammu and Kashmir had been laid.

Drawing blood

Sheikh Abdullah passed away on September 8, 1982. He had long been ailing; unable to campaign in the elections of 1977, audiences had to content themselves with listening to his taped speeches. A year before his death, Abdullah had designated his son, Farooq Abdullah, his political heir. At the time, he had made no secret of the value of the gift: the "crown I am placing on your head", Sheikh Abdullah told his son, "is made of thorns".[131]

Since the 1977 elections, Jammu and Kashmir had been in a state of constant disquiet. One particular concern was the growth of the Islamists. Notwithstanding the defeat of the Jamaat-e-Islami in the 1977 elections, pan-Islamic forces had registered considerable growth. The Iranian Revolution of February 1979 had provided considerable impetus to these forces. So, too, had the United States-backed jihad against the Soviet Union in Afghanistan, a war in which Pakistan had a frontline role. While the effect of these events in the covert war in Jammu and Kashmir was yet to manifest itself, at least some participants were drawing lessons. "A small nation with a small population, with limited resources and weapons rose in revolt against the Soviet onslaught", Altaf Khan told Victoria Schofield in a 1994 interview, "to the extent that the Soviet Union ultimately disintegrated into fragments".[132] "So we got inspired", he proceeded, "if they could offer tough resistance to a super power in the east, we too could fight India".

These events, of course, were some distance in the future, but Altaf Khan's turn from pessimism to renewed hope was mirrored in the attitudes of many young Islamists at the time. In 1979, the Jamaat-e-Islami's youth wing, the Islami Jamiat-e-Tulaba (IJT: Islamic Students Union), inspired by the students of Teheran, launched an agitation demanding compulsory Islamic education for Muslim students of government schools in Jammu and Kashmir.[133] Later that year, the IJT organized an international convention in Srinagar for representatives of several Islamist movements; it was later granted membership in the far-right World Association of Muslim Youth and the International Islamic Federation of Students Organizations, both of which provided access to patronage from the affluent clerical orders of Saudi Arabia. Then, in 1980, the Jamaat-e-Islami and IJT jointly organized another international Islamic conference in Srinagar. The event drew several luminaries, including the Imam of the shrines at Mecca and Medina, Sheikh Abdullah bin Sabil, and the muezzin of the Prophet's Mosque at Medina, Sheikh Qari Khalil. Islamists within Jammu and Kashmir, it seems reasonable to speculate, would have used the opportunity to build contacts with figures central to the Afghanistan jihad, a connection that would prove fateful in years to come. It is possible that Indira Gandhi, who had returned to power in the 1980 General Election, saw the Islamists as a means of containing Sheikh Abdullah. New Delhi, notably, made no effort to restrict the Islamists' activities, even by the simple expedient of denying visas to the Jamaat-e-Islami's foreign guests.

Challenged by the growing influence and religious legitimacy of the Jamaat-e-Islami, Abdullah used a combination of repression and competitive communalism. He banned, for example, a planned World Youth Islamic Council the IJT was scheduled to hold in August 1980. More important, though, he sought to construct a Kashmiri Islamic identity, in contrast with the global Islamic enterprise represented by the Jamaat-e-Islami. He sought to expand his support base into what was described as "Greater Kashmir", that is the Kashmir valley and the adjoining Muslim-majority areas of Ladakh and Jammu, bound together not by "culture or language but their *religious affinity*".[134] In 1979, the district of Kargil was carved out of Ladakh, thus providing administrative institutionalization of the fault-line between the region's Buddhist and Muslim-majority zones. Inevitably, Sheikh Abdullah's "obsession with empowering the *majority community*, the [ethnic] Kashmiris, at the cost of the minorities in Jammu and Ladakh led to growing discontentment".[135] By 1979, Jammu was the scene of a large-scale movement for greater regional autonomy, sparked of by police firing on a student protest in Poonch in the December of the previous year. Ladakh took up the regional autonomy demand where Jammu left off. Through 1982, there were violent clashes between police and protestors in Ladakh. Inevitably, these regional protests fuelled and fed communal anxieties.

It is not, perhaps, coincidental that the JKLF set about planning its first act of war as this unstable political theatre unfolded. According to Hashim Qureshi's insider account, the JKLF first considered bombing the March 1983 conference of foreign ministers of non-aligned countries at New Delhi. For a variety of reasons, the idea was dropped. Instead, the organization planned hijacking an aircraft at the same time. Like the bombing, the hijacking turned out to be a fiasco. Four JKLF cadre – two from Paris and two from London – did indeed board a New Delhi bound airliner but, untrained, unarmed and panicky, decided at the boarding gates not to execute the hijack. In an April 5, 1983 letter to

Qureshi, Amanullah Khan expressed relief at its failure, pointing to United States hostility to the non-aligned movement, and arguing that had

> our mission gone the way we had planned, there was a possibility of linking it to [the] CIA's [Central Intelligence Agency's] doing. That would have done us immense damage which we could never set right just as in the case of [the] Ganga hijacking, there is the allegation that it was a conspiracy hatched by India to divide Pakistan.[136]

Blood was, however, soon to be drawn by the National Liberation Army. Hashim Qureshi arrived in the United Kingdom on January 19, 1984, traveling on a legitimate visa, to discuss the JKLF's future course of action with Amanullah Khan. It was finally decided that the hijacking would now be executed under Qureshi's own command. A sum of UK£6000 was assigned for the operation, Qureshi recorded, and detailed discussions were held on exactly how weapons would be smuggled on board the aircraft. "I had strong reasons", Qureshi recorded,

> to join the mission and take its command in my hands. In my opinion there was no better and more effective a plan than that of hijacking in order to seek Butt Sahib's [Maqbool Butt's] release from the Indian prison. The Indian government would not have run the risk of imperiling [the] lives of 350 passengers on board by refusing to release Maqbool Butt.[137]

Along with key JKLF functionaries Afzal Tahir and Malik Ejaz, Hashim Qureshi spent the night of February 2 watching hired movies. "I wanted to acquaint myself", he wrote, in words that illustrate not just the poor tradecraft and technical resources of the JKLF but also its naivety, "with the new techniques in hijacking".[138]

Unknown to Qureshi, he would never have the opportunity to see whether reel-life hijacking techniques worked in the real world as well. On February 3, even as Qureshi engaged in a political discussion with several sympathizers in the town of Luton, India's Assistant High Commissioner in Birmingham, Ravindra Mhatre, was kidnapped by members of the National Liberation Army. The kidnappers soon issued a letter demanding Maqbool Butt's release, and named Amanullah Khan as their interlocutor of choice. Before serious negotiations could begin, however, the kidnappers panicked. By Qureshi's account, JKLF operative Massrat Iqbal Malik believed that a police raid on the safehouse where Mhatre had been kept was imminent, and asked Amanullah Khan for instructions. "Without waiting for a minute", Qureshi recorded, "Amanullah Khan said, 'shoot him and throw away the dead body somewhere. If we are caught it will mean a disaster and all of us will be implicated.' "[139] Qureshi claims to have prophesied that this course of action would lead India to execute Maqbool Butt in retaliation; Khan, in response, argued that India would not do so because such an act of revenge would "bring defamation to India".[140]

Mhatre's dead body was discovered on February 5, on the grounds of a farm in Leicestershire.[141] He had been shot three times, in the head, neck and chest.[142] India executed Maqbool Butt six days later.[143] He was buried on the grounds of Tihar jail. Two JKLF cadre were found guilty by a British court of having carried out Mhatre's execution, and received life sentences in February 1985. A third suspect, Mohammad Aslam, succeeded in evading the authorities until July 2003, when he was detained in the United States for staying in the country after the expiry of his 90-day visa. Police authorities in the United States subsequently discovered that his fingerprints matched those of a man wanted in the United Kingdom for Mhatre's murder. By then 49 years old, Aslam had abandoned politics, married and spent his time managing an apartment complex in Pottsville, Pennsylvania. He had last been to Jammu and Kashmir in 1985; at the time of writing, he was fighting extradition to the United Kingdom.[144] Both Hashim Qureshi and Amanullah Khan were detained by British authorities but had to be released because of a lack of evidence. Both, however, were expelled from the United Kingdom and made their way to Pakistan. Amanullah Khan would make his peace with the establishment there; Qureshi refused to do so, and was to spend the next decade and a half of his life in Holland, before finally returning to Jammu and Kashmir to face the welter of criminal charges pending against him in Indian courts.

Crossing the line

"In the early 1970s", wrote the would-have-been JKLF operative Nazir Gilani, "the crossing of [the] LoC was as mystical for a Kashmiri youth as the Eve St. Agnes to a virgin." His compatriots, Gilani noted, "seemed mesmerized by a belief that a solution to all their ills on the Indian side of Kashmir lay on the Pakistani side of Kashmir".[145]

To students of the long jihad in Jammu and Kashmir, the case of the NLF–JKLF is of particular interest because for much of its existence, Pakistan was not the promised land. Unlike the operatives of the Master Cell or al-Fatah, the JKLF's members for the most part had to rely on their own resources. Their plans were drawn without the assistance of the ISI, and sometimes lay in direct opposition to Pakistan's tactical interests. A second feature of the NLF–JKLF is also remarkable. While the Master Cell or al-Fatah had drawn their ranks from anti-India political forces within Jammu and Kashmir, the NLF–JKLF's political life was entwined with the lives of the Mirpuri community in Pakistan-administered Kashmir and in the larger Kashmiri diaspora. Maqbool Butt may have emerged from a National Conference background, but neither he nor his organization built upon this political foundation. Both the JKLF's politics and its praxis represent a considerable departure from the models which had so far been used by the organizations which fought the jihad.

Could the reasons for the NLF–JKLF's particularities lie in the composition of its membership? Interestingly, the men who joined the NLF–JKLF were not dissimilar to those who served al-Fatah or the Master Cell. Most of the NLF–JKLF's cadre, as we have seen, came from relatively affluent backgrounds and almost all seem to have been beneficiaries of the new order established after Indian Independence. Butt, for example, was subjected to considerable hardship and deprivation under feudal rule but he was able, after Independence, to gain access to a quality education. Gilani, similarly, was a successful student at Kashmir University prior to his decision to cross the LoC; he had, indeed, won two national scholarships on the strength of his academic achievements.[146] Altaf Khan, likewise, had an exceptional academic record.[147] Like many other NLF cadre, he came from a middle-class home; his father was a government employee who worked as a forest ranger. Neither economic issues nor lack of access to opportunity, then, underpinned the desire of this group of young people to turn to the NLF–JKLF.

Ideology, by contrast, played a key role: "Kashmiri pride", Altaf Khan has called it. But what exactly was this "Kashmiri pride"? Butt, certainly, was no Islamist. Nowhere does invective against Hindus figure in his writing. In a letter from prison to a political activist in Srinagar, he wrote:

> Nations survive because of that spirit, that abiding passion for liberty, which according to our Prophet (praise be upon him), emboldens one to recite the Kalima-e-Tawheed [the call for truth] before a tyrant, and that too with the conviction that this is the greatest jihad.[148]

He drew on other examples, though, to illustrate the ethical basis of his resistance to India:

> Think of it; did Aristotle not have to drink poison? Did not the Prophet of God jump into the fire of Nimrod? Did Jesus not have to kiss the cross erected by the rulers of his time? Did our Prophet (praise be upon him) show despair when he was bloodied because of the stones that were thrown in the markets of Taif? Did Gautam Buddha ever compromise with Brahminic exploitation?
>
> Did Martin Luther King, or Marx and Engels, who gave the ideas of anti-colonialism and national liberation ever compromise with the rulers of their times[149]

Yet, others in the NLF–JKLF hierarchy from the outset rejected the principles of the man they revered. Altaf Khan came from a religious background. His early activism was with an expressly Islamist organization, the Muslim Liberation Front, and his work after Sheikh Abdullah's death with another, the Islamic Students' Organization. He was an ardent admirer, as Victoria Schofield has recorded, of the jihad in Afghanistan, the crucible in which modern Islamist reaction was forged. He would act as a mentor to many of the JKLF's younger leaders during the 1980s, as we shall see in the next chapter. It is also worth noting that Butt's own liberation ideology was profoundly flawed. If he felt any sympathy for other nationalist movements elsewhere in South Asia, we have no record of it. None the NLF–JKLF membership seems to have expressed solidarity with the Bangladesh movement, for example, or to have protested against the carnage unleashed during Operation Searchlight. Nor did the NLF–JKLF make common cause with other nationalist movements, for example that of Balochistan.

We can only guess what Pakistan's intelligence services made of the opportunity they were presented with when Amanullah Khan returned to Pakistan, and once again began the business of reconstructing his organization. India's successful offensive in Bangladesh had demonstrated that the fundamental premises of Operation Gibraltar had been sound. In the right circumstances, and with the right resources, an unconventional war could bring a state, however militarily strong, to its knees. The fact that the NLF-JKLF had succeeded in continuing anti-India activity despite the events of 1971 showed, moreover, that a reservoir of discontent existed in Jammu and Kashmir. If adequate resources could be channelled to the JKLF, it would be possible, at least in principle, to escalate the jihad to levels where it would hurt India significantly. The problem for strategists considering what to do with the NLF–JKLF, however, would have been that India was certain to respond to provocation as it had in 1965. After the war of 1971, pursuing an aggressive policy in Jammu and Kashmir would have appeared about as attractive to Pakistani leaders as a cleaver must seem to the chicken whose neck it is about to sever.

But, as Altaf Khan had realized, changed historical circumstances were creating new opportunities. In the early 1980s, Pakistan would be able to use a variety of circumstances to deter India from responding to sub-conventional war with a nation-threatening escalation. In the years to come, Amanullah Khan and a new generation of JKLF cadre would spearhead a wave of Pakistan-backed violence that would at one stage threaten to sweep the Indian state out of Jammu and Kashmir. In this enterprise, they would have the support of the Islamists. In the cause of defeating India, the alliance that began to emerge after Sheikh Abdullah's death suggested, Islamist and nationalists could co-exist. Butt's ideological heirs, though, would not reap the harvest of neither the seeds he had planted nor of their own struggles. The real beneficiaries, as well shall see, would be the Islamists – Islamists who made clear he was not their hero. In 1984, the Jamaat-e-Islami's house journal, the *Azaan*, carried an obituary for Butt, notably omitting the customary suffix *shaheed*, or martyr, from his name:

> He had entered that world of emotions, where a person like him, burdened by the overwhelming force of emotion carries on without making any distinction between the bitterness and sweetness of life, losing the capacity to distinguish between wrong and right. . . . Be that as it may, [we are] greatly saddened by the death of those youngsters

> who, despite having been so capable, become victims of their emotions, instead of facing the massive boulders in their path with determination and courage.[150]

The NLF–JKLF leaders would have opportunity to ponder the implications of this reading of events in the years after the jihad they had initiated reached full flower. While the contours of a map for reviving the jihad in Jammu and Kashmir had started to become apparent, there was still little sense of just how perilous a business crossing the line would prove.

6

THE WAR OF MANY FRONTS

> all of a sudden a huge procession came. They were raising anti-national slogans along with the usual fundamentalist ones like '*Kashmir mein agar rehna hoga Allah-ho-Akbar kehna hoga*' (If you wish to live in Kashmir you have to say 'Allah ho Akbar') and '*Dil mein rakho Allah ka khauf haath mein rakho Kalashnikov*' [keep the fear of Allah in your heart and a Kalashnikov in hand].
>
> Seema Raina, a nurse, describing a rally 1989 by Islamist groups in Srinagar[1]

LATE on the night of September 18, 1988, a single Jammu and Kashmir Liberation Front operative named Riyaz Ahmad Sheikh arrived outside the gate of Deputy Inspector General of Police Ali Mohammad Watali – one of the police personnel who, in earlier years, had played a key role in dismantling al-Fatah and the Master Cell. Brandishing an assault rifle, he loudly demanded that the officer be awakened to meet his fate. Alarmed, the sentry at Watali's gate put his .303 Lee-Enfield rifle to use. Sheikh thus became the first terrorist to give his life for the ongoing phase of the long jihad in Jammu and Kashmir.

Just a month before the attempted assassination of Watali, a bomb had exploded inside the Telegraph Office in downtown Srinagar.[2] Another explosive went off at the Srinagar Club, then a favoured retreat of the city's elite. The Jammu and Kashmir Liberation Front claim the attacks to be the beginning of the war which, at the time this book is being written, has claimed well over 40,000 lives. To contemporaries, however, the bombing of the Telegraph Office may not have had the same symbolic and portentous significance it now bears. Earlier that year, helicopter skiing had been introduced in Jammu and Kashmir, and major new investments had been made in the power sector.[3] Despite what had happened in front of Watali's home, or at the Telegraph Office, there was no obvious reason to believe the year 1988 would be of specific significance. Many residents of Srinagar had witnessed the informal war, and the rise and fall of the Master Cell, al-Fatah and the National Liberation Front. No one had been killed, after all, in what would then have appeared to be just more punctuation marks in the story of the jihad. Nor had any great damage been caused: the Telegraph Office still stands and functions in the languid manner that characterises most Indian government offices.

For reasons I hope shall become clear as this chapter unfolds, it is not my intent to provide a detailed account of the current phase of the jihad in Jammu and Kashmir here. Nor is it my intention to analyse what some believe to be the root causes of the violence, if indeed the unfolding jihad can be distilled to yield single elements. Many authoritative histories have been written about these events. Manoj Joshi's *The Lost Rebellion* provides a superb account of the key players in the conflict, almost baroque in its detail, as does Mohammad Amir Rana's encyclopedic *A–Z of Jihadi Organizations in Pakistan*; Pradeep Thakur's *Militant Monologues* offers some of those players' own accounts of events. For those interested in the historical context and political dynamics of the ongoing conflict, the work of Victoria Schofield, Sumit Ganguly, and many others have provided contrasting, but nuanced and finely wrought accounts. The proliferation of jihadist groups from 1991 onwards can be seen as the dense branches at the top of a tree, whose roots lie in the Partition of India and even earlier. If the previous chapters of this book have provided an understanding of the trunk of the tree, I shall seek here to inspect the structure of the foliage: an account of the ways in which the secret India–Pakistan war over Jammu and Kashmir, having grown steadily to 1991, finally exploded into full public view.

Jammu and Kashmir is not, however, the only theme in this chapter, even if it lies at the core of its narrative. During this period, it was just one of several fronts on which this war was waged. Pakistan's retreat after the war of 1971, which had led to its dismemberment, would be checked by an unexpected gift: the flow of Western and West Asian aid that streamed in after the Soviet Union invaded Afghanistan. Among other things, the Afghan jihad led the United States to overlook Pakistan's pursuit of nuclear weapons, an effort to offset the conventional military inferiority that was so brutally exposed in 1971. By the early 1980s, Pakistan would test the new balance of power in South Asia by backing terrorist groups fighting to carve Khalistan, a Sikh theocratic state, out of the Indian state of Punjab. This would lead both states to the edge of war before India, confronted with the risk that the fighting might just lead to a nuclear exchange, blinked. By the late 1980s, with terrorist groups having plunged Punjab into a conflict that, at its peak, was considerably more brutal than anything Jammu and Kashmir has so far witnessed, Pakistan was able to turn its eye to the main prize. The jihad in Jammu and Kashmir was, thus, just one theatre of a war with many fronts.

The rise of Zia

In meetings with foreign dignitaries, President Zulfikar Ali Bhutto would sometimes describe the unctuous officer he had appointed Pakistan's Chief of Army Staff as his "monkey".[4] If General Zia-ul-Haq resented the description, he never revealed it, at least in public. Prior to his appointment as Pakistan's army chief, he had on one occasion ordered his subordinates, along with their wives and children, to line the road along which Bhutto was to drive for an official visit. After taking command of the Pakistan Army, the General waited in full ceremonial dress outside a conference room where Bhutto and his staff were meeting, claiming it was his duty to guard his President.[5] Bhutto was to learn at the cost of his life that while Zia-ul-Haq was exceptionally skilled at playing the time-honored role of the *darbari* or professional courtier, the General was nobody's flunkey.

For much of his career, Zia-ul-Haq showed few of the ideological leanings that would lead to his support of an Islamist occupation of the core of both Pakistani society and the state. Indeed, he recommended the removal of an Islamist officer, Major-General Aslam Zuberi, on the grounds that his political views made him a threat to Bhutto's regime.[6] Born to conservative lower middle-class parents who had arrived in Pakistan at Partition, Zia was, in Hassan Abbas' words, "a practicing Muslim more due to force of habit than to temperament".[7] Nor was the General an archetypical warrior. He had not been on the frontlines in the wars of either 1965 or 1971, and his combat experience was restricted to having aided the Jordanian campaign against the Palestinian al-Fateh – ironically enough, at about the same time Pakistan's covert services were helping run an organization in Jammu and Kashmir that shared the same name.

Understanding Zia's rise to power requires an engagement with the multiple crises that confronted the Bhutto regime after 1973. Faith, as I have noted in the last chapter, was one key faultline. Pakistan's Islamists had proven remarkably resistant to Bhutto's efforts to appease them. While Bhutto's Constitution had declared Islam the state religion, committed the state to teach the principles of the faith and the Quran in schools and set up a Council of Islamic Ideology to bring secular laws in line with the *Shariat*, Pakistan's clerics were soon asking for more. By 1974, Bhutto was facing widespread riots against the heterodox Ahmedi sect, led by the Jamaat-e-Islami's student wing, the Islami Jamaat-e-Tulba. Islamists successfully used the agitation to expand their hold both among university students and among followers of the mainstream-conservative Barelwi sect in rural areas and small towns. Bhutto at first sought to contain the agitation by arresting some of the protestors and their leaders, but seeing that this failed to contain the Islamists, he caved in. The Ahmadis, although they had long backed Bhutto and his party, were declared unbelievers, outside the pale of Islam.[8]

Sub-nationalist assertion was another major problem. Since Pakistan's independence in 1947, the Baloch tribes had fought three small wars against the state, two lasting just a few months in 1948 and 1958, and another that ran for six years between 1962 and 1968. The conflict was the consequence, Stephen Cohen has noted, of an illiberal attempt to "modernize a very conservative tribal society whose leaders are very autocratic, but whose authority rests upon the authority of tradition".[9] In the wake of the birth of Bangladesh, however, Baloch separatism flared up again. In 1972, a coalition made up of the National Awami Party, which represented traditional tribal power-structures, and the Islamist Jamaat Ullema

Islam came to power in Balochistan and the North-West Frontier Province. Ten months later, Bhutto sacked the governors of the two provinces, claiming that the National Awami Party was not only resisting his programme of modernization, but also preparing its followers for a war to secede from Pakistan. Fighting broke out. Bhutto's repression of the fourth Baloch rebellion, which began in 1973, was spectacularly brutal. An estimated 80,000 army and paramilitary personnel were committed against ill-armed Baloch irregular forces numbering around 1,000. Helicopter gunships, armor and mortar provided by Iran were used to crush the rebellion, and Pakistan's new elite forces, the Special Security Group, were committed to combat for the first time after their formation.[10] It was of little use, though: the fighting would only end after Bhutto was deposed, and the Pakistan Army secured a political arrangement with the separatists.

Despite these many crises, Bhutto had reason for some optimism going into the 1977 elections. Despite the global economic shock caused by rising oil prices after 1973, the loss of East Pakistan in 1971 and a series of crop failures and floods, Pakistan's economy had registered reasonable industrial growth.[11] Moreover, Bhutto was the only credible democratic leader in South Asia at the time. Battered by the growth of her domestic political opponents, Indira Gandhi had declared a state of Emergency in 1975, suspending the constitutional rights of India's citizens. In Bangladesh, General Zia-ur-Rahman had staged a coup, heralding the coming of a long period of authoritarian rule in that country. Paradoxically, the survival of democracy in Pakistan had thrown up credible challenges to Bhutto's authority, expressed through the medium of a vibrant opposition. Part of the reason for the growing opposition to the Pakistan People's Party [PPP] regime was that Bhutto's promises of radical reforms to improve the conditions of Pakistan's working poor had proved short-lived. As the scholar Hamza Alvi has pointed out, a regime which had come to power riding on the back of working-class militancy soon made its peace with the elite:

> The radical rank and file of the PPP was repressed by its own government. 'Marxist' ministers were sacked. Some of the worst reactionary elements in the country took their place.[12]

Opposition to Bhutto was spearheaded by the Pakistan National Alliance, an Islamist coalition which campaigned for an Islamic form of government. True to form, Bhutto attempted to appease the Islamists. The sale of alcohol was banned, as were horse racing and gambling. Friday was declared the weekly holiday in place of Sunday. Bhutto won the elections, taking 155 seats of 200. The elections were widely believed to have been marred by massive poll fraud, a proposition affirmed by the fact that the official voter turnout figure, 80 per cent, was implausibly high. One credible commentator has estimated that between 35 and 40 seats were stolen from the opposition, but it is worth noting this figure would still have left Bhutto with a majority in the national assembly.[13] Nonetheless, the clerics who formed the core of the PNA were able to use anger against electoral malpractice to devastating effect. Unarmed Islamist cadres took on the police and, after dozens of civilian deaths failed to deter their protests, martial law was imposed in Lahore, Karachi and Hyderabad. It soon became apparent that Bhutto was losing the battle. In Lahore, for example, Brigadier Niaz Ahmad flatly refused orders to fire on civilian demonstrators; two officers posted to replace him followed suit.[14] Bhutto now attempted to negotiate a truce with the PNA. It was too late: power had been gifted to the Army which, sidelined in Pakistani politics ever since 1971, was not about to hand it back.

In March 1977, Bhutto and his ministers were arrested on Zia-ul-Haq's orders, and martial law was imposed across the country. On April 4, 1978, shortly after the Supreme Court of Pakistan rejected an appeal, Bhutto was hanged. Naseem Hassan Shah, one of the judges who rejected Bhutto's final appeal, made it clear years later what had happened: "The higher courts faced the threat of complete closure in the event of a decision against the will of the Martial Law regime."[15] Pakistan's Supreme Court judges were not the only ones who were overawed. Whereas massive crowds had come out to greet Bhutto after his release from jail, none gathered to protest his judicially sanctioned assassination. Yet, as Seyyed Vali Reza Nasr has noted, Bhutto's demise was, in some senses, his own making. Writing of the first retreat executed by Bhutto's regime, the declaration that the Ahmadi were apostates, Nasr wrote:

> The polity, which had only five years earlier been overwhelmingly in support of populism and socialist idealism, had once again exposed itself to manipulation by Islamic symbols. The return of Islam to centre stage was now complete. The fact that all this happened under the aegis of Pakistan's most popular government to date, one which had a strong ideological basis of its own, only attested to the incomparable influence of Islam on the life and thought of Pakistanis. The seemingly implausible resurgence of Islam in lieu of socialism during the Bhutto era meant total victory for Islam and confirmed its central role in Pakistani politics. As populism lost its momentum to Islam, the fate of Bhutto's government was sealed long before Islam actually pulled down the People's Party and its populist government.[16]

Zia-ul-Haq understood which way the storm was blowing and would, as the years went by, prove to be extraordinarily adept at riding it. His abiding contribution to the secret India–Pakistan war would be twofold. He would, as we shall see, energetically pursue Pakistan's nuclear weapons programme, creating a shield behind which the jihad in Jammu and Kashmir could be pursued without inviting Indian retaliation. As important, Zia-ul-Haq placed Islam at the core of the Pakistan Army's *raison d'etre*. Its involvement in Islamist causes was no longer merely tactical; it was an ideological imperative. This use of Islam to legitimize the state's use of coercive means was not, of course, new. General Yahya Khan had, at a late stage of the East Pakistan crisis, described enemy forces as *kaffirs*, or unbelievers, and his own troops as *mujahideen*, or soldiers of Islam. The notion of jihad, similarly, fed and informed each successive phase of covert warfare in Jammu and Kashmir. Now, however, Islam was no longer merely a faith, appeals to which could be made for opportunistic purposes. It was firmly enshrined as an institutional ideology. Courses in Islam became part of the curriculum at officer-training colleges. Soon, religion began to express itself as a key component of military doctrine. Terrorism was now seen as having religious sanction. "This position may not be publicly flaunted", Steven Cohen has noted, "but it is widely held in the army".[17] An opportunity to put the belief into practice was not long in coming.

The Afghan Jihad

On Christmas eve in 1979, Soviet troops landed at Kabul's international airport. Within a few hours, the Afghan President, Hafizullah Amin, had been shot dead by commandos from the Committee for State Security, the Soviet covert service known all over the world as the KGB. Tanks and troops rolled across the Amu Darya river, which marked the border between the Soviet Union and Afghanistan, the next morning. A war that would have consequences of then unimagined magnitude had begun. In the years to come, the jihad in Afghanistan would set off a whirlwind that would turn that country into a wasteland, transfigure Pakistan, bring down the Soviet Union, and turn the Islamists who had been armed and funded by the United States to bring about the defeat of its Cold War adversaries full circle, against their one-time masters. For India, too, the Afghan war would have profound consequences: this war was the crucible in which a new phase of Pakistan's jihad in Jammu and Kashmir would be forged.

For Zia-ul-Haq, the invasion of Afghanistan was a gift, an act of divine provenance as it were. Already tainted by the stigma of having hanged its President, the General's first months in office had proved difficult. By late November 1979, the relationship between the United States and Pakistan had reached an all-time low, a situation that cash-strapped Islamabad could ill-afford. On November 21 that year, a radical Islamist group from Iran occupied the Kaaba, the shrine at the heart of the city of Mecca in Saudi Arabia. Intended to embarrass the pro-United States government of Saudi Arabia, the occupation of the Kaaba coincided with a cycle-borne tour of Rawalpindi by Zia-ul-Haq to popularize the energy-efficient mode of transport. For reasons which are still not clear, Zia-ul-Haq informed the assembled crowd that international radio broadcasts suggested that the United States had engineered the occupation. It was a bizarre suggestion – and would have bizarre consequences. Crowds, led by cadre of the Jamaat-e-Tulba, marched on the United States' Embassy in Islamabad and proceeded to set it on fire.[18]

Within a month of the Embassy siege, however, the Soviet Union's invasion of Afghanistan compelled the United States to energetically woo Pakistan's pariah regime. Challenges to the Kabul regime had been mounting ever since the autumn of 1978, when a rebellion broke out in the remote Hazarajat region. Its efforts to bring about land reform, ensure girls received an education, grant women the right to marry by choice and proscribe the payment of dowries had incensed the traditional leadership of the region's deeply conservative communities. Soon after the Hazarajat rebellion, in March 1979, a massive revolt erupted in the town of Herat, led by junior officers of the Afghan army's 17th Division, including Ismail Khan, Alauddin Khan and Abdul Ahad. More than a dozen Soviet advisers posted in Herat and members of their families were hacked to death by Ismail Khan's forces. Afghan forces responded with massive force, using their Soviet-provided aircraft to bomb Herat. Their efforts, however, proved futile. Ismail Khan's counter-revolution spread to Jalalabad and its surrounding countryside. Islamists were at war with the only communist regime in South Asia. One side had the backing of the Soviet Union, and the other would soon have the support of the other great Cold War power. Soon after the Jalalabad revolt, the Central Intelligence Agency had proposed to the White House that covert military aid be funnelled to the Islamist groups fighting the Kabul regime.[19] Carter chose not to act on this advice, but Iran and Pakistan started making military supplies available to Islamists in Afghanistan, albeit on a limited scale.

At first, aid began to flow from the conservative Gulf monarchies and Saudi Arabia, which hoped to deflect the energies of the many Islamist groups that had sprung up in the wake of the Iranian revolution. The monarchies feared that the new religious right would turn on the regimes of their own countries. Ultra-reactionary groups like the Muslim Brotherhood or the Ahl-e-Hadis were at the principal recipients of these fund flows, often with the approval of the Saudi government. No official data on the scale of these flows is available, but one credible commentator has estimated that the Saudi Arabia-based World Muslim League alone funnelled several million dollars a year to the Afghan jihadists.[20] In the summer of 1979, the White House began to support the Islamist resistance in Afghanistan and authorized the supply of non-military aid to the jihad. Soon after the Soviet military crossed the Amu Darya, weapons began to flow to mujahideen as well. In a memo to President Carter written hours after Soviet troops entered Afghanistan, his national security adviser, Zbigniew Brzezinski, urged that the United States abandon its non-proliferation objectives in Pakistan, and act in concert with West Asian states and China to force the withdrawal of Soviet troops from Afghanistan:

> Anti-Soviet fever swept Washington, arousing support for a new phase of close alliance between the United States and Pakistan. Together, they would challenge the Soviets across the Khyber Pass, much as the British had challenged czarist Russia on the same Afghan ground a century before.[21]

Pakistan was well-placed to act as a base for this war, not just because of its geographical location, but also because of the long-standing relationship of its covert services with Afghan Islamists. In July, 1973, a military-led coup had brought the regime of Mohammad Daud Khan to power in Kabul. Its relations with Pakistan were poor from the outset. Daud asserted Afghanistan's claims to the ethnic-Pashtun enclaves of north-western Pakistan; in response, the regime of Zulfikar Ali Bhutto backed Islamist groups who were facing the wrath of the new political order in Kabul. By 1974, several Islamists who would become prominent in the course of the anti-Soviet jihad, notably Gulbuddin Hekmatyar, Burhanuddin Rabbani and Ahmed Shah Masud, had taken refuge in Pakistan. A political movement made up of Islamists and clerics soon came together. With the assistance of Pakistan's ISI, which wished to retaliate against Afghan support for Baloch and Pashtoon nationalists, the Islamists were encouraged to plan a coup. In July 1975, small groups trained, armed and funded by Pakistan's covert services infiltrated into Afghanistan's provinces hoping to overwhelm the regime.[22] Perhaps unsurprisingly, given its limited resources, the attempted coup of 1974 soon degenerated into a rout. Giles Dorronsoro, however, has suggested a more cynical explanation:

> The coup had no chance of success, as the Pakistani military men who helped lay the groundwork for it must have been aware. It might therefore be supposed that the Islamists had been manipulated by the Inter Service [s] Intelligence, who intention was to deliver a warning to Daud; according to some sources, the ISI may even have tipped Daud off in advance.[23]

This time around, Zia-ul-Haq proved remarkably adept at extracting what could be had from the situation in Afghanistan. As Hassan Abbas has observed, "he was quite clear that he too was opting to be a tool of the United States in its proxy war in Afghanistan".[24] America's use of Pakistan, though, came at a price. Sensitive to the fact that the political climate was turning against President Carter, Zia-ul-Haq dismissed successive offers of military and civilian aid, amounting first to US$ 150 million and then US$ 400 million.[25] His reading of the situation proved correct. Carter's deeply anti-communist successor, President Ronald Reagan, hiked the aid on offer to US$ 3.2 billion, and threw in a generous component of F-16 strike aircraft, weapons that were of no use to the jihad in Afghanistan but did constitute a serious threat to India. Pakistan also succeeded in securing guarantees that the United States would abandon efforts against its nuclear programme, as long as no actual test of a weapon was carried out, and that pressure to democratize would cease. Most important of all, the United States agreed that all military assistance to the Afghan jihadists would be routed through Pakistan's ISI. In practice, this meant that while the United States would pay for the Islamist kite flying in Afghanistan, Pakistan would hold the string – and, if the Soviet Union were indeed forced to withdraw, would control the levers of power in Kabul.

For the moment, though, Pakistan's energies were firmly focused on making sure the jihadist army it had been paid to raise and run inflicted as much pain on

Soviet forces as was possible. Its initial operations, possibly as a consequence of Zia-ul-Haq's desire not to provoke the Soviet Union into direct reprisal against Pakistan, were limited. By 1983, only some 3,000 mujahideen had been trained at the two major camps which had been set up within Pakistan. Those numbers, however, soon rose sharply. By 1987, seven training camps were operating in Pakistan, four near Peshawar and three near Quetta. In 1984, some 20,000 mujahideen were trained at the camps, which operated under the overall command of the ISI Director-General Akhtar Abdul Rehman Malik. Numbers close to that level went through the ISI mill in subsequent years, too. "It is no exaggeration to say", Brigadier Mohammad Yousaf, General Malik's subordinate in charge of Afghan operations has written, "that by the time I left the ISI in 1987, at least 80,000 Mujahideen had received training in Pakistan over a four-year period, and many thousands more had done so in Afghanistan".[26] In that year, 65,000 tons of arms and munitions, including high-technology anti-armor and anti-aircraft missiles, were provided to the jihadist forces, as was some of the most sophisticated communications equipment and technical intelligence in the world.[27]

What happened in subsequent years in Afghanistan is well known; I shall not trace its course here. Steve Coll's magnificent telling of the jihad, *Ghost Wars*, makes clear that none of those involved in that most brutal of conflicts would walked away from it unscathed. The Soviet Union would, of course, pay for its intervention with its own demise; but a price would be paid, too, by the United States, which would witness the Islamist armies it had raised and funded turn on it with spectacular effect. West Asia would, similarly, face savage retribution at the hands of those very groups Saudi Arabia and the Gulf monarchies had patronized. Afghanistan itself, of course, would suffer the most horrific fate of all. In 1987, however, none of this was evident. To Pakistan's strategic establishment, the Afghan jihad would have held out two signal lessons. One, an irregular army, with the right kinds of support, could bleed a superpower at relatively little cost to its patron-state. Two, proxy war could be calibrated to a point where it was not worth the while of an adversary to punish the sponsor-state by going to war. "The water in Afghanistan", Zia-ul-Haq had told his spymaster in December 1979, "must boil at the right temperature".[28] Could the water be heated up to a similar point elsewhere as well? Even as the war in Afghanistan proceeded apace, General Akhtar Malik began applying his mind to a question that, for Pakistan, was not just a question of tactical opportunity: its unceasing war with India.

As a result of events in Afghanistan, Pakistan's strategic establishment would soon arrive at a fateful conclusion: that it could do to India what it had done to the Soviet Union in Afghanistan.

The Khalistan Front

From the early 1980s, a movement of the Sikh religious right, calling for the creation of a theocratic state called Khalistan, had been gathering momentum in the Indian state of Punjab. A variety of complex causes underpinned the growing crisis in Punjab: Prime Minister Indira Gandhi's use of Sikh fundamentalists to marginalize her centre-right opponents, the efforts of her opponents to undermine this enterprise by competitive religious chauvinism, anxieties among clerical elites about the impact of modernization and frustrations generated by the tapering-off of several decades of rapid economic growth.[29] These layers of contestation constituted the political terrain on which the first front of the new phase of covert war for Jammu and Kashmir would be fought.

The ideological genesis of the Khalistan movement, like that of Islamists in Jammu and Kashmir, can be traced among other things to competing religious chauvinisms – in this case Sikh and Hindu – in the early twentieth century. While both Hindu and Sikh chauvinist organizations continued to remain active after the Independence of India, neither commanded any great political influence. Indira Gandhi's party, the Indian National Congress, wielded power in Punjab immediately after Independence. Over the next two decades, however, the Congress' domination came under steady assault from a faction-ridden but none the less powerful opposition, the Shiromani Akali Dal. In an effort to undermine the Shiromani Akali Dal's claims to be the sole spokesperson for Punjab's Sikhs, Indira Gandhi turned to a far-right preacher, Jarnail Singh Bhindranwale. As with Bhutto and the Islamists, the fundamentalist tiger nurtured by Indira Gandhi soon grew large enough to break the chains that contained him:

> In 1980, [Bhindranwale] campaigned actively campaigned for the Congress in three Punjab constituencies and the Janata candidate in one even alleged that Indira Gandhi appeared on the same platform as Bhindranwale. But soon after the elections, Bhindranwale declared his independence and refused to be the tool of the Congress.[30]

During the summer of 1984, Bhindranwale had initiated a low-intensity campaign of terror, targeting political opponents across the ideological spectrum. Much of the terrorist campaign had come to be run out of the Golden Temple in Amritsar, the most revered holy site of the Sikh faith, which Bhindranwale had turned into something of a fortress. On June 5, 1984, the Indian Army initiated Operation Bluestar, a military assault intended to flush Bhindranwale's forces out of the temple. Operating under the command of Major-General Shahbeg Singh, a retired Indian Army officer, the terrorists in the temple turned out to be far better armed than anticipated. Confronted with murderous machine-gun fire, the Indian infantry personnel who initiated the attack made relatively little headway. An armoured personnel carrier brought in to provide some cover to the troops was hit by an RPG-7 rocket-propelled grenade, a weapon widely used by mujahideen forces in Afghanistan but one until then not known to have been possessed by Khalistan terrorists. Eventually, on the morning of June 6, Indian commanders were authorized to use their tanks' main 105-millimetre guns to bring down the terrorists' defences, a tactic that inevitably caused considerable damage to the Golden Temple complex itself. Mopping-up operations continued until July 7.[31]

Operation Bluestar was to have several consequences, not the least of them the assassination of Indira Gandhi herself in October 1984, at the hands of two Sikh personnel in her security detail. For the purposes of this narrative, however, one was central: the growing realization amongst Pakistan's covert services that there were significant opportunities in Punjab. By some accounts, Pakistani intelligence had initiated some level of contact with key figures in the Khalistan movement as early as the 1970s. Jagjit Singh Chauhan, an active participant in early secessionist politics in Punjab and the self-proclaimed President of the National Council of Khalistan, admitted to extensive ties to Pakistani officials and intelligence in the decade before Operation Bluestar.[32] In a series of interviews given in London between 1983 and 1986, Chauhan also claimed to have had contacts with several high-level Pakistani leaders, including the prominent politician, Sir Zafarullah Khan, who at the time of this liaison headed the International Court of Justice at The Hague. Zafarullah Khan, however, denied having ever met Chauhan, stating that he was bound by the terms of his service at the International Court of Justice from interfering in the politics of member-states. Whatever the truth, as Paul Wallace and Surendra Chopra have noted, Chauhan was "used by Pakistan to carry on an anti-India propaganda barrage", specifically by calling "upon Sikhs to revolt against the 'Hindu imperialism' ".[33] Notably, the state-run Voice of Pakistan broadcast an interview with Chauhan in March 1986, a sign of official support for his ideological position.[34]

It would, of course, have been professionally remiss of the ISI *not* to have made contact with Bhindranwale in later years. Given its deep commitment in Afghanistan, it was obviously in Pakistan's interests to back efforts that would keep India – closely allied to the Soviet Union at the time – tied down to its east.[35] Indeed, such links to the Khalistan movement seem to have developed soon after the initiation of the war in Afghanistan. Indian intelligence suspected there was some degree of liaison between the ISI and the Khalistan terrorists from at least 1981, when terrorists hijacked an Indian Airlines Delhi–Srinagar flight to Lahore.[36] The hijackers were not extradited, and were allowed to stay in Pakistan after serving their prison terms. After five subsequent hijackings, new evidence emerged that strengthened these suspicions. A report from the West German government revealed that the pistol used by Khalistan terrorists who hijacked an Indian Airlines flight to Lahore in 1984 was part of a consignment they had supplied to the Pakistan government.[37] None the less, such support seems to have been generally low-grade prior to 1984. One plausible explanation for this may have been, as Mark Tully and Satish Jacob have argued, that Zia-ul-Haq maintained a "cautious attitude" towards the situation in Punjab, afraid that Indira Gandhi was "spoiling for a fight".[38]

By 1990, the year when large-scale violence was to break out in Jammu and Kashmir, killings in Punjab had however reached savage levels: 4,263 people lost their lives in the fighting that year, a number that would rise to 5,265 in 1991

Table 6.1 Fatalities in the Khalistan war

	Civilians	*Terrorists*	*Security force*	*Total*
1981	13	14	2	29
1982	13	7	2	22
1983	75	13	20	108
1984	359	77	20	456
1985	63	2	8	73
1986	520	78	38	636
1987	910	328	95	1333
1988	1949	373	110	2432
1989	1168	703	201	2072
1990	2467	1320	476	4263
1991	2591	2177	497	5265
1992	1518	2113	252	3883
1993	48	798	25	871
1994	2	76	0	78
1995	0	11	0	11
1996	0	3	0	3
1997	56	1	2	59
1998	0	0	0	0
1999	0	0	0	0
2000	18	0	0	18
2001	1	0	0	1

Source: The South Asian Terrorism Portal, www.satp.org.

(Table 6.1). Terrorist violence in Punjab had succeeded in tying down ever larger numbers of Indian military and police personnel. How was this brought about? Assault rifles like the AK-47 and AK-56 started appearing immediately after Operation Bluestar.[39] Starting from 1988, attacks on security forces were increasingly carried out with RPG-7 rockets, mimicking mujahideen tactics. Terrorist groups came to be equipped with not only various Kalashnikov variants, but Chinese-made General-Purpose Machine Guns, sophisticated technology such as night-vision equipment, Dragunov sniper rifles and wireless transceiver sets – all equipment supplied to Pakistan's intelligence services for onward distribution in Afghanistan.[40] That Pakistan was providing infrastructure for Khalistan terrorists is underlined by the fact that 26 per cent of the Kalashnikovs seized in Punjab were intercepted along the India–Pakistan border itself and security forces apprehended over 45,000 persons trying to cross the border illegally between 1986 and 1992.[41]

It takes little imagination to see just why this situation would have been greeted with some satisfaction by Pakistan's military and intelligence services. At the time of Partition, Pakistan had protested Imperial Britain's decision to deny it the Muslim-majority district of Gurdaspur, thus giving India a land route into Jammu and Kashmir. As the historian Shereen Ilahi has demonstrated, Pakistani claims that it was denied Jammu and Kashmir as the result of an imperial conspiracy have little empirical foundation, for the road traverses the *tehsil* of Pathankot, a Hindu-majority area which would in any case have gone to India.[42] As a consequence of its relationship with the Khalistan movement, Pakistan was in a position to attempt to deny India use of this strategically vital axis without committing its regular forces. In the war of 1965, India had stunned Pakistan by retaliating for its assault on Jammu and Kashmir by a counter-offensive through Punjab. Now, however, Indian troops were committed to waging a brutal counter-terrorist campaign in the same areas that had been decisive to their victories in 1965: frontier zones like Khem Karan, Tarn Taran and Amritsar.

Pakistan could not have known, of course, that a situation would arise in 1984 that would bring large numbers of cadre to the side of its covert war against India. By the early 1990s, though, Pakistan had become a significant player. Independent interviews of captured terrorists carried out during the Khalistan conflict make clear that many had obtained training and arms in Pakistan before initiating armed operations in India.[43] Considerable evidence also exists of the direct involvement of Pakistani intelligence personnel with the leadership of Khalistani groups. Talwinder Singh Parmar, the principal suspect in the bombing of an Air India jet that claimed the lives of all 329 on board, escaped to Pakistan before he could be arrested by Canadian investigators. Subsequent trial proceedings threw up evidence that Canada's Secret Intelligence Service was aware of the bombing before it took place, but failed to prevent it for fear that sources would be compromised.[44] When Parmar was eventually shot and killed by the Punjab Police in 1992, two Pakistani nationals were also killed in the firefight that claimed his life. According to the journalist Manoj Joshi, Parmar, who founded the Babbar Khalsa International terrorist group, served as the principal link between the Khalistan movement and the ISI.[45] Another key Babbar Khalsa leader, Wassan Singh Zaffarwal, primarily resided in and conducted operations from a safe-house in Pakistan. Much of Zaffarwal's influence came from his connections with the ISI, which enabled him to leverage the flow of weapons and funds to terrorist groups within Punjab.[46]

Inter-Services Intelligence strategists seemed to have used multiple channels to run the covert war in Punjab. One key source was the affluent Sikh diaspora in the United States of America and Europe, part of which was sympathetic to the Khalistani cause. Maninder Singh, a member of the International Sikh Struggle Committee, stated in an interview that his group's main objective was to "collect funds abroad" which were then used "to purchase arms and to smuggle the same across the Pakistan border to the terrorists".[47] For this latter part of the task, Punjab's well-established trafficking routes were put to use. Tara Kartha's thoughtful study of published reports of India's National Crime Records Bureau has shown that the rise of terrorism in Punjab was preceded by a steady rise in narcotics seizures. Operations conducted against traditional gold and silver smugglers undertaken from 1984 provided evidence that these traffickers had started to show a considerable interest in narcotics as well. In essence, traffickers were being allowed to move heroin out of producing areas in Pakistan and

Afghanistan – and to move the acetic anhydrite used to process the narcotic from India – in return for running weapons for the ISI. Kartha has argued that the "Pakistanis were apt to see narcotics as a resource – in fact comparing it to the oil of the Middle East – seeing it as a problem for the 'decadent West' rather than Pakistanis themselves".[48]

India was in serious trouble, and its strategists started applying their minds to what their options were. War with Pakistan was the ace in India's strategic pack but, as the Indian defence establishment was soon to learn, the new rules of the game prohibited its use.

Brasstacks and the bomb

Pakistan's intelligence services were learning several lessons that would be of use in their bid for the main prize, Jammu and Kashmir. One of these would radically transfigure the structures of the jihad. In 1965 and 1971, Pakistan had been forced to calibrate its sub-conventional offensive, due to fears that India would retaliate against greater efforts by attacking with its superior conventional forces. Now, however, Pakistan had something which rendered the old assumptions redundant: a nuclear bomb.

"If India builds the bomb", Zulfikar Ali Bhutto had famously proclaimed in 1965, "we will eat grass or leaves; even go hungry, but we will get one of our own".[49] At around this time, stung by its defeat by China in the war of 1962, India had embarked on a nuclear programme it hoped would deter a second invasion from the east. Although Bhutto repeatedly pointed to the need for Pakistan to match India's emerging nuclear capabilities, there is little evidence that the polemic was matched by action until after the war of 1971. At a meeting of Pakistani nuclear scientists in 1972, two years before India would test a nuclear devise, Bhutto announced that he wanted "fission in three years".[50] He had little luck, however, building the infrastructure needed to produce a nuclear weapon. Finally, in 1975, Bhutto's weapons programme received a break: a young scientist named A.Q. Khan returned to Pakistan, carrying with him blueprints and information stolen from his employers in Holland, documents which helped put that country on the road to acquiring a bomb.[51]

Khan's act of nuclear theft was driven by patriotism. On May 8, 1974, India carried out a nuclear explosion at a test site in Pokhran, a small village in the desert state of Rajasthan. India insisted that it had tested a device, not a nuclear bomb, and that its purpose was peaceful. Despite the rhetoric, India's military compulsions were obvious. It had a powerful nuclear-armed neighbour to its east, China, at the hands of which it had suffered a humiliating military defeat in 1962. Then, although the United States' displays of support for Pakistan during the 1971 war had been largely symbolic, the presence of the USS Enterprise and its escort ships had strengthened the hands of those in India who wished for insulation against threats to the Indian state. Pakistan, more likely than not, understood that it was not the principal target of India's nuclear-weapons programme, but felt it could no longer afford not to have a bomb of its own. As Hassan Abbas has caustically noted, "India may have exploded a mere 'device,' as it claimed, but the fact remained that if such a peaceful device were to be dropped on Islamabad, the city would be no more".[52]

Although Western governments made some attempts to restrict the flow of components to Pakistan's nuclear-weapons programme, history was on A.Q. Khan's side. Bhutto had been able to secure funding for Pakistan's nuclear programme, which he said would produce an "Islamic bomb" by offering technology to Libya.[53] In April 1979, the United States imposed the Symington Amendment, legislation that mandated sanctions if Pakistan was found to be ignoring its non-proliferation objectives. In December of that year, however, the Soviet Union sent its forces into Afghanistan. As we have seen, Pakistan emerged as a frontline state in the war against communism. Among the consequences of Pakistan's role in the jihad in Afghanistan was that the Symington Amendment was lifted. Although it was well known that Pakistan was using a welter of front companies and covert personnel to acquire components to build a nuclear weapon, its assistance in the war against the Soviet Union was seen as a good enough reason to overlook the issue. By 1982, A.Q. Khan and his staff had succeeded in enriching uranium to a level where it could be used to manufacture a weapon. At some point between 1983 and 1984 – accounts vary – Pakistan had a usable nuclear warhead.[54] In 2004, A.Q. Khan would confess to having sold weapons-related technology to Iran. Transfers are also believed to have been made to North Korea, in return for missile know how.[55]

A.Q. Khan's work – his persistence and ingenuity as seen by his supporters in Pakistan, his catastrophe-inducing criminality as seen by his detractors – had dramatic consequences for the India–Pakistan war then underway in Punjab. In 1986, the violence in Punjab escalated to a point where India began signalling its willingness to go to war. Planned by Lieutenant-General K. Sundarji, who went on to become India's Chief of Army Staff, Operation Brasstacks was a massive exercise intended to demonstrate India's ability to engage in a full-blown war against Pakistan. The largest exercise in Indian military history, Operation Brasstacks began in July, 1986. By December of that year, India had deployed a total of thirteen divisions, or some 160,000 troops, in positions that threatened Pakistan's north–south axis of communication. Pakistan, well aware of the fact that manoeuvers had been used to mask offensive operations elsewhere, notably the Egyptian–Syrian attack that opened the 1973 Yom Kippur war, responded by mobilizing its own forces. The Pakistani Army reserves now threatened the Punjab towns of Amritsar, Pathankot and Gurdaspur, as well as India's rail and road links with Jammu and Kashmir. As India's strike formations were too far away to respond to a potential attack in Punjab, Rajiv Gandhi responded by initiating Operation Trident, which involved airlifting more troops to guard the state's sensitive border with Pakistan.

By January 1987, observers on both sides of the India–Pakistan border, as well as in many world capitals, were genuinely concerned that war might break out. Operation Brasstacks, if testimony heard by the Indian Government's Kargil Review Committee was correct, now provoked what could be interpreted as South Asia's first nuclear threat. The Kargil Review Committee, set up to investigate the circumstances that led to the 1999 Kargil war, recorded that this was officially communicated by Pakistan's Minister of State for Foreign Affairs, Zain Noorani, to the Indian Ambassador in Islamabad, S.K. Singh:

> Noorani, just back from a meeting with President Zia, said he had been authorized to convey a message: if India took any action not conducive to its sovereignty and territorial integrity, then Pakistan was "capable of inflicting unacceptable damage" on it. Pakistan's action would not be limited to northern India alone but also to facilities outside the north. When asked whether this implied an attack on Bombay [now Mumbai], the Pakistani Minister replied that "it might be so".[56]

Although Noorani, by the Kargil Review Committee's account, did not actually refer to the use of nuclear weapons, S.K. Singh could have had little doubt about his meaning. A more specific threat was communicated by A.Q. Khan to the Indian journalist Kuldip Nayar on January 28, 1987. Nayar delayed publishing his interview with Khan, and the scientist's threats did not become public knowledge until after the Brasstacks furor had died down. It is not improbable, though, that Indian diplomats in Islamabad would have learned what the journalist had been told. About a year later, in March 1988, Zia gave an interview to the United States-based magazine *Time*, asserting that "Pakistan can build a bomb whenever it wishes". "What is difficult about a bomb", Zia asked polemically? "Once you have acquired the technology, which Pakistan has, you can do whatever you like".[57] The *New York Times* reported that United States officials had by then concluded that Pakistan had enough weapon-grade uranium for building between four and six nuclear weapons.[58] In the wake of the Brasstacks crisis, and provoked in part by India's 1988 test of its Prithvi ballistic missile, Pakistan accelerated its efforts to acquire missile technology, notably from China.

Indian commentators have generally refrained from making any express linkage between events in Punjab and Operation Brasstacks. Most accounts criticize Rajiv Gandhi and Sundarji for their handling of the affair; one recent book, typically, describes it as a "purely military exercise in which politics got mixed" because of "abrasive and flamboyant personalities" and an "ignorant media".[59] In this case, at least, it could be argued that Rajiv Gandhi's conduct, or that of Sundarji, was not as irrational as many have charged. It seems fairly evident, in retrospect, that Operation Brasstacks must have at least in part been intended to signal India's willingness to initiate a conventional response to the sub-conventional war in Punjab. The costs of being perceived as soft on Khalistan terrorism had hit the image of the Indian National Congress hard, and fuelled the growth of the Hindu-nationalist opposition. Moreover, India had precedent to rely on. Confronted with a sub-conventional offensive by Pakistani irregulars in Jammu and Kashmir in

1965, India had responded with full-scale war. This time around, though, India believed Pakistan had the bomb – and might, under certain circumstances, use it.

Put simply, India had threatened to respond to sub-conventional war with conventional means – and Pakistan had called its bluff. For India's political leaders, the cost of the little war in Punjab was not high enough to justify a big war – a calculation which, as we have seen in the previous chapter, had confronted Pakistan in the build-up to the Bangladesh war. As Sundarji was himself to candidly admit later, "because of nuclear deterrence, the menu of Indian responses to Pakistani provocation . . . no longer includes launching a bold offensive across the Punjab border".[60] India's unwillingness, he continued, to take recourse to its "stated, avowed strategy of reacting in the plains conventionally is because of the nuclear equation". All of this had a deeply unsettling impact on the India–Pakistan strategic balance. As Sumit Ganguly and R. Harrison Wagner have pointed out in a seminal paper, nuclear weapons played, and continue to play, a role in South Asia that is fundamentally different from that seen in the Cold War. During that conflict, they noted,

> nuclear weapons were used to compensate for the perceived conventional inferiority of NATO, but the US was not interested in using military force to upset the status quo in Europe. In South Asia, however, nuclear weapons have helped Pakistan compensate for the conventional superiority of India, and Pakistan *has* been interested in using military force to upset the status quo [my emphasis].[61]

Nuclear blackmail, Pakistan understood correctly, worked. Pakistan's nuclear programme, however untested and crude it might have been at that stage, transformed the order of risk involved for India should it chose conventional war as a means to deter Pakistan. As the legal scholar A.G. Noorani has suggested, the decision to train and arm the massive numbers of mujahideen who would fight in Jammu and Kashmir was taken because of Zia-ul-Haq's delight over "the success of his 'low cost, low risk, high return' investment in Punjab".[62] Interestingly, however, Pakistan seems to have been unsure about how the Operation Brasstacks standoff would play out. By way of caution, it seems to have chosen to scale back support to Khalistan groups for some time during this period. In a 1987 interview, a top operative, Harjinder Singh "Jinda," said that while Pakistan had earlier been supportive of Khalistan groups, it had "lately started backing out".[63] Though Jinda himself had never been to Pakistan, he asserted that many of his comrades-in-arms felt that it had decided to leave them "in the lurch" after Operation Brasstacks.[64]

One possible explanation for this is that India had begun to use offensive covert means against Pakistan with some effect. In the mid-1980s, the regime of Prime Minister Rajiv Gandhi set up two offensive desks in the Research and Analysis Wing, CIT-"X" and CIT-"J," both tasked to retaliate against Pakistan's support of terrorism in Punjab. A series of bombings in Lahore and Karachi demonstrated India's ability to meet terrorism with terrorism. Little hard fact has emerged on this campaign, although the former Indian intelligence official B. Raman has pointed to "the role of our covert action capability in putting an end to the ISI's interference in Punjab by making such interference prohibitively costly to Pakistan".[65] Pakistan appears to have been moved by the Indian counter-campaign to scale back its own support to Khalistan groups for at least some time. Lieutenant-General Hamid Gul, the Director General of the ISI who succeeded General Rehman, met his counterpart in India's Research and Analysis Wing, A.K. Verma, to discuss rules for the war in Khalistan. Brokered through the offices of Jordan's then Crown Prince Hasan bin-Talal, whose wife, Crown Princess Sarvath, is of Pakistani origin, the meetings between the two spymasters were held in Amman and Geneva.[66] While India undertook to end executing reprisal bombings in major Pakistani cities, Gul in turn is thought to have promised that Khalistan groups would be restrained from executing attacks in India's urban centres.

General Gul's promises turned out to be short-lived. India, for its part, resumed its campaign in Pakistan, which attributed two major bombings that took place in Lahore as late as 1990 to an Indian covert operative, Sarabjit Singh.[67] By 1992, with the jihad in Jammu and Kashmir well underway, there is some evidence that the ISI sought to align both theatres of conflict – and to extend them out of their immediate geographical space, in an effort to increase the costs for India as a whole. Waqar Ahmad, an ISI officer believed to have been responsible for handling the Babbar Khalsa International, one of the oldest and most organized Khalistan terrorist groups, is credited with having organized Operation K2. This covert project, which was underway by 1991, sought to link Mumbai-based narcotics traffickers with terrorists in Punjab and Jammu and Kashmir.[68] Operation K2 yielded few results because its key operatives, Manjit Singh and Mohammed Sharif, were arrested in July 1992. By then, however, K2's masterminds had succeeded in recruiting a number of smugglers to move weapons across the Rann of Kutch in Gujarat, paving the way for future operations of the same kind, notably the Mumbai serial bombings of 1993.[69]

If K2 failed, India's counter-terrorism efforts in Punjab had finally begun to yield efforts. Starting from May 1988, India began to build a massive fence along Punjab's border with Pakistan, in an effort to bring an end to cross-border trafficking. Composed of a triple row of barbed wire, the fence was floodlit at night and was also electrified and patrolled. By 2003, it extended from along almost the entire length of the India-Pakistan border, from Gujarat in the south to the northern regions of Jammu and Kashmir.[70] In addition to this physical effort to interdict cross-border traffic, Punjab set out to revitalize the demoralized police force and increase its firepower, mobility and communications capabilities.[71] Led by Director-General of Police K.P.S. Gill, the reinvigorated Punjab Police rapidly demonstrated results. As a result of the police crackdown and heightened border security, terrorist violence began to decline sharply from late 1992. Although fitful terrorist strikes continued – notably the assassination of Punjab Chief Minister Beant Singh in 1995 – the back of the Khalistan movement had been broken by 1993. Pakistan continued its attempts to keep Khalistan forces operation, but to little avail.

For reasons which have never adequately been explained, India learned few doctrinal counter-terrorism lessons from Punjab. The covert desks which carried out retaliatory operations in Pakistan were dismantled by the regime of Prime Minister I.K. Gujral, and no serious effort was made to either fence the Line of Control or upgrade the operational and intelligence capabilities of the Jammu and Kashmir Police. Neither the war in Afghanistan nor the crisis in Punjab seemed, except perhaps to the most perceptive eyes, to have disturbed the lake-like stillness of Jammu and Kashmir's political waters: an error that would cost India dearly.

Troubled waters

Flamboyant, London-educated Farooq Abdullah had succeeded his father, Sheikh Mohammad Abdullah, with an enormous fund of goodwill both within the National Conference and among Jammu and Kashmir's peoples. Among his first actions was to sack several key figures in his father's administration, charging them with corruption. Farooq Abdullah seemed to embody the spirit of the new tourism-driven entrepreneurship visible in urban Kashmir. Fond of driving through Srinagar on his motorcycle, on one famous occasion with the actress Shabana Azmi riding pillion, he was sometimes sourly referred to by his opponents as the "disco Chief Minister." Such commentary had little impact on him. Sheikh Mohammad Abdullah's prophecy, it might have seemed, had been false: his bequest to his son was turning out to resemble a holiday in paradise more closely than "a crown of thorns".[72]

Like Farooq Abdullah's opponents within Jammu and Kashmir, however, New Delhi was beginning to view his government with an increasingly wary eye. Among the first problems he faced was a legacy of his father's rule, the Jammu and Kashmir Grant of Permit for Resettlement Bill. The Bill sought to give Partition refugees who had left the state, as well as their descendants, the right to return to Jammu and Kashmir and claim their property, as long as they were willing to swear allegiance to both its Constitution and that of India. Within Jammu and Kashmir, the legislation caused more than a little concern among Hindus and Sikhs who had arrived from what was now Pakistan-administered Kashmir, and had settled on lands and in homes left behind by Muslims. More important, though, New Delhi believed a state government had no business legislating on matters of citizenship, and was profoundly concerned about the prospect of a large-scale influx of people who it considered to be Pakistani nationals. Farooq Abdullah, however, had little choice but to push the Bill through the two houses of the Jammu and Kashmir legislature; he could not afford for his first major legislative business to consist of a repudiation of his father's legacy.

In the event, a way to avert a showdown was found. Soon after the Resettlement Act was approved by the Jammu and Kashmir Assembly and the Jammu and

Kashmir Legislative Council, the President of India referred the law to the Supreme Court of India, effectively placing it in deep-freeze.[73] A new line of confrontation with New Delhi, however, soon developed. Elections to the Jammu and Kashmir Assembly had been called for in June 1983, and Indira Gandhi pushed hard for the National Conference to contest them in alliance with the Congress. Her concerns were several. The Congress' political position in several key states was less than firm, and the party's interests would have been well served by a victory in Jammu and Kashmir. As important, Indira Gandhi believed that the emerging crisis in Punjab made it imperative for New Delhi to have a direct political role in Jammu and Kashmir's affairs. Sheikh Mohammad Abdullah had resisted such an arrangement, though, and his son followed his political footsteps. In electoral terms, the decision paid off: the National Conference won 46 of 75 seats, to the Congress' 26; the scoreboard for both the Islamist Jamaat-e-Islami and the Hindu-nationalist Bharatiya Janata Party read zero.[74]

Both the National Conference and the Congress, however, had laid the foundations for tragedies to come. While shunning the Congress, Farooq Abdullah had allied with his party's historic rivals, the leadership of the Jamia Masjid in Srinagar. Now led by Mirwaiz Maulvi Farooq, the uncle of the cleric whose 1932 feud with Sheikh Abdullah had been central to the course of the anti-imperialist struggle in Jammu and Kashmir, the double-Farooq alliance had several consequences. It gave space and legitimacy to Islamists in Jammu and Kashmir, for one, and also marked an important symbolic retreat from the National Conference's historical claims to represent all the peoples of the state. From being a secular nationalist formation, however opportunistic its practice of this ideology might have been, the National Conference had made explicit its Muslim and ethnic-Kashmiri character. Congress politicians campaigning in Jammu province jumped at the opportunity to present themselves as guardians of the Hindu interest. The outcome was a remarkably ugly campaign, with issues of ethnic and religious identity at its core.[75] Although the National Conference was able to marginalize the Jamaat-e-Islami in Kashmir province, and the Congress decimated the Bharatiya Janata Party in Jammu, the ethnic-religious faultlines in the state deepened significantly.

Farooq Abdullah did not have long to savor his election triumph. Soon after his victory, he reached out to other regional parties that were seeking greater federal autonomy, notably those that held power in the southern states of Andhra Pradesh, Karnataka and Tamil Nadu, and the Communist-led left-wing alliance in West Bengal. He also held discussions with leaders of the Shiromani Akali Dal in Punjab, a remarkably ill-advised move which was accentuated by his decision to meet Jarnail Singh Bhindranwale himself. Events within Jammu and Kashmir, meanwhile, started to spiral out of hand. A bomb went off at a Srinagar restaurant on August 13, 1983, followed by another at a Srinagar stadium two days later that targeted a celebration of India's Independence. In November that year, an attempt was made to assassinate Justice A.S. Anand, a judge of the Jammu and Kashmir High Court. By far the best publicized of these events, however, was an October 1983 protest by a group of Islamists who waved Pakistani flags and booed the Indian team at a cricket match against the West Indies in Srinagar.[76] By most eyewitness accounts, the disturbances were confined to a small section of the audience in the stadium, but Farooq Abdullah's critics were able to squeeze considerable propagandist advantage from the affair. Soon after the cricket-stadium affair, came the assassination of the diplomat Ravindra Mhatre, and the execution of Maqbool Butt, the high points of the National Liberation Front campaign, which has been discussed in the previous chapter. The following year, 1984, saw a marked escalation in terrorist violence. A bomb exploded at the tourist resort of Nagbal in March, followed by a similar explosion at the Kashmir University library in April. Nilkanth Ganjoo, the judge who had sentenced Maqbool Butt to death, was targeted a few days later, as was the bus stand in the town of Sopore. On May 29, 1984, members of an Islamist political procession attacked Indian military vehicles in Srinagar; five paramilitary personnel were injured. In June, a procession of right-wing Sikhs attacked the offices of the Hindu-chauvinist Arya Samaj and the heterodox Nirankari sect in Jammu.

It appeared that Farooq Abdullah was losing control – *by design*, his critics charged. Abdullah's visit to Pakistan-administered Kashmir, where by Hashim Qureshi's account he shared a platform with the National Liberation Front, became a particular source of embarrassment. Although Abdullah asserted that he had visited Pakistan-administered Kashmir on behalf of Indira Gandhi and his father, so that sentiments there could "be known first hand", few were willing to hear him out.[77] Allegations, never substantiated, were also made that Abdullah had allowed Khalistan terrorist groups to train in Jammu province.[78] Early in the morning of July 2, 1984, Farooq Abdullah was dismissed from office. Weeks of careful machination had preceded the palace coup. Its principal executor was Jammu and Kashmir's Governor Jagmohan Malhotra, a bureaucrat who had earned the trust of Indira Gandhi by effecting some of the worst excesses of the Emergency period. Days before the coup, Jagmohan had received a letter from 13 members of the Jammu and Kashmir Assembly, stating that they had withdrawn their support for the government. Most of the legislators were old-time apparatchiks disgruntled with Farooq Abdullah's efforts to cleanse the party. Among the key figures in this group was Farooq Abdullah's brother-in-law, Gul Mohammad Shah, who had been denied a National Conference ticket in the 1983 elections and evidently had not forgotten the insult. Whatever their motives, though, the 13 rebels reduced the National Conference to a minority in the legislature. Gul Mohammad Shah was rewarded by being made Chief Minister in place of his brother-in-law, with the backing of the Congress.

Indira Gandhi's authorship of the affair was only too obvious. Victoria Schofield has charged that her decision was made "for what essentially were personal reasons".[79] While there is little doubt that the decision to sack Abdullah was a poor one, this assessment seems somewhat excessive. In the somewhat hysterical climate which prevailed after Operation Bluestar, when many in India's security and intelligence establishment believed there was a real threat to the country's existence, it was perhaps inevitable that New Delhi would seek to control the levers of power in its most sensitive frontier state. What followed, however, was equally inevitable. As Balraj Puri has pointed out, the elections signalled that even if Jammu and Kashmir's peoples "wished to remain within India, they would not be free to choose their own government".[80] In office without a popular mandate, the successor-administration of Gul Mohammad Shah was compelled to turn to Islamists and opponents of India, notably Maulana Iftikhar Ansari, Mohammad Shafi Qureshi and Mohiuddin Salati, to win it legitimacy.[81] In February 1986, Jammu and Kashmir saw its first post-Independence communal riots, in the form of attacks on Pandit-owned homes and on Hindu temples in the south Kashmir town of Anantnag. One investigation of the Anantnag riots revealed that members of secular parties – rather than the Jamaat-e-Islami – had played a key role in engineering the violence.[82] In essence, the Islamists who had been decimated in the 1983 elections had been given political space again, this time by secular politicians who, without a popular mandate, sought to obtain legitimacy by tapping religious chauvinist sentiment. Shah attempted to put a lid on the violence by calling in the army, but to little effect. It was part of a pattern of behaviour: 72 of Shah's 90 days in office had seen Jammu and Kashmir's urban residents shuttered inside their homes, earning him the nickname *Gul-e*-Curfew or the Curfew Flower.

New Delhi had realized its chosen nominee had made a mess. In March 1986, Shah's administration was dismissed from office. Jagmohan, as New Delhi's representative in Jammu and Kashmir, now ruled the state directly. As Navnita Chadha Behera has argued, Jagmohan saw Kashmiri identity as a threat, failing to distinguish between its secular forms and its Islamist expression.[83] It was a failing religious chauvinists in Jammu and Kashmir were quick to exploit. Conflicts during Jagmohan's tenure often expressed themselves along explicitly communal lines. A conspicuous fall in the share of Muslims in government employment and in admissions to technical institutions was seen as an outcome of the Governor's Hindu-nationalist sympathies. His decision to proscribe the slaughter on animals on the occasion of the Hindu festival of Janmashtami, similarly, proved a blessing to Jammu and Kashmir's Islamists. The south Kashmir clerical leader, Qazi Nissar, promptly defied the ban. Political contestation was thus being recast as a conflict between "Hindu" New Delhi, and its efforts to impose its will in the state, and "Muslim" Jammu and Kashmir, represented by political Islamists and clerics. Just three years earlier, the political waters of Jammu and Kashmir had appeared still, but no one could now have missed the waves hammering against the shores.

A stolen election?

Farooq Abdullah had stayed distant from the deteriorating situation in Jammu and Kashmir, focusing instead on building allies elsewhere in India. He had the support of the Left, as well as of the regional formations that had dethroned the Congress in southern India. However, none of these allies had the influence to reinstall Abdullah on the political throne in Srinagar – and for power, it turned out, he was not willing to wait. Against the wishes of much of his own party's leadership, Abdullah decided to come to terms with Prime Minister Rajiv Gandhi, who had taken office after his mother's assassination in November 1984. Instead of calling for fresh elections after the government of Gul Mohammad Shah was dismissed, Rajiv Gandhi arranged for the reinstallation of a National Conference–Congress alliance in Jammu and Kashmir. The new Prime Minister had succeeded in obtaining what his mother had sought, without success, in 1983. The deal, Navnita Behera Chadha has observed:

> denuded [the] Kashmiris' secular political identity of its *raison d'etre*. Kashmiris had always taken pride in standing up to political pressures from the Indian State (read central government), but Farooq's deal virtually bartered it away for the prize of power.[84]

A new election was called for 1987. For both the National Conference and the Islamists, it was something of a referendum. With the two major secular, pro-India parties – secular at least in ideology if not, as 1983 had shown, in practice – now in an alliance, the entire oppositional space in the Kashmir valley was vacant for occupation by the Islamists. In this task, they proved adept. Going into the 1987 elections, the Kashmir-based opposition coalesced into the Muslim United Front (MUF), a broad coalition of Islamist parties, notably the Jamaat-e-Islami, Qazi Nissar's Ummat-e-Islami and Maulvi Abbas Ansari's Shia formation, the Anjuman-e-Ittehad-ul-Muslimeen. Most of the organization's constituents were linked to the Jamaat-e-Islami, which also won the right to nominate the bulk of the candidates who fought under the MUF banner.[85] Other political groups, like Abdul Gani Lone's People's Conference and the deposed G.M. Shah's People's National Conference, also backed the MUF, although they were soon to discover, to their dismay, that clerics and not career politicians held the strings of power within the coalition. MUF's constitution specified that it would be "aloof from politics, in so far as it will not involve itself in any non-Muslim political activity", a position that made clear that the religious parties, rather than conservative centre-right opponents of the National Conference like the People's League, would have primacy.[86]

What exactly did the MUF see as permissible activity? From the outset, the MUF campaign focused on Islamic issues: the proliferation of bars in Srinagar was, for example, a major target, on the grounds that this was part of a larger onslaught on Muslim religious practices and culture. For MUF's major constituents, the acquisition of state power was a precursor to Islamist ends. At a March 4, 1987 rally in Srinagar, MUF candidates, clad in the white robes of the Muslim pious, declared variously that Islam could not survive under the authority of a secular state and that Farooq Abdullah was an agent of Hindu imperialism.[87] For the Jamaat-e-Islami leaders, this platform was not new: the organization had long argued that faith made imperative the *Nizam-e-Mustafa*, the state as the Prophet Mohammad had envisaged it. In a broader sense, the Jamaat-e-Islami saw its politics as emerging from the ideological belief that "Kashmiri Muslims need to be converted afresh for accommodating Islamic beliefs in the local framework". As the scholar Mohammad Ishaq Khan has noted, this also was not new; popular religious identity and culture had long been challenged "by the Shariat-oriented culture as generally represented by the Ahl-e-Hadith and the Jamaat".[88] Other constituents in MUF, notably the People's Conference, had a more limited view of the political uses of Islam, which led them to back out of the coalition on the eve of the election. About the use of religion *per se*, though, they had few scruples.

The MUF's formation marked the political coming of age of the *bazaar*, or petty-bourgeois trader class, in Jammu and Kashmir. Its activist cadre came, as the membership of anti-India jihadist groups had historically done, from educational institutions which had been deeply influenced by the Jamaat-e-Islami. Now, however, this activist core had the support of Jammu and Kashmir's traditional elite, particularly urban businessmen and rich peasants closely linked to them; in the main, orchard owners who unlike mainly Hindu feudal proprietors of large rice-field holdings had not been subjected to land reform.[89] These were classes that had long founded their social legitimacy on religion. Backing the high traditions of Islam, as articulated by the Jamaat-e-Islami and Ahl-e-Hadith, was a means of both gaining respectability and possessing a cultural ethos distinct and superior from the syncretic, freewheeling practices of the peasantry. Jammu and Kashmir's dramatic economic growth in the post-Independence period had brought substantial gains for this class, but political power remained firmly in the hands of the peasant-based National Conference. With the emergence of new elites in the 1970s, notably in the tourism sector and in trades linked to the all-India economy, the primacy of the *bazaar* and the orchard owner came under siege. Islam was not just a religious slogan, then: it symbolized the concerns of a social order about the forces of modernity which threatened to obliterate it.

Headed into the 1987 elections, then, a number of oppositions were at play: old elite versus new elite; Islam versus secularism; modernity versus tradition; Kashmiri separatism against the Indian state. It was a bitter contest. Most scholars of the period have condemned the 1987 elections in no uncertain terms, noting that it was marred by electoral malpractice including blatant rigging. Historians of the period have often argued that these events laid the foundations for the terrorist violence that started from 1988: denied power through the ballot box, the argument goes, dissidents in Jammu and Kashmir turned to the Kalashnikov. Several key political figures in Jammu and Kashmir, too, have made much the same argument. Had the 1987 elections not been rigged, the head of the Jamaat-e-Islami, Ghulam Mohammad Bhat, argued in November 1998 that the course of events in Jammu and Kashmir would have been "very different".[90] At the same time, some scholars have noted that the MUF was unlikely to have won even a fair election. Behera, for example, has suggested that in "a fair election, the MUF would have won 10–20 seats at best, and it would not have been able to dislodge Farooq Abdullah".[91] Victoria Schofield, for her part, has found that the MUF had expected to win only "ten out of the forty-four seats they had contested".[92]

While there is little doubt that many of the young cadre of the MUF spearheaded the terrorist violence that followed in the wake of its defeat, these arguments raise a serious question. Was the early terrorist violence representative of a mass rejection of the verdict of the stolen election? Or was it, instead, the work of a political minority, frustrated at the rigging, of course, but also at its failure to acquire mass influence through democratic means? Did the National Conference still have a significant presence amongst its traditional peasant power-base in the Kashmir valley, or had this constituency withered away after Sheikh Abdullah's death? How large were the new class formations represented by the MUF? In some key senses, there is no real way to make confident assertions about these questions. Almost two decades have passed since the event, and any answers respondents would give to a survey on this issue would be profoundly coloured by the events that have passed during this time. Nor does any worthwhile empirical social science research exist for Jammu and Kashmir during this period, which would allow us to engage in informed speculation on these questions. Other means are needed to address these issues. One possible answer may lie within the tainted results of the 1987 elections themselves – an answer that is admittedly very far from satisfactory, but better than no answer at all.

Election malpractices – the capture of polling stations where the opposition is expected to win, the use of coercive and non-coercive means to prevent opposition supporters from turning out to vote, or the use of the state apparatus to alter the results from areas hostile to the ruling party or parties – are, or at least used to be, commonplace in India. Nonetheless, India has never generated the "99 per cent in favor" results often seen in elections held by authoritarian regimes. From first-hand experience, I have found that rigging focuses on specific polling stations or clusters of polling stations, and thus in general has a relatively small impact on the overall voting *numbers*, though it may none the less change *results*. In the election most notorious for rigging that decisively altered its outcome, that of West Bengal in 1972, the Left Front coalition secured 40.43 per cent of the popular vote, to the Congress' 49.08 per cent. We can, for the purposes of this argument, take this to be an index of how much impact rigging might have: that large-scale election fraud would deprive the "real" winner of some 10 percentage points of the popular vote. Using this assumption, we can obtain some idea of what outcomes the 1987 elections, had they been fair, might have yielded. Of the overall popular vote, 10 per cent, 5 per cent and 3 per cent were credited to the leading opposition group in *each* of the Kashmir valley's 42 seats, and the vote-share of the National Conference and Congress in these constituencies debited accordingly. Where the MUF had won, of course, it was allowed to retain the seat (Table 6.2).

Table 6.2 Projected and actual seats in the "stolen election"

Party	*Actual seats, 1987*	*Seats, 3% Swing*	*Seats, 5% Swing*	*Seats, 10% Swing*
National Conference	32	29	25	20
Congress	5	4	4	4
NC + Congress	37	33	29	24
Independents (MUF)	5	9	12	17
Panthers Party	0	0	1	3
Opposition	5	9	13	20

Sources: Derived from Election Commission of India, *Statistical Report on General Election 1987, to Legislative Assembly of Jammu and Kashmir* (New Delhi: Election Commission of India, undated).
Statistical analysis of the raw data was undertaken by Razili Datta, at the USIP.

In each of these three scenarios the National Conference–Congress alliance won a majority of the seats in the Kashmir valley. The 10 per cent scenario yields results roughly in line with the upper limits suggested by Behera; the other two are broadly consistent with what, according to Schofield, the MUF itself had anticipated. However, the data also suggests that the MUF had no real chance of coming to power through democratic means. In the best case scenario from the MUF point of view, it along with the secular Jammu and Kashmir Panthers Party would have had a combined total of 20 seats in the Kashmir valley – and that assuming the latter grouping would have allied with the Islamists. Even then, it would have been four seats short of the National Conference–Congress combine. It has been pointed out, correctly, that the 1987 elections witnessed a dramatic rise in the Islamist vote in the Kashmir valley. The vote-share of the Jamaat-e-Islami rose from 6.6 per cent in 1983 to 31.9 per cent in 1987, while that of the National Conference declined from 59.3 per cent to 49.2 per cent .[93] Had the alliance with Lone stayed in place, MUF may have registered an even stronger showing. It is important to note, though, that the MUF would have had no representation at all in the seats of the Ladakh and Jammu provinces, where the National Conference–Congress alliance faced no real challenge. As such, *even* a victory in the Kashmir valley may have left the MUF short of a majority in the Jammu and Kashmir Assembly.

Put simply, the limits of Islamist expansion through the democratic process had been reached. Islamists in Jammu and Kashmir, and their allies in the Pakistani intelligence establishment, would draw the right conclusions from this affair. There was no way, at least in the foreseeable future, that they would take power in Jammu and Kashmir through their electoral influence alone, and be able to use its democratic institutions to press for accession to Pakistan. However, for the first time since 1947, it seemed that Jammu and Kashmir did have a genuine mass constituency for the Islamists, hostile both to the National Conference and to New Delhi. Where earlier phases of the jihad had failed precisely because of the absence of such a constituency and a political organization to represent it, the conditions now seemed right to make another attempt. Flush with military resources diverted from the Afghan jihad, and having tested Indian responses in Punjab, Pakistan's covert services were ready to initiate a third front in Jammu and Kashmir. What was to follow has been claimed to be something of a revolution. Given its mass base, it might be more accurate to describe it as an attempted putsch: a bid by a defeated social class to seize power after their attempt to acquire it through the formal processes of democracy had been thwarted.

A failed putsch

In late December 1989, television audiences in Srinagar, as in much of the world, watched the disintegration of Romania's Communist regime and the execution of its hated leader, Nicolae Ceausescu. The event had come about with little apparent effort: the Romanian state had collapsed, it seemed, and for no other reason than that a large mass of people had willed it so. Could the same be attempted in Jammu and Kashmir?

Pakistan's ISI appears to have been bracing itself for a renewal of the jihad in Jammu and Kashmir ever since 1984, around the same time when it had escalated its support to Khalistan terrorist groups. According to an authorized biography of General Akhtar Abdul Rehman, the ISI Director-General who commanded its campaign in Afghanistan, General Zia-ul-Haq entrusted the task of planning the jihad to the *amir* of the Jamaat-e-Islami of Jammu and Kashmir, Maulana Sadruddin. Other accounts accord centrality to the Pakistan-administered Kashmir wing of the Jamaat-e-Islami, and in particular the agency of its *amir*, Maulana Abdul Bari. A third version suggests that meetings were held between the ISI and JKLF leaders in Pakistan, notably Amanullah Khan and Hashim Qureshi, again in the summer of 1984.[94] By Qureshi's account, he refused to cooperate with the ISI, after which it ensured Amanullah Khan took charge of the organization. Qureshi, in this narration an opponent of a jihad he believed would do nothing for Jammu and Kashmir, was then forced into exile in Holland.

Other key figures in the jihad, however, have a somewhat different story to tell. Released from jail after his brief flirtation with the National Liberation Front, Mohammad Altaf Khan – who, operating under the alias "Azam Inqilabi" we have encountered in the previous chapter – had rejoined political life. In 1974, he, along with Ayub Thakur, had helped form the Islamic Students Organization, which busied itself in opposing the accord between Sheikh Abdullah and Indira Gandhi. In October 1982, Altaf Khan was again arrested, this time for having called for the liberation of Jammu and Kashmir in the wake of Sheikh Abdullah's death. In the summer of 1983, again out of jail, Altaf Khan crossed the border into Pakistan-administered Kashmir to ask for help with the revitalization of armed struggle. He was to be disappointed. While the ISI provided him with considerable hospitality, Altaf Khan later recalled, he was told that "the country was not yet ready to support a covert operation against

India".[95] He was ordered to return to India and focus his energies on political mobilization, until the time came for armed struggle.

Whatever the truth – and it is possible all these accounts are true, the participants in each organization being unaware of the contacts of the ISI with others – progress towards realizing Pakistan's battle plans was slow. It was only in 1987 that both the JKLF and the Jamaat-e-Islami received authorization for an offensive with the full support of Pakistan.[96] The timing was, quite obviously, linked to the political chaos sparked off by the stolen election. Among the first JKLF cadre to make his way across the border was Abdul Hamid Sheikh, who received a basic course in the use of assault rifles and grenades at an ISI-run facility between February and April 1988. Back in Srinagar, Sheikh organized the nucleus of the JKLF's army-in-formation, along with Yasin Malik, Ashfaq Majid Wani and Javed "Nalqa" Mir Altaf Khan claims to have been responsible for turning this group, made up of supporters of Islamist candidates who fought the stolen election, in the direction of the JKLF. If so, it would illustrate not only the ways in which the experience of the NLF and earlier groups laid the foundations for a new generation of terrorists, but also how opportunity, rather than ideology, shaped organizational affiliations during this period.[97]

At about the same time the JKLF was assembling its army, a number of other groups also scrambled to obtain training and weapons from Pakistan. Many of these had more expressly Islamist affiliations than the JKLF. Fazl-ul-Haq Qureshi, the veteran of the struggles of the Master Cell and al-Fatah, emerged from political hibernation to found the Armed Reserve Force, along with his close friend Abdul Majid Dar, who would go on to occupy a senior position in the Hizb-ul-Mujahideen.[98] Abdul Ahmad Waza of the People's Conference also set up an armed organization. As time progressed, a welter of smaller organizations also asked for assistance. MUF's constituents mirrored this process with energetic political mobilization in Srinagar and other major urban centers. On August 15, 1988, the anniversary of India's Independence, clashes between protestors and police in old-city Srinagar resulted in several injuries on both sides. Two days later, the death of General Zia-ul-Haq and General Akhtar Rehman in an air crash again brought violent crowds on to the streets. This time, four protestors were killed. Demonstrators demanding a ban on Salman Rushdie's *Satanic Verses*, which had been declared heretic by clerical authorities in Iran, clashed with the police in February, 1989.

Inter-Services Intelligence strategists had no shortage of infrastructure or resources with which to outfit their growing corpus of military allies in Jammu and Kashmir. With the Soviet Union in the last stages of its occupation of Afghanistan – its troops had begun to withdraw from that country in May 1988 – diverting some quantities of material for use in Jammu and Kashmir was not a problem. During the first moments of the ongoing phase of the jihad, though, these resources were not needed: its objective was not to confront the state apparatus, but instead to strip it of authority. Of 390 violent incidents in 1988, just six involved attacks on Indian police and paramilitary forces; in 1989, there were 2,154, only 49 of them on security forces. Yet, the few targets singled out for attack were chosen with great care, and hammered home the point that India could no longer protect its own in Jammu and Kashmir. Nilkanth Ganjoo, the judge who had sentenced Maqbool Butt to death and survived an earlier assassination attempt, was among the first to die. Indian intelligence capabilities, too, have been hit hard by the elimination of four key operatives of the Intelligence Bureau, R.N.P. Singh, Kishen Gopal, M.L. Bhan and T.K. Razdan, and the consequent degradation of their network of informers. Jammu and Kashmir's political system was the second major target of jihadist assault. Starting from August 1989, National Conference cadre were warned to disassociate themselves from the party. Several leaders who defied this threat were singled out for assassination. Jihadi groups proceeded to prohibit voters from participating in bye-elections to India's Parliament, which were held in November 1989. Outside one polling station in Baramulla, in northern Kashmir, Islamists placed a coffin, with a sign proclaiming that it was meant for the first person who cast his vote. Such threats succeeded in reducing voter turnout to negligible levels.

While the actual military impact of early jihadist activity was limited, then, it did serve to illustrate in the most theatrical ways possible that Indian authority in Jammu and Kashmir was collapsing. Soon after the terror-scarred national elections of 1989, which saw the defeat of Rajiv Gandhi and the installation of the administration of Prime Minister V.P. Singh, the JKLF scored another significant victory. In December of that year, the organization kidnapped the daughter of Mufti Mohammad Sayeed, a veteran Congress politician from southern Kashmir who had been appointed India's Union Home Minister. Rubaiya Sayeed's captors demanded the release of five JKLF cadre held on terrorism charges in exchange for her release. Despite Farooq Abdullah's warnings that prisoners-for-hostage swap would strengthen the position of the jihadists, the new government in New Delhi pressured the Jammu and Kashmir administration to make the deal. As Abdullah had predicted, the message it sent out was that the Indian state could be bowled over with ease. By early 1990, the flow of young people to training camps in Pakistan had increased to several hundred a month. Capitalizing on the situation, Union Home Minister Sayeed, Abdullah's long-standing opponent in state politics, ensured the appointment of Jagmohan as Governor of Jammu and Kashmir for the second time. Incensed that New Delhi had appointed the very man who had dethroned him in 1984, Abdullah promptly resigned.

Jagmohan had a one-point agenda: the restoration of New Delhi's authority, by any means necessary. Convinced that no option remained, he ordered the commander of the Indian Army's 15 Corps, Lieutenant-General Mohammed Ahmed Zaki, to intervene in support of the state's authority. On the morning of January 21, 1991, massive crowds of protestors, with large groups of armed men among them, challenged Jagmohan's new, hardline approach. Such demonstrations had taken place in previous days and months, but this time the consequences would be horrific. Who fired at whom first remains disputed, but at the end of the protests, at least twelve unarmed people had been shot and killed. What has come to be known as the Gawkadal Massacre still occupies a central place in the consciousness of the anti-India movement in Jammu and Kashmir. From a purely pragmatic point of view, though, it served Jagmohan's purpose well. Four days later, on the eve of the anniversary of India's emergence as a sovereign republic, jihadist groups used Srinagar's mosques to broadcast calls for crowds to emerge and march in protest on the anniversary of India's emergence as an independent republic. It was, Indian authorities believed, to be the moment of the attempted political coup that had been brewing ever since 1987: a declaration of independence would be made in Srinagar, and the flag of Islam hoisted in the place of the Indian national flag. A curfew was declared – and was, with the memories of Gawkadal Massacre still fresh, obeyed.

Where the Kalashnikov of the jihadists had been writing history since 1988, the full coercive force of the Indian state had now joined the contest. Writing of the wave of assassinations of politicians and government officials, Manoj Joshi has correctly noted that by "physically liquidating those who were deemed structures of Indian rule, the militants wanted to create a vacuum that they alone could fill".[99] If that was indeed their hope, the plan failed.

The warriors

Who were the young men who attempted to break free the Indian state in 1988, and who laid the foundations for a grinding sub-conventional conflict that brought two nuclear-weapon states to the edge of full-blown war on three occasions? What were their beliefs? What social and political forces did they represent?

For much of this early period, as I have noted earlier, the JKLF was without dispute the largest organ of the jihad in Jammu and Kashmir. Its stated politics would have been familiar to its founder, Maqbool Butt. Its official manifesto demands that the state of Jammu and Kashmir as it existed prior to 1947 be united as "one fully independent and truly democratic state". It advocates "equal political, economic, religious and social rights" for all citizens "irrespective of race, religion, region, culture and sex".[100] This position, broadcast through the media, has led to the formation of a consensus that the organization is essentially secular-democratic in character, a representative of ethnic Kashmiri nationalism playing a role not wholly dissimilar to that of the National Conference before the accession of Jammu and Kashmir to India. What this argument misses is that the dividing lines between Kashmiri nationalism and religious fundamentalism – as, of course, with other South Asian nationalist movements – has been exceedingly thin. An ideological commitment to secular nationalism, as the political history of Jammu and Kashmir illustrates, can coexist only too easily with the practice of communal chauvinism.

It is to the JKLF's *practice* of politics that we must look for answers – and this practice was far from secular. Farooq Ahmad Dar, widely known by his *nom de guerre* Bitta Karate, was implicated in a series of assassinations directed at the Kashmiri Pandit community, leading to their near-complete exodus by the autumn of 1990. In a television interview carried out shortly after his arrest, he asserted that several of those killings had been carried out on the orders of the JKLF leader Ashfaq Majid Wani.[101] The JKLF also made available its organizational resources to Jamaat-e-Islami cadre. In the words of one commentator, "Jamaat activists went through the JKLF mill".[102] Independence and Islam, notably, were interchangeable slogans:

> the JKLF group decided to raise their profile in Srinagar and disprove charges made by the National Conference activists that they were Congress agents. The forum chosen for the action was the Friday *namaz* at the Jama Masjid, where on any similar occasion 30–40,000 people gather. As soon as Maulvi Farooq finished the prayers, the JKLF boys dispersed amongst the crowd and raised slogans – Islam *zindabad* [long live Islam] and *Hum chahten hein azadi, azadi* [we want freedom; italics added].[103]

I have been unable to find any express JKLF condemnation of the welter of rapes and killings carried out on supposedly Islamic grounds during this phase of terrorism. These included attacks on stores stocking liquor and bars, a ban on beauty parlors, the prohibition of cinema, as well as the throwing of acid on women who did not wear veils. Such attacks have, since they first took place in 1989, been a recurrent motif of the violence in Jammu and Kashmir.[104] Punishment for defiant women then, as now, has included being shot in the legs. Entertainment believed to be immoral was also targeted, noticeably cable television.[105] One particularly gruesome killing was the 1993 murder of Shamima Parveen, the first woman to perform in the traditional Kashmiri satirical dance–drama form, the Bhaand Paather. Parveen was sexually abused and tortured before being shot for her refusal to abandon her theatrical work on television.[106] In 2002, terrorists have insisted that women students in the border district of Rajouri wear veils or else stop attending school.[107]

At least one insider account, authored by Hashim Qureshi, has claimed that these outrages were official Pakistan policy, endorsed not just by the Islamic Right but by the supposedly secular Amanullah Khan faction of the JKLF. Writing of the assassination of Maulvi Mohammad Farooq, the clerical leader who played a key role in organizing the early anti-India protests but was assassinated in May 1990, by the expressly Islamist Hizb-ul-Mujahideen, Qureshi has argued that he:

> was martyred just because he preached three things (a) now when you have taken up the gun raise only the slogan of independence (b) you will not get international support if you raise the slogan of accession to Pakistan and of Nizam-e-Mustafa (c) protect minorities especially the Kashmiri Pandits so that the movement does not get a communal color. Maulavi [*sic*, Maulvi] Farooq was assassinated for these reasons. *The ISI [Inter-Services Intelligence] ran this movement on communal lines*

> *right from the beginning and for that Amanullah and his underlings became its agents* [italics added].[108]

Whether one wishes to take Qureshi at face value or not, the continuity of terrorist ideology and practice across time and organizations is unmistakable. Each periodic assault on women's rights, over the coming decades, provoked considerable national and international outrage, but there was little realization that these were the outcome of the organic ideology of the jihad, not the aberrant actions of marginal groups. Many women's accounts of events during the period of terrorist violence seem to suggest that the levels of terror were actually higher when the supposedly moderate JKLF dominated the landscape, than when pro-Taliban groups took charge of the jihad in the late -1990s. One journalist's account of the views of women bureaucrats and doctors in Jammu and Kashmir is illuminative:

> In the worst days of turmoil, militants dictated much of what happened in civil life. Says Tanvir Jehan, the first and only female District Commissioner in the state, "Till 1995, I too would do exactly what they dictated."
>
> Fundamentalist groups imposed 'rules'—women were pushed into purdah (the veil), deprived of access to contraception and abortion, and prevented from moving freely. And, tragically, no voices were raised within the state establishment to dispute these.
>
> Dr Asma Khan, one of the senior gynecologists [*sic*] at the Lal Ded Maternity Hospital, the only functional gynecological [*sic*] government hospital in Srinagar, says, "Before this problem, there was a growing awareness of contraception in the state, and vasectomies and tubectomies were routine. But for several years now no vasectomy has been performed; tubectomies have been attempted only in cases where another pregnancy could be life-threatening."[109]

One explanation for the ideological basis of such actions may lie in the class background of these early jihadists, linked to the search of the new elites they represented for legitimacy through a particularly chauvinist reading of Islam. There is, sadly, almost no literature on the precise social composition of the first terrorists to cross the LoC for training in Pakistan during this phase of the jihad in Jammu and Kashmir. One author has described them as "in the main young students, youth without jobs, or under-employed youth, mainly city-bred and educated".[110] Yet, several key members of the jihadist leadership of 1987–1990 defy this description. Mushtaq Ahmad Zargar, released from prison in the Indian Airlines Flight 814 hostages-for-prisoners hijack-swap of 1999, was the son of a goldsmith who went on to run his own copper and brass utensil manufacturing store.[111] Of the trained terrorists, 68 per cent, a 1994 study reported, earned between Rs. 500 and Rs. 1,000, and another 11 per cent between Rs. 1,000 and

2,000; although the study omits to record the periodicity of this income, but it seems reasonable to assume it was monthly. Only 19 per cent were classified as dependants.[112] Forty-two per cent of them were engaged in some form of labour-for-wages, as against 19 per cent who were students, 22 per cent who were farmers and 14 per cent who were petty traders. These wages, although modest, would have placed the cadre of the terrorist groups well outside the category of the absolute poor, for annual per-capita income in Jammu and Kashmir for 1979-1980 stood at Rs. 1301.[113]

It is interesting that much of the first-generation leadership of the terrorist groups, like Zargar, came from the deeply conservative petty bourgeoisie of urban Kashmir, a class made up in the main of business families but also mid- and low-level bureaucrats and the rural elite. The JKLF's Mohammad Yasin Malik was the son of a bus driver; his aide Javed Mir a plumber. The *amir* of the Hizb-ul-Mujahideen, Mohammad Yusuf Shah, better known by his somewhat vain *nom de guerre* Syed Salahuddin, came from a family with orchard interests. This was the case, of course, that had backed the MUF, and many of the early jihadists came from that formation's ranks. Malik, for example, compaigned for the Hizb-ul-Mujahideen's Shah in the 1987 elections. As we have seen, the MUF was deeply involved in communal mobilizations, which may be one reason why the practices of the JKLF took the shape they did. In time, perhaps unsurprisingly, a cycle would develop where new formations of Islamists would displace the old, much as the old had eliminated the older political order in Jammu and Kashmir. Maulvi Farooq, as we have noted, was eliminated by jihadist guns; so too was Qazi Nissar, the south Kashmir cleric who had initiated the MUF's first assaults on the National Conference regime on religious issues.[114]

With the wisdom of hindsight, 1990 marked the end of a phase of the jihad in Jammu and Kashmir, the very year it exploded on the world's consciousness. From this point on, the jihad would be given substance not by the anti-India forces which had grouped together in the MUF, but by the ISI and the jihadist groups it sponsored. The time had come for what Manoj Joshi has evocatively described as "the long insurgency".[115]

Towards years of terror

As with many aspects of the history of the jihad in Jammu and Kashmir, the answers to many key questions lie locked away in archives in Pakistan, and in the minds of those in its covert services who served in policy-making positions during this time. Did the ISI intend for Kashmiri Islamists to make a bid to bring down Indian rule in 1990? Or did they, so to speak, jump the gun, taking Pakistan by surprise? Did Pakistan have defined political objectives in Jammu and Kashmir at this stage, or was it merely responding to the flood of cadre that arrived at the ISI's doorsteps after 1987? Was there an end-game shaping policy, or was Jammu and Kashmir merely an opportunity to do to India what it had done in Bangladesh in 1971?

In the absence of adequate empirical material, attempting to answer these questions is a hazardous business. Nonetheless, the circumstances within which the events of 1979–1991 played themselves out are well known, and provide some tentative insight. Sub-conventional warfare, as we have seen, was both a long-standing practice and intellectual concern of the Pakistan Army. Over the coming decades, in part as a consequence of Pakistan's growing military relationship with the United States, these ideas grew more refined. Pakistani strategists paid close attention to debates in the United States on guerilla warfare and Maoist notions of people's war. Where the United States sought to prevent such wars, Steven Cohen has perceptively pointed out, Pakistan studied these "in terms of *launching* a people's war against India, or developing a people's army as a second line of defense".[116] To Pakistani strategists, all the necessary conditions for a people's war to succeed existed in Kashmir: in Cohen's words, "a worthy cause; difficult terrain; a determined, warlike people (the Pakistanis); a sympathetic local population (the Kashmiris); the availability of weapons and equipment; and a high degree of leadership and discipline to prevent (the guerillas) from degenerating into banditry".[117]

Afghanistan would have seemed to have demonstrated the robustness of this model – an epistemological error of great consequence. First, the apparent victory of the mujahideen in Afghanistan was not nearly as dramatic as it seemed. As Giles Dorronsoro has pointed out, the Soviet Union committed relatively modest resources in Afghanistan. In an area considerably larger than Indian-administered Kashmir, the Soviet Union's troop strength at no stage exceeded 100,000 by any great amount.[118] As such, the Soviet Union simply did not have the men needed to hold ground, and often ceded control of large areas, restricting its presence merely to areas of strategic significance and major towns. India, by contrast, would demonstrate its willingness and ability to commit enormous military resources to fighting jihadist groups in Jammu and Kashmir; its forces would fight with numbers overwhelmingly greater to those of their adversaries. Second, the success of the Afghan mujahideen had been enabled by their unfettered access to the military resources of the world's principal power. As Brigadier Mohammad Yusuf had candidly noted in the wake of the Soviet withdrawal, "without full US support, the Jehad [*sic*] did not and still cannot succeed".[119] He continued:

> Without the backing of the US and Saudi Arabia, the Soviets would still be entrenched in that country. Without the intelligence provided by the CIA, many battles would have been lost, and without the CIA's training of our Pakistani instructors, the Mujahideen would have been fearfully ill-equipped to face, and ultimately defeat, a superpower.[120]

From Punjab, similarly, the ISI learned the wrong lessons. At first glance, the situation in Punjab in the late-1980s and early-1990s vindicated Pakistan's belief that its support of terrorism in India was free of cost. Indian security forces were tied down in an endless spiral of violence, with no apparent end in sight, and unable to retaliate against Pakistan for fear of escalation. Yet, beyond the fact that the conflict imposed costs on India, the Khalistan movement did little to advance Pakistan's *own* interests. It would also impose considerable costs on Pakistan, in both conventional and strategic military terms, in ways the strategists who backed the Khalistan movement did not consider. As we shall see in the next chapter, the fact that both the principal states in South Asia possessed nuclear weapons did indeed restrict India's freedom to manoeuver. However, it would be repeatedly demonstrated that it *also* constrained the extent to which Pakistan could escalate the covert war in Jammu and Kashmir. A nuclear war in South Asia, after all, would hurt Pakistan at least as much as it would India, a fact policy-makers on either side of the border turned out to be well aware of. If India did not see the level of Pakistani support for covert warfare in Punjab as justification for high-risk and potentially low-yield war, it would react very differently in Jammu and Kashmir.

All this, however, lay in the future. Despite the defeat of political putsch of 1990, Pakistan's military and covert services had reason to congratulate themselves. After decades spent in retreat, Pakistan was resurgent. Victory seemed within grasp in Afghanistan – and not much farther away in Jammu and Kashmir.

7
THE NUCLEAR JIHAD

> Terror struck into the hearts of the enemies is not only a means, it is the end in itself. Once a condition of terror in the opponent's heart is obtained, hardly anything is left to be achieved. It is the point where the means and the end meet and merge. Terror is not a means of imposing [a] decision upon the enemy; it is *the decision* we wish to impose upon him.
>
> Brigadier S.K. Malik, The Quranic Conception of War.[1]

Mirwaiz Mohammad Farooq's body rests in a corner of the graveyard near the Idgah in Srinagar, where so many of those who lost their lives in the long jihad against India are buried. Mirwaiz Farooq's story has figured at many points in this book, as he was for decades an emblem of secessionist sentiment in urban Jammu and Kashmir, a thorn in the side of the Indian state. Buried nearby is the body of Mohammad Abdullah Bangroo, the Hizb-ul-Mujahideen terrorist who assassinated him. To the faithful, both the assassin and his victim are martyrs; they are martyrs, moreover, for exactly the same cause.

Making sense of the house-of-horrors that Jammu and Kashmir has been reduced to through almost two decades of war – a low-intensity war, it is true, but one that has claimed far more lives than many full-blown ones – is a difficult enterprise. The intensity and violence of what is on display that constitute it make it near-impossible to provide a useful guide to lead us through the exhibits. My account of the jihad of 1990–2001 is not intended to be a detailed rendering of the many and infinitely complex events of those years. My intent is, rather, to provide an overview of this period – a sense of the final flowering of Pakistan's covert war in Jammu and Kashmir, which had steadily grown since 1947–1948 and to provide some understanding of the circumstances that paved the way for its subsequent withering.

One aspect of this phase of the covert war is, to my mind, of critical relevance: it was a *nuclear* jihad, in the sense that it would have been unsustainable, even inconceivable, without Pakistan's acquisition of the Bomb. As we have seen in the previous chapter, the terms and structures of this phase of Pakistani's secret war against India were shaped by its possession of nuclear weapons. India would more likely than not have gone to war when faced with Khalistan terrorism or the putsch of 1988–1989, were it not for fear that a conventional conflict with Pakistan might escalate to catastrophic levels. As we shall see, the architecture of the nuclear jihad – its peaks and troughs – would be underpinned by the new, nuclear reality in South Asia. Full-scale India–Pakistan war, a perpetual threat in each phase of the covert war, would haunt the nuclear jihad. For all the blood spilt during these years, and the constant risk of a cataclysmic confrontation, it is remarkable how little is changed. In *political* terms, that is in the relative influence and weaknesses of the parties who had been engaged in combat through the course of the endless war in Jammu and Kashmir, 2002 was much the same as 1990.

The story of the nuclear jihad, then, illustrates not only the horrors war can inflict, but also how little purpose it can serve when waged by adversaries who cannot be overwhelmed.

The 1990 crisis

In late 1989, even as the Islamist uprising in Srinagar began to reach its climax, Pakistan had begun a series of integrated air–land defence exercises, code-named *Zarb-i-Momin* [Strike of the Believers]. Conceived of by General Mohammad Zia-ul-Haq's successor as Pakistan's Chief of Army Staff, General Mirza Aslam Beg, *Zarb-i-Momin* was in essence a response to India's earlier exercise, Brasstacks. By the beginning of 1990, however, New Delhi had imposed direct rule in Jammu and Kashmir in an effort to put down the near-successful Islamist *putsch*. On January 20, 1990, news of a massacre in Srinagar's Gawkadal area – where Indian troops had fired on Islamist demonstrators – broke as Pakistan's Prime Minister' Benazir Bhutto was presiding over a meeting on that country's Jammu and Kashmir policy. Unable to resist pressure from the right, Bhutto responded by declaring Pakistan's support for the Kashmiris' right of self-determination.

What had been a "mainly Indian affair", Sumit Ganguly and Devin Hagerty have perceptively noted, now began to turn into a "renewed Indo-Pakistani conflict".[2] A day after Bhutto's Jammu and Kashmir policy meeting, her Foreign Minister Sahibzada Yakub Khan delivered what India interpreted as a warning of possible war. India's Prime Minister V.P. Singh responded by asking his Foreign Minister I.K. Gujral to tell Khan that it would be unwise for Pakistan to "mistake our kind words for weakness".[3] By March, 1990, the war-polemic between India and Pakistan escalated. At a meeting that month in Pakistan-administered Kashmir, Benazir Bhutto promised a "thousand-year war" in support of the terrorists operating across the LoC.[4] V.P. Singh responded promptly, telling India's Parliament that such a Pakistani "misadventure would not be without cost".[5] At this stage, however, there were few purely military signs of a war-like build-up. While India had begun to pump large numbers of troops into

Kashmir, both from the Army and the paramilitary Border Security Force, there was no sign of the kinds of movement of armour and artillery that would have preceded offensive action. Two Indian tank units had also been deployed for winter training exercises into the Mahajan ranges in Rajasthan. While this would have reduced India's war-mobilization time in the sector somewhat, it was not in itself of great significance. Pakistan, for its part, had not mobilized its principal strike formation, the Mangala-based 1 Corps, and its air force had not opened its forward operating bases.

By mid-April, however, it appeared that matters were beginning to spiral out of hand. Prime Minister Singh asserted that Pakistan was preparing for war, and that an attack across India's western frontier was likely. He argued that his extraordinary public pronouncement was necessary because differences between Pakistan's President Ghulam Ishaq Khan, its military and Prime Minister Bhutto left it unclear just who was in charge of the country's affairs. Others in the Indian establishment took a similar position. Union Home Minister Mufti Mohammad Sayeed, whose poor handling of his daughter's kidnapping the previous year had left him open to political assault, argued that war "would be fully justified if the objective of freeing Jammu and Kashmir from the stranglehold of the secessionists was achieved".[6] Underpinning these sentiments was the stark fact of escalating violence in Jammu and Kashmir. Indian security force fatalities rose more than tenfold in 1990, to 132 from 13 in 1989. Whereas no terrorists had been killed in action in 1989, 552 lost their lives in 1990, along with 756 civilians, up from just 79 in 1989. For the first time, fatalities in the Jammu and Kashmir jihad had crossed 1,000 a year, a widely accepted benchmark for what constitutes a war (Table 7.1). In addition, India was still facing a gruelling war in Punjab, where the Khalistan campaign was approaching its climax. Much of the Indian Army's energies there were devoted to securing the border against infiltration, and backing the Punjab Police's counter-terrorist operations.

In the United States of America, officials drew the obvious conclusions from this scenario – and responded with alarm. In mid-May 1990, the United States' National Security Advisor, Robert Gates, arrived in South Asia with a simple message for Pakistan. During meetings with President Ghulam Ishaq Khan and General Mirza Aslam Beg, Gates made it clear that each of Washington's war-game exercises had resulted in military defeat for Pakistan. In the event Pakistan saw fit to try its luck in the battlefield regardless, he said, it ought not to expect any assistance from the United States. If Pakistan on the other hand wished to avoid a war, Gates suggested, it needed to stop supporting terrorism in Jammu and Kashmir. His message to New Delhi was along similar lines. While India might win a war with Pakistan, Gates pointed out, its long-term costs would most likely exceed whatever short-term gains it might yield in Jammu and Kashmir. More important, it was at least likely that a war could escalate into nuclear exchange, with calamitous outcomes for both countries. Within two weeks of the Gates mission, India announced that it was withdrawing the armour it had sent to the Mahajan ranges.

Table 7.1 Violence in Jammu and Kashmir

	Attacks		*Fatalities*									*Cross-LoC firing*		
	Attacks on Indian forces	*Attacks on others*	*Indian forces*	*Hindu civilians*	*Muslim civilians*	*Sikh civilians*	*Other civilians*	*Pro-India militia members*	*Terrorists (including foreign)*	*Political activists killed*	*Foreign terrorists killed*	*Exchanges of fire*	*Indian forces killed*	*Civilians killed*
1988	6	1	1	0	29	0	0	0	1	0	0			
1989	49	73	13	6	73	0	0	0	0	0	0			
1990	1098	485	132	177	679	6	0	0	552	13	0			
1991	1999	321	185	34	549	6	5	0	1016	8	3			
1992	3413	507	177	67	747	10	35	0	991	11	19			
1993	2573	539	216	88	891	4	40	0	1584	0	97			
1994	2693	619	220	104	835	11	62	16	1818	7	125			
1995	2253	647	258	97	1013	4	47	39	1545	15	119			
1996	1432	1025	241	114	1175	2	42	131	1313	69	194			
1997	1116	616	203	64	717	1	58	139	1282	48	258			
1998	1211	619	230	159	678	1	39	74	1111	41	394	4314	78	78
1999	1390	660	387	98	684	0	17	104	1184	49	348	4631	315	58
2000	1560	493	499	132	661	40	9	69	1808	28	403	5653	114	36
2001	1994	611	577	105	848	18	0	79	2119	49	488	4153	36	17
2002	1654	679	445	155	803	8	0	16	1747	99	516	5767	81	74
2003	1407	477	303	89	712	5	1	20	1526	34	554	2841	29	38
2004	986	539	302	37	657	1	1	14	964	40	286	0	0	0
2005	806	432	226	57	494	1	0	18	920	39	239	0	0	0

Source: These figures are derived from internal material compiled by India's Union Ministry of Home Affairs, obtained through private sources. It varies marginally from the data published in the Ministry's Annual Reports and the Jammu and Kashmir Government. From experience, I have found this data to be more accurate.

Did nuclear weapons avert a war? Some analysts have suggested that neither India nor Pakistan at this point had capable nuclear arsenals that could be used against their adversaries with any measure of confidence. This, however, misses the point. As Ganguly and Hagerty have argued, building on concepts first developed by McGeorge Bundy,

> nuclear weapons deter war not through the classical modalities identified by deterrence theorists – relative capabilities, demonstrated resolve, nuclear doctrines, escalation dominance and pointed threats – but through the fact of their existence and the accompanying possibility that they might be used at all.[7]

Pakistan may or may not have used nuclear weapons in a war – but India was deterred by the mere prospect that something of the sort might have materialized. Put simply, the cost of the nuclear jihad in Jammu and Kashmir was preferable to that of an outright war, especially a nuclear war. Pakistan, for its part, did not heed Gates' calls to end its support for the nuclear jihad. Nonetheless, it was forced to fight with one hand tied behind its back. Unlike in Afghanistan, mujahideen forces were never provided with equipment that would allow them to pose a serious threat to Indian logistics or military infrastructure. Just one surface-to-air missile ever made its way to Jammu and Kashmir, for example, and that a non-functioning model. Nor did terrorist groups seek to destroy the Jawahar Tunnel on the Pir Panjal range, a critical choke-point for road communications between the Indian plains and the Jammu and Kashmir valley.

Nuclear weapons, then, circumscribed the terms of the nuclear jihad in Jammu and Kashmir. During the crisis of 1990, both sides established lines that could not be crossed. India and Pakistan now geared up for a murderous war of attrition: "our aim is to prick and bleed India", a senior ISI official would succinctly tell the leaders of major jihadist groups, "not to prick India so hard that it declares war against Pakistan".[8] India could not be thrown out of Jammu and Kashmir, it seemed, but the nuclear jihad could impose costs so high that it would force concessions that could not be won on the battlefield. This vision, history would demonstrate, was unadulterated fantasy.

The elephant and the mosquito

"India", wrote the journalist Manoj Joshi in his rich account of the jihad in Jammu and Kashmir, "has often been compared to an elephant. The lumbering vegetarian pachyderm is not a natural predator, yet, in fear or anger, it can and does kill, often with the greatest brutality."[9]

Joshi's metaphor would have been familiar to an earlier generation of participants in the Kashmir jihad. Al-Fatah's Ghulam Rasool Zahgir, writing during the build-up to the war of 1971, had suggested that the Jammu and Kashmir jihad needed to employ methods similar to those "a mosquito does while fighting with an elephant".[10] Now, India was confronted with swarms far larger than those it had ever witnessed before; swarms, moreover, with the ability to bite particularly hard. For six years from 1990, the elephant would wage an inelegant but merciless battle, swatting its enemies with all the resources at its command.

From the spring of 1990, the thousands of young people who had crossed the LoC the previous summer began to return from training camps in Pakistan, transformed by professional military training and modern weapons into a formidable irregular army. We know relatively little, as I have noted in the previous chapter, about who these young men were, in particular their ideological and class background. Many accounts of the period have, indeed, represented these recruits in terms suggesting that they had no specific *class* moorings, and were instead the outcome of a *national* sentiment that cut across social boundaries:

> The women would put *mehendi* (henna) on their sons going to Pakistan and daughters gave them a hero's welcome on return. Having a mujahideen in the family became a status symbol. The families and the peer groups competed for crossing the border to get arms training. Children carried placards saying "Indian dogs go Home" or "*Mujahideen Qaum Zindaba*" [Long Live the Warrior Race]. A new vocabulary of violence depicted militants as "freedom fighters" and the security personnel as "occupation forces." Curfews were described as "martial law" and martyrs' graveyards became "places of pilgrimage."[11]

Yet, there is little empirical evidence to support this thesis. From my work as a journalist, I know of entire villages where not one single person joined the ranks of the jihadists; conversely, there are others where a majority of young men did so. Within the limited context of Srinagar, some neighbourhoods contributed huge numbers of cadre; other communities, like the *Hanjis*, almost none. On why this is so, there is no literature, and I shall not hazard guesses here. What we do know of recruits, as I have argued in the previous chapter, is that they were in the main urban, at least during the first phase of the jihad, and from specific economic and social categories. Clearly, a large gap exists in social science research that needs to be filled in before any firm conclusions can be drawn.

Somewhat better research exists on how those who did volunteer to serve in the jihad were armed and trained. Ejaz Ashraf, who later served as a senior leader in the terrorist organization al-Jihad, travelled across the LoC with some 80 other recruits in December 1989, for a six month course of military instruction.[12] According to Ashraf's account, by this time there was already a fairly elaborate system to funnel recruits to the camps, composed of guides who knew their way across the mountain passes, as well as supporters who provided food, shelter and snow-boots. Once in Pakistan-administered Kashmir, these groups reported to the nearest Pakistan Army outposts, from where they were despatched to training facilities further away from the frontier. Mohammad Altaf Khan – who using the pseudonym "Azam Inqilabi" or Great Revolutionary we have encountered as a member of al-Fatah – ran the network used by Ashraf. Together, many such networks are believed to have funnelled up to 20,000 cadres into camps by the mid-1990s.

Of these networks, that of the JKLF was without dispute the largest at the beginning of 1990. Built from the political cadre who had backed the MUF in the election of 1987, the JKLF also had access to the considerable resources Pakistan's ISI had placed at the disposal of its Muzaffarabad-based leader, Amanullah Khan. By mid-1990, however, other political groupings that had operated under the MUF umbrella were also positioning themselves for a role in the nuclear jihad. Perhaps the largest of these was the Jamaat-e-Islami. Interestingly, the Jamaat-e-Islami had maintained a distance from the attempted putsch of 1988–1989, perhaps understanding that its chances of success were minimal. It was only in April 1990 that the Jamaat-e-Islami leader Syed Ali Shah Geelani, who had been elected to the Jammu and Kashmir legislature as an MUF-backed candidate in 1987, first made express his support for the jihad.[13] With its formidable network of cadre across Jammu and Kashmir, however, the Jamaat-e-Islami soon became the patron of the largest jihadist group, the Hizb-ul-Mujahideen. A plethora of other groups emerged, most of them, like the Hizb-ul-Mujahideen, with expressly Islamist sympathies and committed not to the independence of Jammu and Kashmir, but its accession to Pakistan.

Most of these groups also had one form of affiliation or the other with MUF's constituent political bodies. Al-Barq, for example, was closely allied to Abdul Gani Lone's People's Conference; al-Umar to the Mirwaiz of Srinagar. Thus, by the summer of 1991, perhaps apprehensive because of their experience with al-Fatah that these quasi-independent groups would be vulnerable to deal-making with the Indian state, Pakistan's covert services threw their weight behind the Hizb-ul-Mujahideen. Aided by some numbers of Pakistani and West Asian nationals who had gained combat experience in Afghanistan, the Hizb-ul-Mujahideen was soon engaged in a full-scale war with the JKLF. On several occasions, JKLF leaders have charged that the decimation of their organization in 1992–1993 was engineered by a tacit alliance between Hizb-ul-Mujahideen and Indian forces. As early as 1991, for example, Amanullah Khan asserted that Hizb-ul-Mujahideen treachery had ensured that "500 important commanders of the JKLF have been martyred to date".[14] Several commentators have suggested that this conflict was the outcome of the fissures between Kashmiri nationalists and pro-Pakistan parties, and between secular-nationalists and Islamists. This line of argument, however, ignores the empirically well-documented fact that the Hizb-ul-Mujahideen had little hesitation about taking on competing pro-Pakistan Islamist groups, either. Ikhwan-ul-Muslimeen cadre were decimated by the Hizb-ul-Mujahideen in Anantnag, Bandipora and Badgam in 1992; clashes between the organization against both al-Barq and al-Jihad were also reported.[15]

Ideology may indeed have been one reason for Pakistan's growing support of the Hizb-ul-Mujahideen, then, but it was not perhaps the sole one. Pakistan's concerns at this stage were several. First, the proliferation of jihadist groups, representing diverse power centers, not only undermined efficient military operations, but also imposed political costs on the jihad. A plethora of competing platforms did little for the claim that there was such a singular entity as the Kashmiri liberation movement, which India would have to come to terms with at some stage. As important, the stakes for Pakistan itself transcended the jihad in Jammu and Kashmir. Uncalibrated warfare in Jammu and Kashmir, the experience of 1990 would have made it clear, held out the risk of war with India. Brigadier Mohammad Yousaf's account of his management of the anti-Soviet jihad in Afghanistan makes it clear just how worried Pakistan had been that the conflict could escalate into a war that "would have been the end of Pakistan".[16] Through the Afghan jihad, Brigadier Yousaf records, Pakistan had learned the dangers of "allowing the water to come perilously close to the boil". His organization, the ISI, would most likely have had the same concerns in Jammu and Kashmir. The war needed to be carefully managed and the Hizb-ul-Mujahideen, given its links to the Jamaat-e-Islami in both Pakistan and Jammu and Kashmir, would have seemed a more reliable ally than the other prospective groups. Moreover, the ISI had come to believe in Afghanistan that hardline Islamists made more efficient fighters than their more moderate counterparts, an institutional mentality that may have played a role in Jammu and Kashmir as well.[17]

Caught off guard in 1989, Indian troops had been ill-prepared for to deal with the forces arrayed against them. By 1992, however, the rudiments of a counter-terrorist grid were in place; to use Sumit Ganguly felicitous phrase, India succeeded "in bringing about a degree of order, if not law, to Kashmir".[18] Old-city areas in Srinagar, Baramulla and Sopore, which in 1991–1992 had in effect become no-go zones for Indian forces, once again came under the state's authority. To ISI strategists, it must have been evident by mid-1990 that India was winning the war of the elephant and the mosquito: it was simply too large an adversary to be brought to its knees by pinprick-like bites, no matter how infuriating and painful. Moreover, Indian intelligence and military organizations had succeeded in creating counter-terrorist militia groups from the organizations which had been at the receiving end of the Hizb-ul-Mujahideen campaign two years earlier, notably the Ikhwan-ul-Muslimeen. In Srinagar, Anantnag and Bandipora, militia organizations several hundreds strong, led by Mohammad "Kukka" Parrey, Liaqat Ali Khan, Usman Majid and Javed Ahmad Shah, inflicted havoc on both the Hizb-ul-Mujahideen and its overground political patrons in the Jamaat-e-Islami.[19] "The war between police and terrorists", Eric Hobsbawm has noted, "is one of nerves as well as guns".[20] No great effort was needed to see, in the summer of 1994, which side was under greater strain.

Pakistani strategists sought to respond to this situation by pumping in grow
numbers of battle-hardened personnel, often its own nationals and those
West Asia. A welter of Pakistan-based organizations, such as the Lashkar-
Harkat-ul-Jihad Islami, Harkat-ul-Ansar and Harkat-ul-Mujahideen
play an ever-greater role in the nuclear jihad. Armed with a
brutality – these groups were responsible not just for a series

of Hindu communities in rural Jammu and Kashmir, but also for the brutal executions of Muslims believed to be hostile to their ultra-reactionary version of Islam – the new cadre were greeted by Islamist political leaders as *mehmaan* mujahideen or guest fighters. The kidnapping of five Western tourists in 1995 brought one of these organizations, the Harkat-ul-Ansar, considerable international notoriety. In purely military terms, however, their induction served some purpose. Killings of Indian troops and police personnel began to rise again from 1995, a clear sign that the enemy was now better armed and trained than in the past. Even worse, from India's point of view, the ratio of personnel it lost for each mujahideen killed, one important index of the success of counter-terrorist operations, declined steadily. Killings of civilians by terrorists grew steadily as well, as new organizations sought to impose their authority on the ground and show that the forces fighting the nuclear jihad had not yet lost their authority over civil society.

India viewed these developments with concern, but not alarm. The induction of the *mehmaan* mujahideen made it unlikely that the nuclear jihad would come to an end just yet. But that a few more mosquitoes, however large, would bring the elephant to its knees seemed profoundly unlikely.

New armies

For the new jihadist organizations operating in Jammu and Kashmir, Kashmir was just one battlefield in a larger war between Islam and *kufr* or unbelief. Pakistan-based jihadist groups found that they had surplus resources after the mujahideen captured Kabul in the spring of 1992, to achieve their larger objectives, both global and local.

We have no real way of gauging the numbers of *mehmaan* mujahideen committed to the jihad in Jammu and Kashmir. What the available data does make clear, however, is their growing importance in the fighting. In 1994, 119 of 1,545 terrorists killed by Indian forces in Jammu and Kashmir, or less than one in ten, were believed to be foreigners. By 1998, that figure was up to 398 of 1,111 – an index of both the diminishing enthusiasm of Kashmiri recruits for the nuclear jihad and the growing importance of the new jihadist groups to keeping it
... by no means all, of these organizations' cadre were drawn
...aries run by organizations of the Islamist
...akistani journalist Mohammad Amir Rana,
...t on jihadist organizations in the region,
...en motivated by their religious education.
...r, also seem to have played a significant
...pect of liberation from a repressive social
...v opportunities for self-realization. One of
...nted to the role of Indian popular film in
...noting that these "promoted heroism and
...rtunity to become a hero".[21]

Opportunities for adventure were, indeed, considerable. The Harkat-ul-Mujahideen, for example, began to transform itself into an organization of global reach after the fall of Kabul, funnelling cadre not just into Jammu and Kashmir, but Egypt, Tunisia, Algeria, Bosnia, Chechnya, Tajikistan, Myanmar and the Philippines.[22] Its name changed from time to time – from 1993 to 1997, for example, it operated as the Harkat-ul-Ansar, after merging with the Harkat-ul-Jihad Islami, and then reverted back to its original nomenclature after the United States declared it a terrorist organization – but the networks that underpinned its functioning remained constant. The Harkat-ul-Mujahideen's overall leader Maulana Fazl-ur-Rehman Khalil used radical elements drawn from among the Tabligh-i-Jamaat, a proselytizing organization that in itself claims to remain above politics, but has had considerable influence in Pakistani life after it was patronized by the regime of General Zia-ul-Haq. Harkat-affiliated figures frequently used these networks to fight for domestic objectives within Pakistan. In September 1995, for example, several senior Pakistani military officials personally linked with the Tabligh-i-Jamaat and officially involved in supplying weapons to the Harkat-ul-Mujahideen were charged with attempting to stage a coup against the government of Prime Minister Benazir Bhutto. One of the key suspects, Major-General Zaheer-ul-Islam Abbasi, had in 1989 been expelled from New Delhi, where he was serving on a diplomatic assignment, on espionage charges. Pakistani newspapers reported that the abortive coup was intended to re-establish direct military control over the jihad in Jammu and Kashmir jihad.[23]

A similar combination of both global and Pakistan-specific ambitions can be seen in the case of another major jihadist group, the Lashkar-e-Taiba. In 1987, Hafiz Mohammad Saeed, an academic at the University of Engineering and Technology at Lahore, set up the Markaz Dawa wal'Irshad [Centre for Proselytization and Preaching] with the support of a colleague, Zafar Iqbal, and Osama bin-Laden's key lieutenant, Abdullah Azam. With support both from the regime of General Zia-ul-Haq and al-Qaeda, the Markaz flourished. By 2001, it occupied over 190 acres of land at Murdike, near Lahore, and included a number of education facilities, a mosque, businesses, a swimming pool and three housing complexes. The Lashkar-e-Taiba was founded as the armed wing of the Markaz, committed to fighting for the establishment of its version of an Islamist state. At a November 1997 conference held by the Markaz Dawa wal'Irshad, Saeed called for an end to democracy in Pakistan, arguing that "the notion of the sovereignty of the people is anti-Islamic".[24] Pakistani newspapers noted that the venue was festooned with signboards proclaiming that the appropriate response to democracy was through grenade and bomb explosions ["*jamhooriyat ka jawaab*, grenade *aur* blast"].[25] The Markaz's hostility to democracy made it a natural ally of Pakistan's military establishment: notwithstanding its designation as a terrorist organization by the United States, the Director-General of the ISI, Lieutenant-General Mahmood Ahmad, attended its April 2001 convention, where a resolution was passed calling on cadre "to capture Hindu temples, destroy the idols and then hoist the flag of Islam on them".[26]

By the late 1990s, a growing effort was being made to give a formal architecture to the alliance between the jihad in Jammu and Kashmir and its global counterparts. In February, 1998, al-Qaida's Osama bin Laden sponsored a conference in Khost, Afghanistan, which was attended by representatives of several jihadi groups active in Jammu and Kashmir, including the Harkat-ul-Jihad Islami, the Harkat-ul-Mujahideen, the Lashkar-e-Taiba and al-Badr. The conference led to the formation of the International Islamic Front Against Jews and Crusaders. Among other things, the International Islamic Front decided to commit greater resources to creating an Islamic emirate in Jammu and Kashmir which could serve as an alternate base to Afghanistan if the need arose.[27] Although there is no evidence that al-Qaeda ever actually operated directly in Jammu and Kashmir, the formation of the International Islamic Front had several visible manifestations. Its constituents renewed their efforts to ensure compliance with their version of Islamic law, notably by ordering women to wear the enveloping *burkha*, and through proscribing television and other pop-culture media. It is important to note, though, that these campaigns were not entirely new. In 2001, for example, the Lashkar-e-Taiba had opposed a campaign to immunize children against polio in Pakistan-administered Kashmir, claiming that it was a cover for a birth-control project which would render children "unable to reproduce".[28] Similar campaigns, as we have seen in the previous chapter, were conducted in the first stages of the Jammu and Kashmir jihad.

Some commentators have sought to make sharp ideological distinctions between the political sponsor of the Hizb-ul-Mujahideen, the Jammu and Kashmir Jamaat-e-Islami, and those of new formations like the Lashkar-e-Taiba. In a 2002 essay, Yoginder Sikand argued that while the Jamaat-e-Islami sought to place the war in Jammu and Kashmir within an Islamist discursive framework, of a war between disbelief and Islam, it was not explicitly communal in character:

> Three features are of particular importance in Geelani's description of the jihad. Firstly, the jihad is seen as directed against the Indian state and its agents, not against Hindus or Indians as such. Secondly, the jihad has the limited goal of freeing Kashmir from Indian control. Thirdly, the mujihadin have no intention of intervening in Indian internal affairs after the liberation of Kashmir. Once the Kashmir issue is solved by freeing the territory from Indian control and merging it with Pakistan, the two countries, Geelani writes, will be able to establish peaceful and cordial relations with each other, for the root cause of the tensions between the two countries is the dispute over the issue of Kashmir.[29]

While valuable, such distinctions run the risk of losing sight of the wood while identifying the trees. Much of the ideological content of the new jihadist groups marked a *development* of the ideas of the organizations of the first phase of the struggle, not a *discontinuity*. Even the JKLF, during the early 1990s, cast its agenda in expressly Islamist terms. Despite its ostensible commitment to the creation of an independent, secular–democratic state, on ground its cadre called for the creation of an Islamic democracy, an economic system they called Islamic socialism, and the protection of minority rights as prescribed by the Quran and religious tradition.[30] Much of this language, particularly that on the rights of minorities, would have been wholly acceptable to organizations further to the right; the Taliban in Afghanistan, after all, claimed that their fascist policies on Hindu and Sikh citizens had the sanction of the Quran. Indeed, as Navnita Chadha Behera has pointed out, Islam, rather than Kashmiri national history, was the very *moral basis* for the armed struggle.[31] Even the JKLF's notion of independence was theocratic in nature: one popular slogan from early 1990s, Behera records, was "*Azadi ka matlab kya? La Illahi il-Allah!*" [What is the meaning of freedom? It means there is no god but God]. Similarly, as I have noted in the previous chapter, the notion of *azadi*, or independence, was often twinned with that of the *Nizam-e-Mustafa* or Islamic state.

In practice, too, the mujahideen sponsored by the Jamaat-e-Islami or the JKLF often anticipated the practices of the mujahideen sponsored by the Jaish-e-Mohammad or Lashkar-e-Taiba. In the late 1990s, for example, the Lashkar-e-Taiba involved itself in a succession of large-scale killings of Hindus in both the Kashmir valley and rural Jammu. It is worth noting, however, that the Hizb-ul-Mujahideen actively participated in some such atrocities, in particular leading the massacre of Kashmir Pandits at Wandhama in 1998.[32] While it is true the Jamaat-e-Islami avoids anti-Hindu polemic of the Lashkar-e-Taiba variety, there is no great ideological distance between its aspirations and those of Islamist groups even further to the Right. Geelani's ideas on the very *basis* for Kashmiri accession to Pakistan are deeply similar to those of groups like the Lashkar-e-Taiba. In matters of faith, belief and customs, Geelani argues, Hindus and Muslims are set irrevocably apart, as they are divided by such matters as food, clothing and lifestyles. He describes it as being as difficult for Muslims to live in a Hindu milieu as "for a fish to stay alive in a desert". Muslims, he argues, cannot live harmoniously with a Hindu majority without their own religion and traditions coming under a grave threat, one major factor being what he perceives to be Hinduism's capacity to assimilate other religions. For Islam to be preserved and promoted in Jammu and Kashmir, in Geelani's conception, it is necessary for it to be separated from India.[33]

Geelani's is not impolite enough to say just why Hindus and Muslims may not coexist. Nor does he move on to the rational corollary of his argument – that Indian Muslims cannot live as citizens of secular India, either. The Markaz has shown no such squeamishness, candidly asserting that "the Hindus have no compassion in their religion".[34] In another article, Hafiz Mohammad Saeed wrote that "the Hindu is a mean enemy and the proper way to deal with him is the one adopted by our forefathers . . . who crushed them by force".[35] Sikand has argued that this "sort of anti-Hindu rhetoric is not a prominent feature in Geelani's writings, and thus represents a further radicalization of the Kashmiri jihadist discourse".[36] My suggestion here is that while this is indeed so, there is in fact marked *continuity* of thought and practice. Terrorist groups in Jammu and Kashmir, starting with the JKLF and running onwards to Lashkar and Jaish-e-Mohammad, span a relatively small part of a black–white ideological spectrum. Ideologues of the Islamist Right locate the conflict in Jammu and Kashmir within the paradigm of a larger South Asian communal conflict. For Saeed, Partition was merely punctuation in an yet unfinished battle between "Hindu" India and Islam, a battle which is in turn part of a larger *jihad* between the true faith and unbelievers everywhere. His colleague, the Harkat-ul-Mujahideen's Fazl-ur-Rahman argued that, "Delhi, Calcutta, Mumbai and Washington are the real targets of Militants. Muslims should co-operate with militants for dominance of Islam in the world."[37]

Over the coming years, the evolution of the jihad in precisely this direction would have definitive consequences for South Asia, and the world. For the moment, however, the fact was that the polemic had altered little. The influx of the new Islamist groups had raised the costs of conflict for India somewhat, but not enough to push it any closer to making concessions on Jammu and Kashmir than it had been at any point since 1947–1948. Within Jammu and Kashmir, the declining enthusiasm of local cadre for the jihad was becoming painfully apparent, illustrated in their declining share of battlefield fatalities. Something new was desperately needed if the nuclear jihad was to live up to its promise of being something other than just a minor inconvenience for the elephant.

The Kargil war

In late April 1999, three shepherds from the small frontier hamlet of Batalik left for the high ranges with their flocks, as a part of the grand summer movement that takes place across the great Himalayan ranges.[38] Tashi Namgyal, Ali Raza Stanba and Morup Tsering were among the first shepherds to leave for the high pastures that spring, perhaps hoping to have the opportunity to bag some of the region's famous *Tahr*, or mountain goats, a pastime that has proved resistant to official efforts to protect the species. On the morning of May 3, Namgyal discovered an altogether unexpected variety of wildlife: a group of men clad in *Pathan* suits, clothing favoured by both Pakistani irregulars and troops, digging trenches along the Jubbar *Langpa*, a stream that runs down from glaciers along the LoC towards Batalik. He promptly reported the discovery to the Indian Army. Despite some skepticism about Namgyal's claims, the 121 Independent Infantry Brigade sent out a small patrol into the area.[39] It was ambushed, as were subsequent Indian probes into the area.

India's fourth war with Pakistan had begun. Before it ended, the armies of both countries would have together lost over a thousand dead on what was perhaps the highest battlefield in the world.

To understand the events of the Kargil war and its relationship with the jihad in Jammu and Kashmir, it is necessary to go back in time. Recent Pakistani scholarship suggests that its military repeatedly considered the contours of a limited war against India more than once since the mid-1980s. Kargil, given its strategic location between the Kashmir valley and Ladakh, the poor road infrastructure offering access to the region and the difficult terrain on which an Indian counter-offensive would have to fight, was an ideal location for such an enterprise. According to Hassan Abbas, Pakistan's Military Operations Directorate had outlined a detailed plan for war in Kargil during the regime of General Zia-ul-Haq, but Pakistan's military ruler had rejected it on the grounds that it could "lead us into a full scale war with India".[40] Such plans, given Pakistan's renewed sub-conventional offensive against India during the Zia-ul-Haq years, made eminent sense. So, too, did Zia-ul-Haq's concerns. In the late 1990s, however, the Kargil idea was revived. It gained the support of powerful patrons in Pakistan's military, notably Lieutenant-General Mohammad Aziz Khan, an ethnic Kashmiri with a deep ideological commitment to the Islamist right, Lieutenant-General Mahmood Ahmad, the commander of Pakistan's 10 Corps, and Major-General Javed Hassan, the commander of the Pakistani troops in the Northern Areas, the very men who Tashi Namgyal had spotted in the Batalik heights.

As the decade drew to a close, the need for dramatic action in Kashmir would have seemed ever-greater to Pakistan's military establishment. In the summer of 1996, India had managed to hold elections to Jammu and Kashmir's six seats in the national Parliament. Although the elections were characterized by both high levels of violence and allegations of malpractice, the significance of the fact that they could be held at all was lost on no one.[41] Moreover, charges that Indian forces and their allies in the counter-terror militias had corralled hesitant voters to voting centers did not obscure the fact that there seemed to be a genuine constituency for change in Jammu and Kashmir. Notably, there was no clear correlation between allegations that the state had coerced voters and the actual turnout seen on ground. Voter turnout was *low* in some areas where intense coercion was alleged, like Anantnag and Sopore; it was, conversely, *high* in others where there appeared to be no sign of state pressure to participate in the elections.[42] Politicians read the signs correctly: after years of direct central rule, a large mass of people within Jammu and Kashmir wanted a representative political order sensitive to their everyday concerns. Later that year, in September, elections to the Jammu and Kashmir Assembly brought a National Conference government, led by Farooq Abdullah, back to power. The National Conference picked up 37.98 per cent of the popular vote in the seats it contested, and 57 of 81 seats.[43]

In some senses, then, Jammu and Kashmir was back to where it had been in 1987. Given time, it seemed probable that the new regime would be able to reconstruct those systems of patronage and influence which had sustained mainstream politics in the state prior to the Islamist onslaught of 1988–1989 and the nuclear jihad. In any case, the fact that an election had been held meant that the Hurriyat Conference could no longer claim to be the sole spokesperson for *all* opinion in Jammu and Kashmir. Aware of the risks the new situation posed to their position, at least some in the Hurriyat, as we shall see, began to consider the prospect of a dialogue with New Delhi. All the circumstances seemed favourable. The Congress-led regime of Prime Minister Narasimha Rao had been replaced by a centre-left coalition. The new Prime Minister, I.K. Gujral, had opened peace talks with Pakistan, and was also a strong advocate of a dialogue within Jammu and Kashmir, too. Little tangible was achieved in these early moves towards détente, but the fact that it was taking place at all could not have been welcomed by the architects of the nuclear jihad in Pakistan's military establishment. In the winter of 1997, for example, Pakistani forces shelled the Indian town of Kargil, causing some civilian casualties in what appeared to be an effort to sour relations. It was a sideshow – but one, it would turn out, of some significance.

India's decision to test its nuclear arsenal in May 1998, a decision that took the world by surprise, rudely interrupted the slow movement towards peace seen during the previous years. An intense debate has taken place over just why India chose to test its weapons at this time, but it seems probable that a combination of factors, including the Bharatiya Janata Party's (BJP) domestic political compulsions, played a role. It is also possible that the new government in New Delhi believed that bringing its nuclear weapons out of the closet would help stabilize its relationship with Pakistan, and force an end to the nuclear jihad, and its periodic explosion into crisis and war. Speaking to India's Parliament in 1999, Prime Minister Vajpayee suggested that the nuclear bomb is "the kind of weapon that helps in preserving the peace".[44] "If in the days of the Cold War there was no use of force", he argued, "it was because of the balance of terror". Pakistan promptly responded with nuclear tests of its own. To its military establishment, however, the new situation in South Asia represented an opportunity to escalate the jihad in Jammu and Kashmir, rather than one to strengthen the status-quo. With both nations' nuclear weapons out in the open, Pakistani strategists believed, India would be more reluctant than ever before to risk the prospect of a low-intensity conflict escalating into a calamitous exchange.[45] General Zia-ul-Haq's concern that a limited incursion into Kargil could spark off a full-blown war, Pakistan's military establishment correctly understood, was less pressing now than a decade earlier.

In February 1999, Prime Minister Vajpayee travelled across the India–Pakistan border for a renewed attempt at peacemaking. Even as the two leaders issued what is known as the Lahore Declaration, a commitment binding both countries to nuclear confidence-building measures and reiterating their willingness to solve the Jammu and Kashmir dispute through peaceful means, Pakistani troops had initiated a series of incursions across the LoC in Kargil. Nawaz Sharif was later to claim that the offensive was initiated without his assent. Most Pakistani commentators, however, believe he was aware of at least the contours of the enterprise, if not its entire import, by April 1999.[46] Pakistan's objectives in initiating the Kargil offensive were, by most accounts, simple: it hoped to draw international attention to the conflict in Jammu and Kashmir, specifically the risks it held out in the transfigured strategic context of South Asia, and to revitalize the jihad India had succeeded in wearing down.

In the event, Pakistan's offensive in Kargil did not yield the results its strategic establishment had anticipated. While initial Indian military responses were often confused and disorganized, the consequence of senior and mid-level commanders seeking to cover up the failures that had allowed Pakistan to capture ground in the first place, the tide began to turn by June 1999.[47] Aided by air strikes and massive artillery support, Indian soldiers were able to slowly overwhelm the multiple complexes of high-altitude bunkers Pakistani troops had succeeded in erecting through the winter and spring of 1999. Nor did the international situation evolve as Pakistan's military had hoped. In late June 1999, the commander-in-chief of the United States Central Command categorically told Prime Minister Sharif that Pakistan would have to withdraw its troops from the Indian side of the LoC. More important, General Zinni made it clear that Pakistan ought not to expect any political concessions on Jammu and Kashmir in return for restoring the status -quo.[48] Under intense pressure, Prime Minister Sharif made a visit to Washington, DC on July 4, hoping to secure at least some face-saving formula. He received an icy welcome, and was bluntly told to ensure that Pakistani troops withdrew immediately. On July 12, Prime Minister Sharif formally called on the mujahideen his government continued to claim were on the Kargil heights to pull back. Two days later, the first sets of Pakistani troops began pulling back from their positions, ceding them to advancing Indian troops.

Several key questions about the Kargil conflict remain unanswered, not in the least because no credible account of its strategic ambitions and reasoning has yet emerged from Pakistan. The war was, however, to set off a series of complex events. Relations between Prime Minister Sharif and his hand-picked Chief of Army Staff General Pervez Musharraf deteriorated dramatically. Sharif's efforts to sack General Musharraf eventually culminated in a military coup. While General Musharraf's rise to power was to have several fateful consequences for Pakistan itself, it was also to give a new form to the nuclear jihad. Like much of Pakistan's military establishment, General Musharraf believed that the reverses in Kargil notwithstanding, Pakistan could continue to escalate the jihad within Jammu and Kashmir. As Ashley Tellis, C. Christine Fair and Jamison Jo Medby have noted in a perceptive monograph, the Pakistani establishment "concluded that the use of *Pakistani* troops in Kargil invited political failure", but did not derive from this the further conclusion "that other forms of violence are either illegitimate or ineffective for altering the status quo".[49] What was needed was a successful calibration of the levels of violence within Jammu and Kashmir: its expansion to new theatres, and its ratcheting-up to a point where it would inflict an altogether new level of pain on India.

Talking to terror

Kargil, General Pervez Musharraf reminisced some years after he usurped power in an October 1999 military coup, had "proved a lesson to the Indians and a rude awakening to the world of the reality of Kashmir".[50] Many dismissed the observation as the ill-founded polemic of a military leader who could not accept that his cherished war plan had led to disaster. In one key sense, though, General Musharraf was right. To India, Pakistan and the world, it was evident that the nuclear jihad had reached a decisive phase. The outcome of this contestation, though, was to be far from what General Musharraf had perhaps expected.

To all the principal players in the covert war in Jammu and Kashmir, the post-Kargil stakes were high. Although the National Democratic Alliance government in New Delhi had been re-elected to power in the national elections held in 1999, it had seen its vote-share diminish, despite its war record. One statistical analysis has shown that had the opposition Congress party not suffered a split in the single state of Maharashtra, it – and not the BJP – would most likely have led whatever coalition in New Delhi.[51] To Prime Minister Atal Behari Vajpayee, the fact that Pakistan had apparently been tamed – and was, moreover, in a considerable state of domestic ferment following the October 1999 coup – seemed to offer the opportunity to secure an historic outcome in Jammu and Kashmir. If New Delhi could secure a settlement with its principal opponents in Jammu and Kashmir, the National Democratic Alliance would have been able not just to retrieve lost political ground, but to establish itself as the centre of gravity in Indian politics. For General Musharraf, domestic political compulsions were just as strong. Internationally isolated by the coup and the subsequent exile of both Nawaz Sharif and Benazir Bhutto and facing considerable resentment within the military over the outcome of his Kargil adventure, his sole allies in Pakistani politics were on the Islamist far-right. Escalating the nuclear jihad would not only secure their support, but demonstrate Musharraf's commitment to a cause which many in Pakistan saw as central to their core national identity.

India's post-Kargil response to the continuing war in Jammu and Kashmir can be likened to a two-legged stool. The first pillar of this process was New Delhi's effort to engage the largest terrorist group in Jammu and Kashmir, the Hizb-ul-Mujahideen, in dialogue. This was manifested in a brief declaration of a ceasefire by the Hizb-ul-Mujahideen in 2000, and its five-month long reciprocation by India. Officially described as a Non-Initiation of Combat Operations, India advertised the ceasefire as a gesture made in observance of the month of Ramzan, a sacred time for Muslims around the world. The second pillar of Indian strategy sought to give the Ramzan ceasefire political meaning by engaging political secessionists within Jammu and Kashmir. Together, these two elements constituted what I call the Ramzan *Process*. What the Ramzan Process lacked, however, was the proverbial third leg of the stool – a means to deal with Pakistan; more specifically, Pakistan's willingness to escalate violence in Jammu and Kashmir to dangerous and possibly war-inducing levels, despite its recent defeat in Kargil. This failure was eventually to drive India and Pakistan to the edge of another war: a near-war that would, as we shall see, transfigure the course of the nuclear jihad.

By the account of the friends of the Hizb-ul-Mujahideen commander Abdul Majid Dar, the Ramzan ceasefire of 2000 was of divine provenance. Standing before the *Kaaba*, the black rock that forms the centrepiece the Haj pilgrimage, Majid Dar had a vision of the suffering that a decade of terror had inflicted on Jammu and Kashmir, and was as a result moved to work towards bringing peace.[52] For those dissatisfied with this "god-did-it" narration of events, little is on offer about the precise sequence of events leading to the Ramzan ceasefire. None of the key participants have either confirmed or denied the accounts of its genesis that have appeared in media accounts. Its broad contours, however, seem clear. In early 2000, Ghulam Mohammad Bhat, the *amir* of the Jamaat-e-Islami Jammu and Kashmir, made contact with Majid Dar. The two conducted their dialogue both in the United Arab Emirates, where Majid Dar's wife worked as a doctor, and in Saudi Arabia, where they used the Haj pilgrimage as cover for their covert contact. Soon after, G.M. Bhat and intermediaries from the ethnic-Kashmiri diaspora initiated contact with the Government of India. A.S. Dulat, the then head of India's external intelligence service, the Research and Analysis Wing, who had served in the Intelligence Bureau during the Islamist putsch attempt of 1988, played a key role in this early dialogue, along with Brajesh Mishra, the principal secretary to Prime Minister Vajpayee.

Events moved rapidly and, in April 2000, Majid Dar flew into Jammu and Kashmir through Kathmandu and New Delhi, after guarantees of safe passage and protection were provided by the Government of India. Back in the Kashmir valley after several years, Dar set about making allies at two distinct levels. First, he approached key Hizb-ul-Mujahideen field commanders to inquire how they would respond to a political engagement with the Government of India. Two powerful commanders, Masood Tantrey and Khurshid Ahmad Zargar, were particularly receptive to the idea; none voiced outright opposition to it. At the same time, key figures in the All Parties Hurriyat Conference, a platform for several major secessionist organizations, were sounded out on their position. Two centrists in the Hurriyat, Abdul Gani Bhat and Abdul Gani Lone, proved supportive of the proposal for a ceasefire. Syed Ali Shah Geelani, the hardline Islamist who served as the Jamaat-e-Islami's representative in the Hurriyat, was less enthusiastic. After some persuasion, however, Geelani agreed to go along with Majid Dar's proposals, although it is unclear if he knew of his *amir*'s role in shaping them.

On July 24, 2000 Majid Dar summoned a small group of journalists to a safehouse in Srinagar, and announced that the Hizb-ul-Mujahideen would observe a three-month unilateral ceasefire. He said the organization had decided to do so to "dispel Indian propaganda that we are terrorists, rather than a people fighting for our birthright, freedom".[53] He laid down a few pre-conditions: the ceasefire was subject to the cessation of Indian violence against civilians and political activists; the use of the ceasefire by India as a "tactical weapon" for propaganda, he added, would subvert its purpose. Significantly, Majid Dar let it be known that the Hizb-ul-Mujahideen was open to the Hurriyat engaging New Delhi in direct dialogue. "Let them talk to anybody", he said, "the aim of the exercise should be to resolve the issue amicably, through a dialogue without preconditions". The Hizb-ul-Mujahideen itself, Dar continued, would encourage politicians from

India and abroad to visit Jammu and Kashmir, and begin a dialogue with its people. Conscious of the reaction his statement was likely to provoke from jihadist groups with large numbers of Pakistani nationals among their ranks, Dar described their cadres as "our brothers who have come to our help". "Once the problem is resolved amicably and peace is restored", Dar concluded, "they will return peacefully".

Majid Dar, it seems likely, aimed to force both Pakistan and the jihadist groups directed by its intelligence services into accepting a ceasefire without India first agreeing to their presence in the negotiation room. While there is no hard evidence to support the proposition that Pakistan was taken by surprise, it seems to be borne out by subsequent events. Majid Dar's announcement and Shah's subsequent endorsement of the ceasefire were blacked out on Pakistani television. The United Jihad Council, a coalition of 14 Pakistan-based terrorist groups operating in Jammu and Kashmir, moreover, promptly removed Shah from his position as its chief, and demanded that the Hizb-ul-Mujahideen immediately withdraw the ceasefire. Shah was deemed a traitor to the cause, and widely condemned in Pakistan.[54] Jihadist reaction to the ceasefire was not restricted to polemic. The Jaish-e-Mohammad, the Jamait-ul-Mujahideen and al-Umar Mujahideen, all members of the United Jihad Council, jointly claimed credit for a series of six bomb blasts in Srinagar, which they said had been executed to protest the ceasefire.[55]

Still in the diplomatic doghouse after Kargil, Pakistan was in no position to oppose the ceasefire in public. Just in March 2000, after all, it had been subjected to the unpleasant experience of President Bill Clinton spending five days in India on an official visit, and stopping in Pakistan for just a few hours on his way home. Pakistan would, however, fight by proxy. In days to come, jihadist attacks on the ceasefire would gain in intensity. Starting from the night of August 1, jihadist groups carried out a series of mass killings intended to force India to resume offensive operations against the Hizb-ul-Mujahideen. One hundred civilians, mostly members of religious minorities in Jammu and Kashmir but also some Muslims, were killed in the first wave of attacks.[56] Such attacks had taken place with depressing regularity in Jammu and Kashmir, but in their sheer scale and brutality, the massacres of August 1 were unprecedented. Other outrages soon followed. After a series of killings in the mountain districts, particularly in Doda, the Indian government was forced to impose the Disturbed Areas Act, a legislation that gives the armed forces extensive special powers to the provinces of Jammu and Ladakh.[57] While the legislative measure had limited ground-level impact, since armed forces had long operated against terrorists in these areas, it did serve to illustrate just how much pressure jihadi groups had been able to mount – and how difficult it was becoming for the National Democratic Alliance government in New Delhi to continue with a peace process in the face of this unrelenting assault.

For the moment, however, New Delhi chose to ride out the jihadi offensive, and continued to voice its commitment to the Ramzan ceasefire. The Hurriyat, however, displayed less conviction, and soon backed out of an agreement its leaders had themselves endorsed – albeit not in public. On July 28, just days after the United Jihad Council voiced its ire, the Hurriyat put out a press release describing the ceasefire as "a step taken in haste".[58] "The Hizb leadership", it argued, echoing the language used by the rejectionist constituents of the United Jihad Council "has also failed to perceive the Indian machinations and cunning behaviour that has always been there to divide Kashmiri opinion on issues like this". At the same time, however, the Hurriyat insisted that the dispute on Jammu and Kashmir "should be resolved through peaceful means, to ensure the prosperity of the region". This last formulation points us to the twin meanings of thc Hurriyat position. If the rejectionists grouped around Geelani objected to the Ramzan ceasefire on first principles, the realists had their own set of concerns, notably that a dialogue carried out in these circumstances would accord primacy to the Hizb-ul-Mujahideen itself, and not to the politicians who claimed to speak for the armed struggle.

From the point of view of both the Hurriyat's realists and the central Hizb-ul-Mujahideen command, Majid Dar's choice of Fazl-ul-Haq Qureshi, his old comrade in arms in Tehreek Jihad-e-Islami, as his chosen mediator with New Delhi, posed a particular problem. A long-standing participant in anti-India terrorist activity in Jammu and Kashmir, Qureshi had served as a member of the Master Cell and al-Fatah. In September 1974, he, along with Farooq Rehmani and Nazir Ahmad Wani, formed the People's League to continue the anti-India struggle. Bruised by its defeat in the India–Pakistan war of 1971, Pakistan's covert services were less than enthused by this enterprise, whose leaders they believed to be *agent provocateurs*. As early as 1979, the People's League's leadership had formulated a three-year plan for an uprising against Indian rule in Jammu and Kashmir. Pakistan, its attention focused on the growing anti-Soviet Islamist campaign in Afghanistan, remained sceptical. In 1988, the then People's League chief Abdul Aziz Sheikh finally returned to Jammu and Kashmir from Pakistan, and began raising cadre for armed action. Later the same year, however, the League broke into two units, with a one-time Hurriyat executive member, Shabbir Shah, and S. Hamid forming the now-defunct Muslim Janbaaz Force. Sheikh and Mohammad Farooq Rehmani, for their part, set up the Tehreek Jihad-e-Islami. It was an ill-fated move, for Pakistan threw its weight behind the Hizb-ul-Mujahideen, and both Shah and Rehmani found themselves militarily marginalized. Under similar pressure from Pakistan's covert services, much of the Tehreek Jihad-e-Islami cadre joined the Hizb-ul-Mujahideen; Qureshi and Rchmani stoically distanced themselves from the proceedings.

Given his political roots, then, Qureshi was viewed with suspicion by both the rejectionists and the realists, to both of whom he was a direct political competitor. Given his decision to keep his distance from the Hizb-ul-Mujahideen, it seems reasonable to speculate that Qureshi had few friends in Pakistan's Jammu and Kashmir-policy establishment, either. Majid Dar's decision not to conduct his negotiations through the Hurriyat made matters all the more difficult, since Pakistan at least had some leverage over that organization – but none over Qureshi. All of this sharpened the dilemma for the Pakistani strategists of the jihad in Jammu and Kashmir. While Pakistan did not wish to be seen as a spoiler, it was precipitously close to losing all control over the war in which it had invested so much. When the August 1 massacres failed to derail the ceasefire, however, direct Pakistani intervention became inevitable. Even as the Hizb-ul-Mujahideen delegation met India's Home Secretary, Kamal Pande, for talks in Srinagar on August 3, its *amir*, Shah, was being pressured to announce an August 8 deadline for the inclusion of Pakistan in the dialogue. Majid Dar, significantly, stayed away from this meeting; he perhaps understood that a crisis was imminent, and did not wish to be a part to what would follow.

Ultimately, little took place at the meeting – the only public engagement of jihadist groups with the Government of India. Both sides' representatives agreed that the Hizb-ul-Mujahideen demand for the release of prisoners, as well as a cutback in search-and-cordon operations, would be considered by their principals before a second round of talks was held on August 7. Even as the negotiators were leaving the meeting, however, the Hizb-ul-Mujahideen's central command made public its August 8 deadline for the inclusion of Pakistan in the talks. Political dialogue, Mohammad Yusuf Shah said, executing a neat volte-face, had to precede an end to hostilities. Qureshi attempted to persuade Shah to extend the deadline, but to little effect. Meanwhile, under pressure from the right-wing in his own party, which was incensed by the August 1 massacres, Prime Minister Vajpayee was forced to announce in Parliament that any negotiations would be held within the framework of the Indian constitution – something which ruled out even the theoretical possibility of a territorial compromise. Unsurprisingly, the second round of talks was never held. At 5:35 PM on August 8, 2000, Indian signals intelligence began jamming the half a dozen frequencies used by the Hizb-ul-Mujahideen for routine communications. Five minutes earlier, the Mohammad Yusuf Shah had announced that the ceasefire his organization had announced a fortnight earlier had come to an end.

In purely military terms, India's experience of the ceasefire was far from positive. During its five-month course, the numbers of violent incidents actually increased, in comparison with the same months of 1999–2000 and 1998–1999. Although the overall numbers of killings declined marginally, this was largely a consequence of the fact that Indian forces had been ordered to stop initiating offensive operations. Killings of civilians, a key means through which terrorist groups exercise political authority, actually increased. A measure intended to bring peace to Jammu and Kashmir had, in the short term at least, succeeded in realizing precisely the opposite outcome. Terrorist groups were more than able to compensate for the divisions between the pro-dialogue and rejectionist commanders of the Hizb-ul-Mujahideen – and to thus demonstrate that Pakistan was well equipped to undermine the peace process unless it was present at the table. Three times through the coming year, however, India would renew its commitment not to engage in offensive operations against terrorist groups, hoping against hope that something could be salvaged from the ruins of the Ramzan ceasefire.

A peculiar situation had thus arisen: a dialogue intended to end armed violence had ended up convincing political secessionists that the continued use of guns was, in fact, their only guarantee of relevance. Were the jihad's guns to be silenced, both realists and rejectionists would, after all, lose their right to speak for the war.

The war after the war

Unlike the pro-dialogue elements in the Hizb-ul-Mujahideen, jihadists in Pakistan had no intention of silencing their guns. In February 2000, at a rally in Islamabad's Aabpara Square, just a few hundred metres from the headquarters of the ISI, the Lashkar-e-Taiba's overall head Hafiz Muhammad Saeed proclaimed that the jihad in Jammu and Kashmir had reached a new stage. Kargil, he said, had been the first component of this new campaign; the wave of *fidayeen* suicide-squad attacks the organization had unleashed on major security and civilian installations was the second. "Very soon", he promised, "we will be launching a third round".[59] His deputy, Abdul Rahman Makki, announced that the Lashkar would soon initiate operations in Hyderabad, a city claimed by Pakistan's Islamist right-wing to have been seized illegally by Indian forces from its Muslim monarch in 1948.

Indian security analysts took Saeed seriously, and with good reason. Pakistani defeat in the Kargil war had not meant the beginning of peace in Jammu and Kashmir. One dramatic sign of Pakistan's willingness to step up the pace of the nuclear jihad was provided after an Indian Airlines flight was hijacked to Kandhahar in Afghanistan on Christmas Eve in 1999. The three terrorists released by India in return for the safety of the passengers, Maulana Masood Azhar, Syed Omar Sheikh and Mushtaq Zargar, soon surfaced in Pakistan, as did the hijackers themselves. Theatrical episodes of this kind amplified a larger reality. Indian security force fatalities – a category which includes regular soldiers, paramilitary personnel, police and irregulars fighting, as it were, on the same team – had been in decline ever since 1996. The year 1999, however, had witnessed the highest levels of Indian force fatalities ever seen in the course of the Jammu and Kashmir conflict, 555 – a figure excluding troops lost in the war itself. Indian security force fatalities rose again in 2000 to 638, and to 706 in 2001. One particular source of concern for Indian military planners was that the ratio of terrorists killed to security force personnel lost fell to a record low in 1999, to just over 2:1. Although this ratio recovered somewhat in subsequent years, to the vicinity of 3:1, this was still lower than in the pre-Kargil period. What the figures meant was simple: India was facing better armed and trained terrorist cadre than had been seen prior to the Kargil war, and in greater numbers.

Unlike terrorists with families and futures on the Indian side of the LoC, the *mehmaan* mujahideen had no interest in a détente process. What evidence is available suggests that their numbers had increased significantly in the wake of the Kargil war. Killings of foreign terrorists by Indian forces registered a significant increase from 1996 onwards. Of the foreign terrorists, 348 died in combat that year, followed by 403 in 2000, 488 in 2001 and 516 in 2002 – a year when Pakistan was to have good reason to believe it could soon be at war in India. Foreign terrorists, a relatively marginal component of terrorist cadre in Jammu and Kashmir in early years, had thus come to constitute over a third of their strength, as measured by relative fatalities. Indian politicians and officials have often exaggerated the foreign component of terrorist cadre active in Jammu and Kashmir; one senior Border Security Force official asserted in 2003, for example, that it stood at 75 per cent.[60] Nonetheless, the demonstrable increase in numbers between 1996 and 2002 casts some light on the increasing ferocity of conflict after the Kargil war.

Policy-makers in New Delhi assumed that a dialogue process would create a polarization between mainly ethnic-Kashmiri organizations, like the Hizb-ul-Mujahideen, and those made up of greater numbers of Pakistani nationals, like the Lashkar-e-Taiba, Jaish-e-Mohammad, Harkat-ul-Mujahideen and Harkat-ul-Jihad Islami. In the real world, the lines between these groups were often diffused. Inayatullah Khan, a Pakistani national who operated using the *nom de guerre* Bilal-e-Habshi, commanded both Lashkar-e-Taiba and Hizb-ul-Mujahideen units in the Budgam–Beerwah area over a period of seven years before his eventual elimination in December 2003.[61] If Inayatullah Khan led mainly ethnic Kashmiris, his one-time comrade in arms, Abdul Hamid Gada, led a group which was mirror-opposite in character. Foreign terrorists played a key role in Gada's operations, notably in the execution of a mass killing of 23 Pandits, nine of them women and four young children, at Wandhama.[62] Pakistan nationals working for groups other than the Hizb-ul-Mujahideen also cooperated closely with its cadre. Mohammad Suhail Malik, a Lashkar-e-Taiba terrorist charged with having participated in the massacre of 36 Sikh villagers at Chattisinghpora, told his interrogators of one earlier joint action with the Hizb-ul-Mujahideen. On that occasion, Malik, along with four other Lashkar terrorists, joined two Pakistan-national members of the Hizb-ul-Mujahideen in the ambush of a civilian bus hired to carry army personnel. Ethnic Kashmiri cadre of the Hizb-ul-Mujahideen were used to follow the movements of the bus and alert the ambush unit of its arrival.

Pakistan had an army, therefore, with which it could undermine India's effort to secure a deal with elements of the Hizb-ul-Mujahideen. The Kargil war's outcome, however, had demoralized the rank- and- file of terrorist groups and means needed to persuade them that the war was still worth fighting. Tactics were soon devised. Starting with an attack on the BSF's sector headquarters at Bandipora in north Kashmir, the Pakistan-based Lashkar-e-Taiba and Jaish-e-Mohammad unleashed a series of *fidayeen* attacks on high-value civilian and military targets. Generally translated as suicide-squad attacks, a partial misnomer since few involved bombings of the kind seen in Sri Lanka or Israel, *fidayeen* units targeted two key symbols of state authority apart from the BSF in 1999: the headquarters of the Indian Army's 15 Corps in Srinagar, and the offices of the crack Special Operations Group of the Jammu and Kashmir Police in the same city. Seventeen security force personnel and five terrorists were killed in these attacks, a relatively trivial number given the overall levels of combat-related fatalities. What the *fidayeen* attacks did do, however, was to signal to cadre demoralized by the outcome of the Kargil war that the larger campaign against India was far from finished.

Both the scale and the frequency of *fidayeen* targets rose steadily in the coming years, although their military utility is unclear. Thirty-eight Indian security force personnel were killed in fidayeen attacks in 2000, for the loss of 18 terrorists, numbers that increased in 2001 to 91 and 36, respectively. It is possible that the wave of *fidayeen* attacks led Indian forces to commit more personnel for defensive purposes but, judging by the steady escalation in the number of terrorists killed from 1999 onwards, this does not seem to have significantly impeded their offensive capabilities. Although it could not have been lost on the leadership of jihadi groups that their losses in *fidayeen* missions represented a neat reversal of the attrition ratio recorded in combat in Jammu and Kashmir, it would also have become clear that each such operation meant the loss of highly motivated personnel. Certainly, by 2003, Indian security forces seemed to have learned to deal with *fidayeen* attacks, sustaining just 23 fatalities while claiming 16 of their attackers. Instead, the real value of the *fidayeen* attacks lay in their propagandistic value, and the fact that they were able to carry the war in Jammu and Kashmir to the Indian state – the very fact, of course, that led India and Pakistan to the edge of war.

We suffer, sadly, from a near-complete absence of information on the decision-making processes at the command levels of both jihadi groups and the Pakistani military-intelligence infrastructure which guided the course of their operations in Jammu and Kashmir during this time. However, it seems safe to surmise that on the eve of the 2001 crisis, some of the key lessons of warfare since 1999 must have become clear. Pakistan could, indeed, escalate warfare within Jammu and Kashmir to unprecedented levels. Indian forces were, however, able to respond to this escalation by simply stepping up their own operations. The war of attrition waged in Jammu and Kashmir since 1988 had simply reached new levels, without giving Pakistan significantly greater political leverage. While it could sabotage Indian efforts at securing a unilateral dialogue with terrorist groups, Pakistan had not yet been able to decisively tip the scales. For India, too, the escalating war held out problems. While it could militarily contain the jihad in Jammu and Kashmir, movement towards peace had become near impossible. High levels of killings were, moreover, politically damaging – as were suicide-squad attacks on symbols of Indian control of Jammu and Kashmir.

All of this might, however, have been tolerated, had it not been for one critical fact. The Lashkar-e-Taiba's Abdul Rahman Makki had meant what he said in February 2000: the jihad in Jammu and Kashmir had begun to expand beyond its traditional geographical limitations.

Crossing the threshold

"The real war", the Lashkar-e-Taiba's Saeed had asserted after the end of the Kargil war, "will be inside [India]". He promised to "unfurl the Islamic flag on the Red Fort".[63] It was a promise of great emotive significance: a manifesto, as it were, for the next phase of the nuclear jihad.

Flying the flag of Islam on the Red Fort in New Delhi has been a long-standing motif in Islamist discourse, as old as Partition itself. In my discussion of the first phase of the Jammu and Kashmir jihad, I noted that as early as April 1948, Kasim Rizvi, an Islamist militia leader who sought to fight off Indian forces that had entered the state of Hyderabad, proclaimed that this was his ultimate objective.[64] Pakistan's history of support to such groups, described earlier in this book, has a pedigree of precisely the same length. One remarkably candid admission has come from Lieutenant-General Gul Hasan Khan, who served as commander-in-chief of the Pakistani armed forces. General Khan's memoirs record that an unnamed "elder statesman" in Pakistan organized the covert supply of weapons to the princely state of Hyderabad in 1948, which was using armed force to resist accession to the Indian Union. According to General Khan, the "elder statesman" organized at least one shipment of .22 pistols on a DC-3 aircraft.[65] Fantasies of conquering India remained strong as the decades went by: having arrived in New Delhi in 1994, the jihadist leader Syed Omar Sheikh was to record in his prison diaries, he had surveyed the city from the point of view of "a future conqueror, as I fondly imagined myself to be".[66]

India's first major terrorist group of the Islamic Right was, however, born neither in Srinagar nor in Hyderabad, not in Karachi nor in Lahore. In 1985, activists of the Jamaat Ahl-e-Hadis' Gorba faction gathered in the western-Indian town of Bhiwandi to speak about the need for armed Muslim resistance to the wave of communal violence India had seen since early that year. Two key figures were present at that meeting: Azam Ghauri, who went on to form a Lashkar-e-Taiba-based unit in Andhra Pradesh and Abdul Karim "Tunda," nicknamed for his deformed arm, who was to go on to become the Lashkar's top operative in India. At the end of the meeting, they formed the Tanzim Islahul Muslimeen [Organization for the Correction of Muslims], committed to the defence of Muslims during communal riots. The Tanzim's early activities were mildly farcical, consisting of self-defence drills using bamboo poles and ideological classes, both practices borrowed from an ultra-Right Hindu organization, the Rashtriya Swayamsevak Sangh.[67] Among their most enthusiastic recruits was Jalees Ansari, the son of a textile mill worker who went on to become a medical doctor – and to help set off a series of 43 explosions in Mumbai and Hyderabad and 7 separate explosions on trains on December 6, 1993, the first anniversary of the Babri Masjid's demolition by Hindu fundamentalists.

Ansari had been tasked to execute a second series of explosions on January 26, 1994, 13 days after his arrest. By the time India's federal police organization, the Central Bureau of Investigations, picked him up, however, both Karim and

Ghauri had disappeared. Karim is believed to have travelled to Calcutta and then to Dacca, where he again made contact with the Lashkar-e-Taiba network. The Lashkar-e-Taiba commander then responsible for its Indian operations outside of Jammu and Kashmir, Zaki-ur-Rahman Lakhvi, put him to work running new recruits from the north-Indian Muslim community, like Amir Hashim, who went on to execute a series of bomb attacks in New Delhi, Rohtak and Jalandhar. Ghauri, in turn, first hid out in Andhra Pradesh, and then travelled on a fake passport to Saudi Arabia. In 1995, Saudi national Hamid Bahajib, a key financier of the Lashkar's India activities, arranged for Ghauri to travel to Pakistan. He later returned to Hyderabad, and before he was killed in a shootout with the state police, carried out a series of bombings and assassinations in and around the city.

Pakistan's intelligence services were well poised to take advantage of the growing, if marginal, influence of jihadists among young Muslims across India. Earlier in this book, I have described Operation K2, an abortive efforts to set up a pan-India alliance between Khalistan terrorists and their counterparts in Jammu and Kashmir. Despite the failure of K2, efforts to forge these kinds of alliances proceeded on several different fronts. In January 1994, Mohammad Masood Azhar, who went on to found the Jaish-e-Mohammad in the wake of his release from prison as part of the Indian Airlines hostages-for-prisoners swap of 1999, was dispatched to India. His task was to bring about a reconciliation between the fractious cadre of the Harkat-ul-Mujahideen and the Harkat-ul-Jihad Islami, whose parent organizations had merged to form the Harkat-ul-Ansar. At this time, Azhar described the ideological content of his mission in location-specific terms. The organization's main objective, he told his interrogators, was "to liberate Kashmir from Indian rule, and to establish Islamic rule in Kashmir".[68] Before leaving for Srinagar, however, he spent considerable time attempting to network with ultra-conservative theologians in the northern Indian province of Uttar Pradesh. In the course of three days, he travelled between half a dozen cities, covering hundreds of kilometres. He sought and, in some cases, secured meetings with several important clerics of the ultra-conservative Deoband seminary.[69]

Despite the failure of K2 and the arrest of Azhar, the fallout of the demolition of the Babri Masjid seems to have encouraged Pakistan's intelligence services to renew its efforts at forging pan-India alliances. By the end of 1994, the ISI had managed to form the Jammu and Kashmir Islamic Front [JKIF], a body unique at the time for having no affiliation with any secessionist political organization within Jammu and Kashmir. It was believed to have attracted considerable funding from Saudi Arabia-based religious organizations, and drew ideological inspiration from the circle of the revanchist preacher and deputy head of the Lashkar-e-Taiba, Maulana Abdul Rahman Makki. The JKIF's leadership, Sajjad Ahmad Keno, Hilal Ahmad Baig, Bilal Ahmad Baig and Javed Ahmad Krava, were drawn from the Students' Liberation Front, which had broken from the ranks of the JKLF in the early 1990s. Its task was to work together with the mafia figures who had carried out the Mumbai serial bombings of 1993, executed as retaliation for a Hindu fundamentalist pogrom against

Muslims earlier that year.[70] In 1995, the JKIF released a photograph of one of the key accused in the serial bombings, Abdul Razzak "Tiger" Memon, along with Keno. The photograph, it was then claimed, had been taken in Srinagar. One of the participants in the affair, Usman Majid, has since confirmed the long-standing speculation that it was in fact taken at an ISI-run safehouse in Muzaffarabad, Pakistan.[71] Among the JKIF's more murderous acts was a bombing of the busy Lajpat Nagar market in New Delhi in 1996, which claimed a dozen lives.

Although, the JKIF was in near-terminal demise by 1998, the ideas it was founded on continued to flourish. That summer, the Jammu and Kashmir Police's Special Operations Group [SOG] eliminated the Hizb-ul-Mujahideen's top Kashmir valley commander, Ali Mohammad Dar. Better known by his *nom de guerre* Burhanuddin Hijazi, Dar was among the organization's best strategic minds. Dozens of pages of hand-written notes were recovered from Dar's temporary Srinagar hideout. Page 66 in one of the Dar diaries suggests new courses of action on an all-India basis. "Ways and means should be found", it records, "to launch the movement in India on [a] priority basis".[72] This can be achieved by "above all, a system of launching and logistics working to push through in a better way". To do this, he suggests a broad linkage with criminal organizations elsewhere in the country. "Kingpins of the underworld [should] be contacted", Dar advocated, "to have the weapons and ammunition launched for us through other possible ways". "A cell of three persons" would work "to develop relations with underworld beings [*sic*] like Dawood Ibrahim and trying [*sic*] to have a project of counterfeit currency". A year earlier, in December 1998, the Hizb-ul-Mujahideen had promised to take the "war against India outside Jammu and Kashmir", and threatened a "move towards Delhi". In addition, the December 1998 issue of *Majallah al-Dawa*, the in-house magazine of the Lashkar's political and financial patron, the Markaz Dawa wal'Irshad, reported the organization's belief that its campaign in Jammu and Kashmir was "just the beginning" and described its plans to extend its activities through India.

Pakistani nationals came to play an increasingly direct role in these activities. Ghauri's elimination was preceded, in July 1998, by the arrest in Hyderabad of top Lashkar activist Mohammad Salim Junaid, a resident of Kala Gujran village in Pakistan's Jhelum district. Junaid had begun his career with the Lashkar-e-Taiba in 1991 as a foot soldier for the jihad in Jammu and Kashmir, rising rapidly through the organization's hierarchy as a protégé of Azam Cheema, who was in charge of the trans-border movements of Lashkar-e-Taiba's operatives. Wasim Akbar, shot dead by the Jammu and Kashmir Police in 2001, is believed to have been responsible for a bomb explosion in Jalandhar, Punjab. In May 1998, another key Lashkar-e-Taiba member active in Uttar Pradesh, known only by his alias Abu Talha, was killed in an encounter with the SOG in Srinagar. Then, on July 30, 1998, the Delhi Police arrested three other members of the "Tunda" cell, who were led by Abdul Sattar, a resident of Islamnagar in Pakistan's Faislabad district. With his colleagues Shoaib Alam and Mohammad Faisal Hussain, Sattar had put together a base in the famous pottery town of Khurja, in the north Indian state of Uttar Pradesh. The group had built a bunker under a pottery kiln for the storage of explosives. There is considerable evidence that groups like the Lashkar-e-Taiba have been able to set up a wide pan-India support network through which operatives manage to obtain cover identities. Junaid, for example, had married a Hyderabadi woman and set up a spare-parts export enterprise. Similarly, Lashkar operative Zahid Hussain, tried to set up a business after being tasked to develop bases outside Jammu and Kashmir.

All of this had been seen both in India and Pakistan as part of the business of jihad as usual. While Pakistan's covert services had managed to extend the reach of the jihad in Jammu and Kashmir beyond the state, they had not yet posed a serious threat to India. Now, however, the landscape was to be irrevocably transfigured. In a December 1999 interview, the Harkat-ul-Jihad Islami *amir* Maulana Fazl-ur-Rahman Khalil threatened that if India did not immediately withdraw from territories claimed by Pakistan, "all of its states will become Kashmir".[73] The all-India jihad was nowhere near that point in December 2001, but New Delhi had no intention of allowing matters to drift until it was.

Towards the near-war

It takes little effort to see that the new wave of pan-India terrorism that broke out after the Kargil war was of a fundamentally different order than the kinds of relatively low-level terrorist activity jihadist groups had engaged in prior to the Kargil war. The political impact of Lashkar-e-Taiba's December 2000 attack on the Red Fort in New Delhi far transcended the damage it caused. Coming just two days after the Government of India announced an extension of the Ramzan ceasefire, the outrage sent out obvious messages. Both to the ruling BJP's core constituency amongst the Hindu right and to voters at large, the attack signalled that the Government of India had in fundamental ways failed to protect Indian sovereignty, notwithstanding its claims of triumph in the Kargil war. In an interview with the Pakistani newspaper *Ausaf*, Hafiz Muhammad Saeed announced that several similar attacks would follow.[74]

On December 13, 2001, a terrorist group stormed India's Parliament building in New Delhi sparking off a massive mobilization of its troops and a stand-off that, on more than one occasion, threatened to escalate into war. I shall deal with this crisis, the role of the tragic events of September 11, 2001, in preparing the ground for it, as well as the consequences of these in the next, concluding chapter of this book. However, some questions need to be addressed at this stage. The first of these is why Pakistan's intelligence services believed that they could escalate the jihad to new levels without risking an Indian response. Kargil had, after all, established that Pakistan's nuclear umbrella was not leak-proof – that, if it were pushed hard enough, New Delhi would still consider a full-scale war. Given the inherent fragility of the coalition government in India, the pressures on New Delhi to respond were all the more enormous: being seen as weak could have cost the BJP not just long-term electoral support, but also the backing of its sometimes fickle partners in the National Democratic Alliance. It is possible, of course, that Pakistani strategists simply did not expect the kind of reaction that followed. India, after all, had not threatened war after the horrific Mumbai serial bombings of 1993. Yet, a nuanced understanding of history ought to have led Pakistan to consider the consequences of its actions with greater care. India had almost gone to war in 1987 as a response to Pakistani support of Khalistan terrorists, and had come very near to doing so again in 1990, after the outbreak of widespread violence in Jammu and Kashmir.

Why then did Pakistan act as it did? Confronted with the prospect of losing control of the string that flew the Hizb-ul-Mujahideen kite, it is possible that Pakistani strategists simply did not game the possible long-term consequences of escalating the nuclear jihad. Among their principal problems was that Majid Dar proved remarkably resistant to calls from the Hizb-ul-Mujahideen's central command to terminate the dialogue process. By July 2001, the Hizb-ul-Mujahideen commander Mohammad Yusuf Shah was sufficiently alarmed by the way events were heading to shake up the organization's field command. He ordered Majid Dar and his associates back to Pakistan, a demand they unceremoniously rejected. Shah's chosen replacement for Majid Dar, the portly 54-year old Ghulam Hassan Khan, who also used the code-names Saif-ul-Islam and Engineer Zamaan, arrived in Jammu and Kashmir in October 2001. By then, the stage had been set for the bitter internecine warfare the Hizb-ul-Mujahideen had tried so hard to avoid. On August 25, 2001, the Baramulla-based pro-Shah division commander Shaqir Ghaznavi had organized the assassination of Dar's key aide, Farooq Sheikh Mirchal. Soon after Khan's arrival, Indian intelligence succeeded in cracking *hawala* funds transfer to several of the new commanders, strangling the resources they needed to establish their authority. Among the first of the seizures were funds intended for Mirchal's successor as Kupwara division head, Javed Ahmad Rather, code-named Zubair-ul-Islam. While no evidence exists on who the informer was, the pro-Shah faction of the Hizb-ul-Mujahideen made the obvious connection. From the outset, Ghulam Hassan Khan lived in fear of betrayal – and would, indeed, eventually be killed in a targeted operation carried out by Indian forces in 2004.

Within Jammu and Kashmir, then, the Hizb-ul-Mujahideen's factions were well and truly at war. Soon, they would have the express support of the centrists in the Hurriyat. In the midst of the crisis within the Hizb-ul-Mujahideen, Lone visited Pakistan to attend the marriage of his elder son with the daughter of the JKLF leader, Amanullah Khan. During a meeting with President Pervez Musharraf in late 2004, the People's Conference leader made clear his support for the ceasefire and for bilateral dialogue with India. In a subsequent interview with *The Washington Post*, Lone said that "the biggest danger now is from the [Islamist] extremists".[75] The far right, he said, "will make serious efforts to undermine the ceasefire". To prevent that outcome, the Union Government offered the Hurriyat realists the opportunity to visit Pakistan to consult with leaders there. The sole condition was that the team should not include Geelani.

While the visit did not materialize, Geelani found himself isolated within the Hurriyat on the issue. Lone, who travelled alone to Pakistan, was among his most bitter critics. "On the one hand", Lone said of the Hurriyat's demand for passports to travel to Pakistan, "we ask for a legal right that stands denied to us. But in the same breath we say that allow us to go to Pakistan, and when we will reach there, we will tell the mujahideen to sharpen their weapons against India. I see no logic in it."

Geelani responded to his marginalization in the Hurriyat executive by mobilizing the Islamist right on the streets. With the support of terrorist groups, he gained no small success. Bhat's enthusiasm for dialogue had dulled considerably after a near-successful February 22, 2001 attempt on his life. Lone had led a stubborn rearguard action, hoping to push the Hurriyat to begin dialogue with the Union Government mediator, K.C. Pant. Terrorist threats, again, ensured he was unable to succeed. The General Council of the Hurriyat rejected the realists' calls after a grenade went off during the meeting called to discuss the issue. At the 2001 remembrance of the assassination of Umar Farooq's father, Mirwaiz Mohammad Farooq's death, armed men gathered around the rostrum shouted Lone down. "*Haath mein haath do, Lashkar ko saath do*" [walk hand in hand with the Lashkar-e-Taiba], went the slogans "*Hurriyat mein rahna hoga to* Pakistan *kehna hoga*" [those who want to stay in the Hurriyat must support Pakistan]. Lone, however, refused to cave in. In mid-April 2001, he and Umar Farooq, now the only two vocal realists in the Hurriyat, were quietly granted permission to travel to Sharjah to hold an extended meeting with Sardar Abdul Qayoom Khan, the head of the Jammu and Kashmir Committee set up by Pakistan's military ruler, President Pervez Musharraf. The meeting was the first in several years between major political figures from both sides of the LoC in Jammu and Kashmir. Pakistan's ISI chief, Ehtaz-ul-Haq, is also believed to have been present on the sidelines of that meeting.

Lone offered little insight into what had been discussed with Khan during the April 17 meeting. He did, however, reiterate his commitment to dialogue. "We will go back and take the ideas we discussed here to our respective governments so that violence can end", he said.[76] "If the [Indian] government is not ready to allow self–determination", Lone continued, "the alternative is that they should be ready to settle the dispute through a meaningful dialogue involving all parties concerned". This in itself was of a piece with stated Hurriyat policy. What was significant, however, was that Lone did not join Khan in attacking India's human rights record in Jammu and Kashmir. Instead, he demanded that jihadi groups "leave us alone", as they were defaming the freedom movement. Meanwhile, Geelani again came under fire from within his own party, the Jamaat-e-Islami, which passed a resolution supporting the "conciliatory stance adopted by Umar Farooq and Abdul Gani Lone".[77] Hospitalized for cancer treatment, the Islamist hardliner found his position challenged by Khaliq Hanif, a one-time ally of the rejectionists. Hanif succeeded in pushing through an unprecedented political resolution in which the Jamaat-e-Islami stated that it would not oppose the coming elections to the Jammu and Kashmir Assembly. The resolution added that should the Hurriyat choose to oppose the elections, the Jamaat-e-Islami, as its largest constituent, would oppose the decision. From the optic of both the Islamists, a break in secessionist ranks had become inevitable. Accommodation between the two groups – and their representatives among the jihadists – was simply no longer possible.

A few column-centimetres of newsprint provoked the final showdown. On May 1, 2002, the Srinagar newspaper *Greater Kashmir* carried an article authored by Hizb-ul-Mujahideen deputy commander-in-chief Abdul Ahmad Bhat, who uses the code-names Moin-ul-Islam and Umar Javed. Bhat stated that if "today India begins a genuine process of settlement and peace, we will not wait till tomorrow. We will give up our defensive [military] operation right now".[78] The Hizb-ul-Mujahideen deputy chief added that if "India takes an initiative with good intentions, she will find us ten steps ahead of her one step. We will at once give up guns and observe real ceasefire so that [a] solution-finding path receives a headway [*sic*]." This was widely interpreted, correctly or otherwise, as an endorsement of efforts by the Indian Prime Minister's Office to bring a coalition of secessionist groupings into the electoral process. Shah, who had appointed Bhat to contain just these kinds of ideas, was infuriated. The expulsions of Dar, his second-in-command Khurshid Ahmad Zargar and their associate, central division commander Zafar Abdul Fateh, followed the day after the article appeared in print. Other mid-level commanders who backed Dar were also removed after they protested the decision; Abdul Ahmad Bhat himself escaped the axe by claiming the article was a hoax.

The wages of the peace process

"Don't shoot", the Hizb-ul-Mujahideen commander Ghulam Rasool Dar had shouted out to journalists on August 3, 2000, just before meeting his Indian interlocutors, "my life is in danger". He was right: the Ramzan détente was to take a terrible toll in lives.

Lone was shot and killed by a jihadist hit-squad, just before the elections that were held in 2002. His death was to have a profound impact on political developments in the Hurriyat, but these would not unfold until well after the Ramzan Process ended. Abdul Majid Dar was killed the following year, not long after the elimination of his deputy, Farooq Mirchal. Neither killing was a surprise, for Shah had expelled the moderates from the Hizb-ul-Mujahideen in May 2002, the equivalent of a death sentence.[79] The doves, too, were to have their vengeance. Majid Dar's supporters were to stage a coup that would end in Shah losing much of his infrastructure and cadre in Muzaffarabad, thus effectively dividing the Hizb-ul-Mujahideen into two.[80] Ghulam Rasool Khan the Hizb-ul-Mujahideen commander sent into to replace Majid Dar was eliminated by Indian forces in April, 2003. His successor Ghulam Rasool Dar, went the same way in early January the next year, as did a number of second-rung commanders who had opposed the ceasefire, notably Shabbir Bhaduri of southern Kashmir. Few key participants in the process are now around to tell the tale. Of those who can, notably Dulat and Brajesh Mishra on the Indian side, Mohammad Yusuf Shah and Zargar in Pakistan-administered Jammu and Kashmir, and the Jamaat-e-Islami's Ghulam Mohammad Bhat in Indian-administered Jammu and Kashmir, none have chosen to do so, at least in public.

In July 2001, Prime Minister Vajpayee and President Musharraf would make an effort to put in place the missing third element of the Ramzan process: a means of providing Pakistan a good reason to go along with India's efforts to bring about an end to the violence in Jammu and Kashmir. Less then six months before the attack on Parliament House, the two leaders met in the city of Agra, hoping to hammer out the framework for an India–Pakistan rapprochement. Despite enormous expectations, the Agra summit, as it came to be known, ended in acrimony with both India and Pakistan accusing each other of unreasonable behaviour. As the commentator B. Raman has pointed out, the fundamental problem was that both sides came to the table with enormously different expectations.[81] Vajpayee hoped for formal assurances that cross-border terrorism would end and that this issue would form part of a future India–Pakistan dialogue on Jammu and Kashmir. Pakistan, on the other hand, believed that India had come to the table because the post-Kargil escalation of the jihad in Jammu and Kashmir had inflicted enough pain for it to seek a solution – and that it could therefore afford to demand concessions on Jammu and Kashmir *before* discussing an end to violence. It was, events would show, a fundamental miscalculation by Musharraf, although South Asia would be brought to the edge of war before it would be learned.

While those events shall be discussed in the next chapter, it is perhaps worth recording at this stage itself that events in Jammu and Kashmir after the end of the near-war of 2001–2002 would take the form that all the key participants in the Ramzan process may have hoped for. Violence declined, and for the first time since 1988–1989 it appeared that an abiding peace in the region was at least a prospect. It is worth considering, however, whether it was necessary for so long to have been taken for this to be realized; whether tens of thousands lives had to have been lost for the lessons of the nuclear jihad to become apparent. Students of nuclear deterrence have long been aware of what is known as the "stability–instability paradox," an idea first developed by the scholar Glenn Snyder. Broadly speaking, Snyder suggested that while the fear of escalation provoked by the possession of nuclear-weapons did bring about some stability between adversaries so armed, they could also create *incentives* for low intensity conflicts to be waged in peripheral areas so long as both sides respected certain thresholds.[82] Since the late 1980s, as Sumit Ganguly has noted, events in South Asia have illustrated the workings of the stability–instability paradox, with Pakistan using its possession of nuclear weapons – however rudimentary, untested or even non-existent these may have actually been at any stage – to shield itself from Indian retaliation while at once pursuing sub-conventional warfare.[83]

Yet, what Pakistani strategists appear not to have comprehended is that the stability–instability paradox also imposed critical restraints on their pursuit of covert war against India. If Pakistan succeeded in deterring a possible attack by India in response to its support of the Khalistan movement, it was in turn severely restricted by the risks of war. Pakistan was unable to go to war in 1990, which would have been a rational course given that India was pinned down in both Jammu and Kashmir and Punjab. It could not do so for fear that the conflict would escalate out of hand. During the course of the nuclear jihad, similarly, Pakistan could not supply jihadist groups the kinds of weapons that could have interdicted Indian air logistics in Jammu and Kashmir, or even posed a significant threat to the Pathankot–Srinagar highway. In 1999, likewise, its responses to the Indian counter-offensive were severely restricted. Pakistan could not use its air assets to take on the jets bombing its troops on the Kargil heights, nor use the jihadist groups in Jammu and Kashmir to act in ways which might have posed serious threat to Indian logistics and communications lines.

Pakistani strategists had long seen the jihad in Jammu and Kashmir as a means to bleed India: a tactic expressed, at various points in its course and by various figures, through the mosquito–elephant metaphor. Yet, the fact is that there are limits to the damage a mosquito can inflict upon en elephant. As the historian Eric Hobsbawm has noted, history illustrates that warfare by irregulars only rarely poses a real threat to modern states.

> The test of a guerilla group comes when it sets itself such ambitious tasks as the overthrow of a political regime or the expulsion of a regular force of occupiers, and especially when it sets out to do this not in some remote corner of a country (the "liberated area") but over an entire national territory. Until the early twentieth century, hardly any guerilla movements faced this test; they operated in extremely inaccessible and marginal regions – mountain country is the commonest example – or opposed relatively primitive and inefficient governments, native or foreign. Guerilla actions have sometimes played an important part in major modern wars, either alone or in exceptionally favourable conditions, as with the Tyrolese against the French in 1809, or more usually, as ancillaries to regular forces – during the Napoleonic wars, for example, or in our century in Spain and Russia. *However, by themselves and for any length of time, they almost certainly had little more than nuisance value* . . . [emphasis added].[84]

We may never know if the ISI strategists who planned the nuclear jihad were students of history: had they been, they might have understood that the war had, in some key senses, been lost even as it began. What was true of the limitations of Tyrolese insurgents operating against Napoleon's forces, it had turned out, also applied to the nuclear jihad. In 1988, Jammu and Kashmir had been at peace and Pakistan had little role in shaping its destiny. India saw no reasons to make concessions, despite the occasional costs of war and the persistent irritant of the long-running jihad. In 2002, despite the loss of some 40,000 lives, a war and several near-wars, nothing had fundamentally changed. Unnoticed by contemporaries, as Indian and Pakistani troops began to mass along their frontiers after the December 13, 2001 attack on the Parliament House in New Delhi, the nuclear jihad had reached its limits. From here on, it would wind down – even though the road towards peace would prove riddled with both political potholes and jihadist land-mines.

8

TOWARDS PEACE?

> Wala ho baghawano, nav baharuk shaan paida kar,
> pholan gul gath karan bulbul, timay saman paida kar
> [Arise, Gardener, and bring on a glorious new spring;
> nightingales hover over roses, bring on a better time]:
> Prime Minister Atal Behari Vajpayee, quoting
> a couplet from the Kashmiri poet Ghulam Ahmad
> Mehjoor, at a rally in Srinagar on
> April 18, 2003.[1]

In the summer of 1989, or so goes the story, Major Mir emerged from the hollowed-out Chinar tree in the small village of Wathora that had been his home for years. "It is time for me to go to war again", the old Indian Army soldier-turned-mystic supposedly declared, "and I do not know whether or not I will return".

Major Mir, some in the village believed, had gone along with a team of his fellow Sufi mystics to do battle against the forces of the devil, whomever they were; other versions have it that Major Mir and the other mystics were the custodians of Kashmir's cultural essence, and will return with it when peace comes again. The stories about Major Mir were, most likely, fantasy; like all such tales, they spoke of both despair and hope. Until 2002, Wathora's small community of folk artists, who were among the last practitioners of the dying traditions of the *Bhaand Paather* dance–drama, were only rarely able to perform. Jihadist groups had deemed their satirical, sometimes bawdy, performances un-Islamic. On occasion, the Sufi shrines where the *Paather* were performed had been attacked.

Late in the summer of 2002, Wathora's *Bhaand Paather* performers resumed regular public performances, at shrines and folk festivals across central Kashmir. A full-blown India–Pakistan war, which had seemed almost unavoidable just months earlier, had been averted. Elections were held that autumn, and were widely hailed as among the fairest Jammu and Kashmir had ever seen. A new coalition government replaced the National Conference, marking a widening of democratic space; New Delhi also opened negotiations with the All Parties

Hurriyat Conference. Most importantly, violence declined steadily. By 2005, killings in Kashmir were at their lowest level since the late-1980s, and most observers concurred that infiltration across the LoC by terrorist groups had declined to a small fraction of their earlier levels. A vigorous India–Pakistan détente was in place, a bus service across the LoC had been launched, which allowed the first direct contact between individuals and families on the two sides of the state since the war of 1947–1948, and politicians were speaking of the prospect of a permanent resolution of the conflict in Kashmir.

To employ Prime Minister Vajpayee's symbolism, spring around the corner? Or, to borrow from the fable of Wathora, is Major Mir preparing to return?

I shall not seek to answer this question here, not least of all because pessimists have a depressing record of being correct in their predictions about the future of Kashmir. This concluding chapter instead provides a brief overview of the circumstances that brought about the decline of the nuclear jihad after 2002, and the challenges that still preclude securing an abiding peace in Jammu and Kashmir.

The 2001–2002 crisis

At 11:40 AM on December 13, 2001, a white Ambassador car – a brand of vehicle commonly used by Indian officials – pulled into the gates of India's Parliament building in New Delhi. It bore a red command light, as do many official vehicles, as well as what appeared from a distance to be a label authorizing entry into the complex. Security guards, well aware of the ire of senior officials who are stopped and searched, allowed the vehicle to pass without even a cursory examination. From the point of view of the five terrorists inside the car, armed with assault rifles, grenades and explosives, things could not have gone better. But they then made a small mistake. The excited driver rammed into the back of a vehicle used by India's then Vice-President Krishan Kant. Its driver protested, and a police guard ordered the white Ambassador to back away.

All five terrorists were killed in the shootout that ensued, along with nine security personnel. But the message was not lost on India's political class, many of whom would have lost their lives before lunchtime that day if fortune had not favoured them. Indian intelligence personnel managed to identify two Pakistan-based jihadist groups as the perpetrators of the attack. Hafiz Mohammad Saeed's Lashkar-e-Taiba, which operated with considerable Pakistani official patronage, and the Jaish-e-Mohammad, founded by Maulana Masood Azhar, who had been released from an Indian jail in a prisoners-for-hostages swap after an Indian Airlines plane had been hijacked in 1999, were held responsible for the strike. Both organizations had carried out dozens of major bombings in Kashmir; the Jaish-e-Mohammad, notably, had successfully stormed the Jammu and Kashmir Legislature building in October 2001, killing several people. Within a day, Indian officials articulated an explicit set of demands for Pakistan: they wanted both organizations proscribed, an end to infiltration across the LoC and the extradition of 20 terror suspects believed to be harbored in Pakistan.

India had made such demands several times in the past, to little effect. This time, however, New Delhi put military muscle behind its demands. India moved its offensive formations to assault positions along its 2,200-kilometre frontier with Pakistan and ordered its air force to prepare for strikes. Prime Minister Vajpayee made clear that all options, including war, were being considered to deter Pakistan's continuing support for jihadist groups.[2] Union Defence Minister George Fernandes went one step further, hinting that India had prepared its nuclear assets for retaliatory use in the event of a Pakistani first-strike.[3] Pakistan scrambled to catch up with the Indian build-up, but by all accounts there was considerable panic in Islamabad, where strategists were well aware that they were in no position to match their adversary's military resources. Although Islamabad, under pressure from the United States, detained some low-level Lashkar-e-Taiba and Jaish-e-Mohammad operatives, this was too little to satisfy the Indians, who continued their build-up apace. Although Prime Minister Vajpayee and General Pervez Musharraf met at a regional summit in January, the atmosphere was frosty, and no measures towards de-escalation took place.

By mid-January, India appeared firm in its resolve to risk a full-scale war unless its demands were met, whatever the consequences. Discussing the prospect of a Pakistani first-strike with journalists on January 11, 2002, India's army chief, General S. Padmanabhan, bluntly asserted that "if anyone uses nuclear weapons against India, Indian forces, Indian assets at sea, Indian economic or human interests, the perpetrators of that particular outrage will be punished so severely that their continuation in any fray will be in doubt".[4] Padmanabhan's uncharacteristically aggressive tone was interpreted, correctly, as a warning to Pakistan of the perils of responding to an Indian conventional offensive with nuclear means. Conscious of the enormous risks that this stand-off posed, the United States began to pressure Pakistan to act against the Lashkar-e-Taiba and Jaish-e-Mohammad. A day after General Padmanabhan's remarks, General Musharraf appeared on Pakistani television, promising to prevent Pakistani soil for use in terrorist activity against India – more or less the position that India had sought at the Agra Summit in July 2001, without success. Both Masood Azhar and Saeed had earlier been detained by Pakistani authorities, along with some low-level organizational functionaries.

In response, India chose to keep up its coercive pressure on Pakistan, hoping to secure further concessions. By mid-summer, however, that strategy appeared to have backfired. On May 14, 2002, the Lashkar-e-Taiba launched an attack on a military barracks outside the town of Jammu, killing 33 people, most of them the wives and children of military personnel deployed elsewhere. A week later, the APHC leader Abdul Gani Lone, as we have seen in the previous chapter was a key advocate of direct talks with New Delhi, was assassinated. Lone's son, Sajjad Lone, blamed the attack on his father's Islamist opponents in general, and the rejectionist leader Sayyid Ali Shah Geelani in particular.[5] New Delhi, for its part, renewed its war threats in response to the strike in Jammu. Several squadrons of Indian Air Force aircraft were moved to forward bases, while the

Indian Navy moved warships, including its aircraft carrier, into positions from which the Pakistani port of Karachi could be blockaded. General Musharraf responded to this renewed build-up with bellicosity, charging New Delhi with terrorism against India's religious minorities, and by threatening to withdraw Pakistani forces fighting the Taliban in Afghanistan for use along its eastern border.

By June 2002, however, the crisis had passed. One decisive factor in the de-escalation was the United States' pressure on Pakistan. At the end of May 2002, President George W. Bush publicly upbraided Musharraf, demanding that he "must stop incursions across the Line of Control".[6] "He said he would do so", Bush continued, "We and others are making it clear to him that he must live up to his word". Although violence within Kashmir had yet to register a significant decline, it was clear by the summer of 2002 that killings were indeed somewhat lower than in the previous year. Since then, that trend has continued: violence in Kashmir has fallen each year since 2002. The numbers of fatalities of both civilians and combatants, as well as the overall numbers of violent acts carried out by jihadist groups, have fallen to levels similar to those seen before the Kargil war of 1999, which had started an escalatory cycle. More important, the numbers of foreign terrorists, mainly Pakistani nationals, killed by Indian forces has dropped to low levels. It is hard to reliably quantify the level of infiltration by jihadi cadre across the LoC with any real accuracy, but the empirically demonstrated fact that fewer numbers are being killed seems to suggest that General Musharraf has at least part-delivered on his 2002 promise to end cross-border terrorism.

The death of the nuclear Jihad?

It is this last outcome of the 2001–2002 stand-off that distinguishes it from the near-wars and crises that both preceded and punctuated the nuclear jihad. At the end of Operation Brasstacks, Pakistan escalated support for Khalistan terrorist groups, and succeeded in significantly ratcheting up the intensity of sub-conventional war in the Indian state of Punjab. While the Indian war threat of 1990 seems to have succeeded in establishing some important red lines for the kinds of assistance and support that Pakistan could and could not provide to the armies of the jihad in Jammu and Kashmir, its end again marked the beginning of another upward phase in the escalatory cycle. Similarly, Pakistan's defeat in the 1999 Kargil war did not deter it from actually *increasing* its support to the jihadist forces operating in Kashmir. In stark contrast, despite apparently calling India's war-bluff during the 2001–2002 stand-off, Pakistan has this time around chosen to back down.

All of which, of course, begs the obvious question: *why*? On the face of it, the 2001–2002 stand-off ought to have persuaded Pakistan of the robustness of its nuclear umbrella. After all, India did not go to war despite the considerable provocation presented by attack on Parliament House because it feared events might spiral out of hand. "The fear of Pakistan's resort to a possible nuclear threat", Sumit Ganguly and Devin Hagerty have argued, "was paramount in the minds of Indian decision-makers, thereby inhibiting a resort to all-out war".[7] If so, and there is no evidence so far that challenges their assessment, why did Pakistan de-escalate the nuclear jihad? In General Pervez Musharraf's telling of events, the crisis was "a trial of who blinked first".[8] "India blinked", Musharraf has argued, after which Pakistan "decided to move our forces back". "A change of heart took place in Indian leadership", he claims, which enabled a détente to begin. Musharraf does not, however, address the fact of diminishing violence in Jammu and Kashmir, nor Pakistan's role in bringing it about.

One answer – and we are still far too close to recent events to do anything other than speculate – may lie in the dramatic fallout from the September 11, 2001, terrorist attacks on New York and Washington, DC. Pakistan, which had long backed the Taliban regime in Afghanistan, and was deeply intertwined with the jihadist networks of which that organization was just a part, faced intense pressure from the United States to sever these linkages and battle extremists operating within and outside its borders. Less than a week after the attacks, Pakistan attempted to persuade the Taliban to hand over the al-Qaeda leader Osama bin-Laden, using the services among others of the Director-General of the ISI, Lieutenant-General Mahmood Ahmad, and Nizamuddin Shamzai, an influential Karachi cleric with deep links both to jihadist organizations like the Jaish-e-Mohammad and to the Taliban's supreme leader, Mullah Mohammad Omar.[9] Shamzai's inclusion in the enterprise was of particular significance, since he had taught Omar at the Binori Town seminary in Karachi, an ideological powerhouse of the Islamist right. The effort, however, failed. Musharraf now had to choose between the Islamists within and outside of the Pakistan Army, who had backed his rise to power, and his nation's historic superpower patron.

Days after the collapse of the effort to broker a deal with the Taliban, Pakistan allied itself with the United States, acceding to demands for the use of its soil for a war in Afghanistan and a volte-face in Pakistani foreign policy. Over the coming months, Islamists in the Army who had been among Musharraf's closest allies, notably Lieutenant-General Mahmood Ahmad and Lieutenant-General Muzaffar Usmani, were eased out of office. Musharraf's own account of the dilemma is instructive:

> I confronted acute challenges on one side but also saw great opportunities on the other. I decided on the route of opportunities. I had to absorb external pressure and mould domestic opinion towards my decision. It was a tough decision to side with the US and the Coalition to fight terrorism. Domestic opinion was divided but I saw that the vast majority of moderates were behind me, while the religious extremists got violently against [*sic*]. We surmounted domestic pressures with courage and perseverance. While the economy kept moving forward (with additional external assistance) the schism between the

> minority of extremists and vast majority moderates kept increasing. The Nation [*sic*] was faced with the challenges of sectarian and religious terrorism. We acquitted ourselves well and continue to do so. Our fight against Al Qaeda in our cities and mountains is leading towards success. Today we face the uphill task of bringing about a societal renaissance by ridding Pakistan of the dominance of an extremist minority and establishing the unequivocal authority of the silent, dormant, vast moderate majority.[10]

If, for the United States, action against jihadist groups in Pakistan was a logical extension of its new global threat-perception, the new situation also represented a significant opportunity for India. In the wake of the Parliament House attack, India was able to argue that if the United States could use force to punish acts of terrorism, it could scarcely criticize others for doing precisely the same thing. United States pressure on Musharraf may thus have been the outcome of its own pragmatic self-interest, and the need to protect itself from charges of supporting double standards by battling terrorism in Afghanistan but not in Pakistan. During the early months of the United States' war against the Taliban, many in India did make precisely that charge. In one interview, Prime Minister Atal Behari Vajpayee demanded that "Pakistan should understand that there can be no double standards on terrorism: it cannot fight terrorism to its West and sponsor it to its East".[11] "Its membership of the international coalition against terrorism", the Indian Prime Minister continued "cannot be used as a cover for terrorist activities directed against India". It takes little to see that the message was directed as much to Washington, DC, as to Islamabad.

Musharraf's responses to the pressures upon him were instructive. While he did indeed come down hard on jihadist groups which were Pakistan-centric such as the Sipah-e-Sahiba Pakistan or the Lashkar-e-Jhangvi, both of which were embroiled in sectarian warfare between members of the country's Sunni majority and Shia minority, his actions on groups spearheading the jihad in Jammu and Kashmir were more nuanced. Some jihadist groups escaped any form of government containment. Maulana Fazl-ur-Rahman Khalil, the head of the Harkat-ul-Mujahideen and one of the signatories of the edict bin-Laden had issued in 1998 calling for war against the United States and Israel, was untouched, as was Qari Saifullah Akhtar, the head of the Harkat-ul-Jihad Islami. While both the Lashkar-e-Taiba and Jaish-e-Mohammad were proscribed in the wake of the 2001–2002 crisis and their leadership arrested, their organizational structures and terrorist training camps were soon up and running again albeit under different names. The Jamaat-ud -Dawa, the post-crisis reincarnation of the Lashkar-e-Taiba's earlier patron, the Markaz Dawa wal'Irshad, was reported to have raised substantial funds through public events carried out on the build-up to the Eid festival in March 2004.[12]

What this suggests is an effort by Musharraf to separate the *state-authorized* jihad, as expressed in Jammu and Kashmir, from the jihad *hostile* to the state – a form most visibly manifested in the three attempts on his life that took place within ten days in December, 2004.[13] While the effort to seek alliances with elements of the religious right while opposing others may appear to be a self-defeating strategy, it is not an unusual one. Despite an abundance of historical experience, South Asian leaders have shown a deep reluctance to learn that animals reared to bite their enemies can also bite their masters. It is also possible that Pakistan's strategic establishment does not wish to altogether abandon its sole source of leverage on Jammu and Kashmir and the source of its ability to influence events in the Indian-administered part of the region. As Stephen Cohen has perceptively noted, within Pakistan's strategic establishment, "the dominant view is that Pakistan can continue to harass 'soft' India".[14] Indeed, the project of the jihad in Kashmir is not merely to inflict costs upon India. It is, instead, an enterprise of great ideological import to the national self-identity of Pakistan; not "an international issue as much as an extension of domestic politics and the remnants of a flawed partition".[15] It is little wonder that Musharraf's handling of jihadist groups in Jammu and Kashmir has been so ambiguous: to use Hassan Abbas's acidic phrase, he has proved to be "a master of half-measures".[16]

No final answers, then, are readily available as to why the outcome of the 2001–2002 crisis was so different to that of its predecessors. Whatever the reason, though, there is no disputing that even the state-authorized jihad has diminished in its scale and intensity. We can only speculate as to the reasons. Pressure from the United States, concerned that a war in south Asia could derail its campaign in Afghanistan and against al-Qaeda worldwide, may be the major restraining force on Pakistan. It is also possible that Pakistan's military establishment may itself have realized that even the *prospect* of war affects its economy more severely than that of India; that the enterprise of creating an economically vibrant and politically stable state at a time of intense internal challenges to its authority cannot proceed if dogged by external crisis as well. If one objective of the long jihad was to bleed India, to secure strategic parity by inflicting costs which would retard its economic growth, the tactic has proved counterproductive. Through 1980–1990, the decade prior to the initiation of the nuclear jihad, the growth of industry and the Gross Domestic Product in Pakistan exceeded that of India. From 1990 to 2000, that situation was reversed. In the decade after 1993, India's growth rate exceeded that of Pakistan in each year; in four of those years, it was twice as high as that of Pakistan. Where Pakistan ranked higher than India and Bangladesh in the United Nations Human Development Index of 1991, it was overtaken by both these countries by 2003 and along with Nepal is now among the only two non-African countries to be classified as a low-human development nation.[17] As the Indian security-affairs commentator B. Raman has pointed out:

> it needs to be underlined that when Pakistan started its proxy war India's foreign exchange reserves, for reasons not connected with Jammu and Kashmir, had touched rock bottom, forcing it to pledge its gold to

> borrow money to repay its debts. Today, despite the proxy war, India's reserves are US $ 40 billion plus, a figure which the Pakistanis cannot hope to see for many generations in their own country. The proxy war is hurting their economy. Not ours.[18]

Both possibilities are, of course, not mutually exclusive. The truth is that reading intentions that underpin the actions of politicians is at best a hazardous business; in situations such as this one, where there are considerable gaps in our knowledge, it is perilously akin to gazing at entrails. Nevertheless, one unequivocal fact stares us in the face: a de-escalation in violence, a *necessary* condition for peace in Kashmir, has taken place. The next question is whether this de-escalation is also a *sufficient* condition for peace.

Days of détente

To attempt what is of necessity a tentative answer to this question, we need to turn our attention to the political events that have emerged from the near-war of 2001–2002.

Late in the autumn of 2002, elections to the Legislative Assembly of Jammu and Kashmir were held for the second time since the beginning of the long jihad. Most observers had expected Chief Minister Farooq Abdullah and the National Conference to be re-elected for a second six-year term in office, but Jammu and Kashmir's voters overthrew conventional wisdom. It was a bloody election – 99 activists of political parties were killed by terrorists that year, a record level of carnage that exceeded even the figure registered in 1996, when the previous elections were held. Much of the terrorist violence was directed at the National Conference, with the Hizb-ul-Mujahideen and Lashkar-e-Taiba directing the electorate to vote against the ruling party. Nonetheless, unlike those held in 1996, the elections were widely deemed to have been free and fair. A coalition between the People's Democratic Party (PDP) and the Congress now took power in Kashmir, backed by several smaller partners including some independents and the Communist Party of India (Marxist).

Several features of the 2002 elections were of considerable significance. First, the election did not mark a complete defeat of the National Conference. In terms of its share of the popular vote, the party remained the largest in both the Kashmir valley and in the state as a whole, polling 28.18 per cent of the vote in all of the state's 87 seats together.[19] By contrast, the PDP took 14.64 per cent of the vote in the seats it contested, and just 9.28 per cent state-wide. For its part, the Congress won 24.24 per cent of the votes in the seats where it contested, and 24.24 per cent state-wide. It took the *combined* forces of the PDP and the Congress to dethrone the National Conference – a reversal of the alliances, it should be noted, of those which had underpinned the "stolen election" of 1987. Significantly, the PDP had self-consciously styled itself on the MUF, borrowing the flag of the Islamist alliance, and often drawing on its propaganda. Most of the new government's Jammu-region seats were won by the Congress, as they had been in 1987; in the Kashmir valley, however, the PDP rose as the principal challengers to National Conference hegemony.[20] As such, these elections marked a coming-together of different oppositional forces from the two regions. Had such an alliance been built prior to the 1987 elections, a MUF–Congress alliance might well have taken power in Jammu and Kashmir 15 years earlier – a lesson, if more were needed, of the centrality of coalition-building to electoral outcomes in India.

A second major feature of the 2002 election was the emergence of an unspoken tactical alliance between elements of the PDP and the Hizb-ul-Mujahideen, one that was directed at overthrowing the National Conference. While this covert alliance did have a formal architecture – Mehbooba Mufti, the daughter of the new Chief Minister, Mufti Mohammad Sayeed, was reported to have met senior Hizb-ul-Mujahideen commanders on several occasions – its working at the local level were of more interest.[21] On several occasions during and after the 2002 elections, elements in the PDP offered financial incentives or other forms of patronage to terrorist groups, in return for political backing and electoral support.[22] In one particularly embarrassing case, a senior PDP leader was found to have links, albeit somewhat remote ones, with a terrorist group involved in an attack on a Hindu temple in the state of Gujarat.[23] Similar alliances had existed between the National Conference and elements of the Hizb-ul-Mujahideen in the past, for example those involving Jammu and Kashmir's one-time Minister of State for Home Mushtaq Lone.[24]

Did these alliances mark an opening of political space? Perhaps so and perhaps not. As the historian Eric Hobsbawm has pointed out, such alliances are endemic in pre-capitalist societies, particularly where state authority has been undermined or otherwise diminished. An element of bandit-like behaviour had existed for a considerable length of time in the long jihad. Extortion was common, as was the use of violence to corner contracts.[25] The Hizb-ul-Mujahideen's Abdul Hamid Gada, for example, had cultivated a considerable Robin Hood reputation, as had the Lashkar-e-Taiba's Badgam-area commander, Inayatullah Khan, widely known by his *nom de guerre* Bilal-e-Habshi.[26] None of this was new, of course. Hayat Mir, the Pakistani covert operative who was affiliated to the Master Cell, had engaged in activities similar to those Inayatullah Khan would adopt several decades later, using social-bandit tactics to win legitimacy for the jihad. It was inevitable that local magnates, whatever their stated ideological affiliations, would seek linkages with jihadist groups operating in their areas of influence:

> The politics of areas ruled by pre-capitalist landowners turn on the rivalries and relationships of the leading landed families and their respective followers and clients. The power and influence of the head of such a family rests, in the final analysis, on the number of men to whom he is patron, offering protection and receiving in turn those services of loyalty

> and dependence which are the measure of his prestige, and consequently of his capacity to make alliances: fighting, voting or whatever else determines local power.[27]

What was new about the post-2002 situation was that these local alliances, a long-standing feature of Jammu and Kashmir's political landscape during the nuclear jihad, began to acquire a political significance that transcended their geographical limitations. In the context of the existing divisions between the Hizb-ul-Mujahideen moderates and the rejectionists, which were highlighted during the Ramzan Process, the alliances meant a sharpening of the fault-lines within anti-India political formations. The first shots in this war within the secessionists were fired shortly before the 2002 elections, when the assassinated centrist leader Abdul Gani Lone's party, the People's Conference, put up proxy candidates in defiance of the All Parties Hurriyat Conference line. Incensed by the Hurriyat's failure to act against the People's Conference, and by the moderates' general unwillingness to campaign against the 2002 elections, the Islamist leader Syed Ali Shah Geelani broke from the organization. In the process, however, he lost the support of his parent organization, the Jamaat-e-Islami, which largely backed the pro-dialogue Hurriyat leadership. Efforts to bring about reconciliation continued fitfully through 2004 and 2005, but to little avail.

In January 2004, the pro-dialogue faction of the All Parties Hurriyat Conference met India's Home Minister and Deputy Prime Minister L.K. Advani. It was the first formal meeting between a high level Indian political functionary and Kashmiri secessionists; the earlier round of engagement with the Hizb-ul-Mujahideen, it may be recalled, was conducted through bureaucrats. Little emerged from this meeting, but episodic movement towards further dialogue continued. In December 2003, a cease-fire went into place along the LoC, putting an end to the regular fire exchanges between India and Pakistani forces – and significantly reducing the level of cover available to jihadist groups engaged in infiltration. India was also able to use the cease-fire to complete construction of a counter-infiltration fence along the LoC. Although power shifted hands in New Delhi in 2004, when the National Democratic Alliance government led by Prime Minister Vajpayee was defeated, and a centre-left coalition brought Prime Minister Manmohan Singh to office, the dialogue process deepened. A succession of meetings between Manmohan Singh and President Pervez Musharraf took place through 2004 and 2005, punctuating a high-level official engagement between India and Pakistan.

By the summer of 2005, the results were there for all to see: despite continuing terrorist action, levels of violence had declined to an all-time low. A bus service had begun between Srinagar, in Indian-administered Jammu and Kashmir, and Muzaffarabad, on its Pakistan-controlled side. Instead of passports and visas, Jammu and Kashmir residents wishing to use this service required only permits, a gesture of significant symbolic concern since it met the sovereignty concerns of both countries. Using this bus service, Hurriyat leaders from the moderate faction travelled to Pakistan, capitalizing on the opportunity to revive their contacts with the jihadist leadership based there and to explore new avenues for political progress. Within Indian-administered Jammu and Kashmir, meanwhile, the wages of peace were becoming evident. Tourist inflows began to improve, and a number of new trade and economic-development projects were initiated. India–Pakistan détente also proceeded apace, with the countries engaging in dialogue on a number of secondary disputes, as well as on strengthening regional cooperation, notably through the construction of a natural gas pipeline ferrying fuel from Iran to India. For the first time since the beginning of the long jihad, peace seems to be a realistic prospect, rather than a pipe-dream.

Challenges ahead

Can an abiding peace in fact be put in place? It seems likely that three major issues will confront the dialogue process as it proceeds. The extent to which political leaders are able to negotiate their ways through the potential pitfalls will determine the degree of failure or the success of the ongoing détente.

First, it is unclear what the political forces arrayed against India in Jammu and Kashmir might stand to gain from a resolution of the conflict. The All Parties Hurriyat Conference emerged, after all, from the ruins of the MUF and, if the People's Conference's experience of the 2002 elections is any guide, there is little reason to believe its constituents have significantly expanded their constituency since then. It is possible, then, that the elements that together constitute the Hurriyat may actually find their role *diminished*, rather than enhanced, by movement towards a democratic resolution of the conflict in Jammu and Kashmir. Second, ever since the second phase of the nuclear jihad, the war in Jammu and Kashmir has been dominated by organizations with no real political stake in the state's future. Jihadist organizations like the Jaish-e-Mohammad and Lashkar-e-Taiba, with their belief in a millenarian war between Islam and its enemies, have little to gain from peace between India and Pakistan. Even the Hizb-ul-Mujahideen's *amir*, Mohammad Yusuf Shah, has described efforts to bring about a ceasefire as a "waste of time", and asserted that, after the détente process, the case for a "massive armed struggle has increased more than ever and the Kashmiri youth need to prepare themselves for fighting in maximum numbers".[28]

Finally, there is Pakistan itself: specifically, its unwillingness to terminate support for the jihad altogether. Terrorism can be distilled to two essential factors: the capabilities required to carry out an act of violence and the intention of executing one. While the substantial decline in violence in Jammu and Kashmir makes it clear that jihadi intentions have been worked on by Pakistan's covert services, there are no signs that their capabilities have been diminished. No leaders of major terrorist groups, including several proscribed by the United States, have been arrested. No operatives of significance have been prosecuted nor weapons seized. Indeed, there were even signs of expansion: the Lashkar-e-Taiba found, for example, to be funnelling some numbers of cadre to fight allied forces in Iraq, although the full scale of these activities is unclear, as is the question of whether they had the sanction of the organization's top leadership.[29] Indian commentators are not alone in noting that jihadist infrastructure remains active in Pakistan. The United States of Federal Bureau of Investigations, for example, arrested four individuals in 2005 who had trained in jihadist camps in 2003–2004, and then set up an al-Qaeda cell in San Francisco.[30] Afghan authorities have also complained bitterly that Pakistan continues to turn a blind eye to Taliban activities conducted from its soil, a charge that acquired urgency after a series of cross-border attacks in the summer of 2005 and the attempted assassination of the United States Ambassador to Kabul, Zalmay Khalilzad.[31]

Many good reasons exist for Musharraf's unwillingness to confront terrorism head-on, not the least of them being the prospect of a destabilizing pitched battle against the tens of thousands of trained cadre that Islamist groups have at their command. Against this, however, the risks of *any* level of terrorist violence, the more so where it involves the patronage of one nuclear-armed state in an adversarial relationship with another, must be considered. Terrorists armed with a gun ultimately decide when to use it – and Pakistan's failure to take away their capabilities could have serious strategic repercussions in the future. In 2005, responding to the lessons of the Kargil war and the 2001–2002 near war, India developed a new cold-start doctrine that envisaged the use of up to eight integrated battle groups capable of fighting short-duration limited wars with an intent to impose costs on Pakistan for its sub-conventional campaign in Jammu and Kashmir. As one thoughtful commentary by the Pakistan Air Force officer Khurshid Khan has noted, the new Indian doctrine envisages a situation where the "Indian Army could decisively degrade Pakistan's military potential without crossing its nuclear threshold and giving the international community the time or opportunity to intercede".[32]

Does all this mean there is nothing for India to do, other than pressure the world community for action against Pakistan or find military means with which to coerce its adversary? Most certainly not. From an early stage of the long jihad, as we have seen, Indian policy-makers variously acquiesced in and engineered appalling abuses of democracy. If the long jihad did not succeed in winning Jammu and Kashmir for India, it succeeded in pushing the Indian state to undermine those very institutions on which the legitimacy of its rule rested. Efforts to give democracy content and meaning are imperative, notably a broad-based dialogue where all political forces in Jammu and Kashmir may deliberate on the state's future. Second, India needs to finds means to assure Jammu and Kashmir's Muslims that their religious rights are secure in what is, after all, a Hindu-majority state. The periodic anti-Muslim pogroms that have scarred India's post-Independence history have done nothing to help India's case in Jammu and Kashmir. Ending the appalling record of impunity Hindu chauvinists implicated in this violence have enjoyed would, I believe, do not a little to mitigate the fears of ordinary people in Jammu and Kashmir. Pan-India communal conflict is more closely linked to Jammu and Kashmir than most people imagine. On a visit to New Delhi shortly after Independence, Sheikh Abdullah pointed to this fact, andtried to explain the psyche of Kashmiri Muslims.

> There isn't a single Muslim in Kapurthala, Alwar or Bharatpur. Some of these had been Muslim majority states. Try to symbiotically [*sic*] understand the Kashmiri Muslims. They are afraid that the same fate lies ahead for them as well.[33]

Yet, a resolution of the multiple crisis and conflicts which together constitute what we call the Kashmir conflict may be years or even decades in the future. Bringing an end to the long jihad offers the best prospect of ensuring that there at least remains a place where a conflict may one day be solved.

NOTES

1 A HOUSE ON A HILL

1 Cited in Prithvi Nath Kaul Bamzai, *Kashmir and Power Politics from Lake Success to Tashkent* (Metropolitan Book Company: New Delhi, 1966), 182.
2 Victoria Schofield, *Kashmir in Conflict: India, Pakistan and the Unending War* (London: I.B. Tauris, 2003).
3 Sumit Ganguly, *The Crisis in Kashmir: Portents of War, Prospects of Peace* (Cambridge and Washington, DC: Cambridge University Press and Woodrow Wilson Center Press, 1997).
4 Navnita Chadha Behera, *State, Identity and Violence: Jammu, Kashmir and Ladakh* (New Delhi: Manohar Books, 2000).
5 Muhammad Amir Rana [trans. Saba Ansari], *A-Z of Jehadi Organizations in Pakistan* (Lahore: Mashal Books, 2004), 83 (Operation Cyclone was the United States of America's covert programme of support for terrorist groups of the Islamist right, opposed to the Soviet Union's intervention in Afghanistan).
6 Chandrashekhar Dasgupta, *War and Diplomacy in Kashmir: 1947–1948* (New Delhi: Sage Publications, 2002), 102.
7 Sheikh Mohammad Abdullah [trans. Khushwant Singh], *Flames of the Chinar: An Autobiography* (New York: Viking, 1993).
8 Hassan Gardezi, "The Resurgence of Islam: Islamic Ideology and Encounters With Imperialism", in Hassan Gardezi and Jamil Rashid (eds), *Pakistan: The Roots of Dictatorship* (London: Zed Press, 1983), 359.
9 Kanchan L., "The Media in Terror: Analysing Reportage From Theatres of Conflict", Faultlines Volume 8, (New Delhi: Institute for Conflict Management, April 2001). Online at www.satp.org/satporgtp/publication/faultlines/volume8/Arti\cle2.htm.
10 Eric Hobsbawm, *Bandits* (New York: The New Press, 1999), 123.
11 Alan Clark, *Barbarossa: The Russian-German Conflict, 1941–1945* (London: Weidenfeld and Nicolson, 1965), 218, cited in Geoffrey Roberts, *Victory at Stalingrad* (Edinburgh and London: Pearson, 2002), 77.
12 Steven P. Cohen, *The Idea of Pakistan* (New Delhi: Oxford University Press, 2005), 193.
13 Hobsbawm, op. cit., 120–121.
14 Mridu Rai, *Hindu Rulers, Muslim Subjects: Islam, Rights and the History of Kashmir* (Princeton University Press: Princeton, 2004), 1.
15 C.A. Bayly, *Empire and Information: Intelligence Gathering and Social Communication in India, 1780–1870* (Cambridge: Cambridge University Press, 1999), 135–136.
16 Peter Hopkirk, *The Great Game: The Struggle for Empire in Central Asia* (Tokyo and New York: Kodansha International, 1994), 90.
17 Hopkirk, op. cit., 329–332. For a thorough – and thoroughly readable – account of the Pundits' path-breaking covert work in the eastern Hi'malayas, see Peter Hopkirk,

Trespassers on the Roof of the World: The Secret Exploration of Tibet (Tokyo and New York: Kodansha International, 1995).
18 Bayly, op. cit., 231–232.
19 Ibid., 140.
20 F.M. Hassnain, *Freedom Struggle in Kashmir* (New Delhi: Rima Publishing House, 1988), 15.
21 Rai, op. cit., 16–17.
22 Hari Om, *Mulism of Jammu & Kashmir: A Study of the Spread of Education and Consciousness, 1857–1925* (New Delhi: Archives Publishers, 1982), 48 and 107–108.
23 Cited in Hassnain, op. cit., 45.
24 Ibid., 58.
25 For an account of the rioting, see Devendra Swarup and Sushil Aggarwal, *The Roots of the Kashmir Problem: The Continuing Battle Between Secularism and Communal Separatism* (New Delhi: Manthan Prakashan, 1992), 72–74.
26 Hassnain, op. cit., 58.
27 Altaf Hussain, *Shabbir Shah: A Living Legend in Kashmir History*, (Srinagar: Noble Publishing House, not dated), 4.
28 Alistair Lamb, *Kashmir: A Disputed Legacy, 1856–1990* (Hertingfordbury: Roxford Books, 1991).
29 For example, see Prem Shankar Jha, *Kashmir, 1947: Rival Versions of History* (New Delhi: Oxford University Press, 1996).
30 Jack Weatherford, *Genghis Khan and the Making of the Modern World* (New York: Three Rivers Press, 2005), 28.

2 THE INFORMAL WAR

1 Prithvi Nath Kaul Bamzai, *Kashmir and Power Politics from Lake Success to Tashkent* (New Delhi: Metropolitan Book Company, 1966), 182.
2 Salman Rushdie, *Midnight's Children* (New York: Penguin, 1991).
3 Tariq Naqash, "Bus Service Thrills Divided Families", *The Dawn* (Karachi), April 5, 2005.
4 C. Dasgupta, *War and Diplomacy in Kashmir: 1947–1948* (New Delhi: Sage Publications, 2002), 39.
5 Victoria Schofield, *Kashmir in Conflict: India, Pakistan and the Unending War* (London: I.B. Tauris, 2003), 41.
6 P.L. Lakhanpal, *Essential Documents and Notes on Kashmir Dispute* (New Delhi: International Books, 1965), 50.
7 Ibid., 54.
8 Ibid., 56.
9 Lieutenant-General L.P. Sen, *Slender Was The Thread* (New Delhi: Orient Longman, 1994), 27–29.
10 Ibid., 30.
11 Shereen Ilahi, "The Radcliffe Boundary Commission and the Fate of Kashmir", *India Review*, Vol. 2, No. 1 (London: Frank Cass, January 2003), 96.
12 Akbar Khan, *Raiders in Kashmir* (Islamabad: National Book Foundation, 1970), 9.
13 ibid., 10.
14 Ibid., 9–10.
15 Ibid., 6.
16 Ibid., 14.
17 Ibid., 12.
18 Ibid., 19.
19 Cited in Bamzai, op. cit., 106.

20 Margaret Bourke-White, *Halfway to Freedom* (Mumbai: Asia Publishing House, 1950), 163.
21 Bamzai, op. cit., 104.
22 Sen, op. cit., 36.
23 Bourke-White, op. cit., 164.
24 Sen, op. cit., 103–104.
25 Bourke-White, op. cit., 164.
26 Dasgupta, op. cit., 48–49.
27 Akbar Khan, op. cit., 63.
28 Ibid., 68–69.
29 Dasgupta, op. cit., 84.
30 Ibid., 84–85.
31 Ibid., 109.
32 Ibid., 111.
33 Francis Tucker, *India's Partition and Human Debasement, Vol. 1* (New Delhi: Akashdeep Publishing House, 1988), 27.
34 Ibid., 26–27.
35 Dasgupta, op. cit., 130–132.
36 Sen, op. cit., 288.
37 Sen, op. cit., 296.
38 Akbar Khan, op. cit., 156.
39 Dasgupta, op. cit., 102.
40 [SECRET] *Report on Pakistani Organized Subversion, Sabotage and Infiltration in Jammu and Kashmir* (Jammu and Kashmir: Criminal Investigation Department, 1966), 16.
41 Sen, op. cit., 19.
42 Ibid., 19.
43 Ibid., 18.
44 Gul Hasan Khan, *Memoirs* (Karachi: Oxford Univeristy Press, 1993), 230–231.
45 Sen, op. cit., 11.
46 Associated Press of India dispatch of April 12, 1948, cited in Sen, op. cit., 13.
47 "Dated March 14, 1951: India will not be bullied," *The Hindu* (Chennai), March 14, 2001. Online at http://www.hinduonnet.com/thehindu/2001/03/14/stories/10141045.htm.
48 Gul Hasan Khan, op. cit., 89–93.
49 [SECRET] *Report on Pakistani Organized Subversion*, op. cit., 17–18.
50 Bamzai, op. cit., 220.
51 Sumit Ganguly, *The Crisis in Kashmir: Portents of War, Signs of Peace* (Cambridge: Cambridge University Press and the Woodrow Wilson Center Press, 1997), 39.
52 Cited in M.J. Akbar, *Kashmir: Behind the Vale* (New Delhi: Viking, 1991), 159.
53 Schofield, op. cit., 75–77.
54 Akbar, op. cit., 139.
55 Prem Nath Bazaz, *Democracy Through Intimidation and Terror* (New Delhi: Heritage, 1978), 15.
56 Cited in ibid., 16.
57 Akbar, op. cit., 138.
58 Bamzai, op. cit., 202.
59 Bamzai, op. cit., 202.
60 Schofield, op. cit., 77–79.
61 Cited in Bamzai, op. cit., 203.
62 *The New York Times*, July 5, 1953, cited in Bamzai, op. cit., 203.
63 Schofield (Schofield, op. cit., 79) has asserted that the National Conference victory was sweeping because the election was "boycotted by the Praja Parishad." The boycott was, in fact, to protest the rejection of its candidates' nomination papers.

64 Navnita Chadha Behera, *State, Identity and Violence: Jammu, Kashmir and Ladakh* (New Delhi: Manohar, 2000), 83.
65 Schofield, op. cit., 79.
66 Ibid., 92.
67 Ajit Bhattacharjee, *Kashmir: The Wounded Valley* (New Delhi: UBSPD, 1994), 192.
68 B.N. Mullik, *Chinese Betrayal: My Years With Nehru* (Bombay: Allied Publishers, 1971), 86.
69 Y.D. Gundevia, "On Sheikh Abdullah", in *The Testament of Sheikh Abdullah* [no author] (New Delhi: Palit and Palit, 1974), 111.
70 Bhattacharjee, op. cit., 192.
71 Gundevia, op. cit., 111.
72 Ibid., 117.
73 Bamzai, op. cit., 238–239.
74 Owen Bennett Jones, *Pakistan: Eye of the Storm* (New Haven and London: Yale University Press, 2003), 225.
75 The text of the Baghdad Pact is available online at http://www.yale.edu/lawweb/avalon/mideast/baghdad.htm.
76 Behera, op. cit., 108–109.
77 Balraj Puri, "Jammu and Kashmir", in Myron Weiner, ed., *State Politics in India* (Princeton: Princeton University Press, 1966), 228.
78 Joseph Korbel, *Danger in Kashmir* (Princeton: Princeton University Press, 1966), 245.
79 Behera, op. cit., 110.
80 Gundevia, op. cit., 117.
81 Schofield, op. cit., 95.
82 Akbar Khan, op. cit., 178.
83 Ibid., 179.
84 Ibid., 180.
85 Ibid., 180–181.
86 Ibid., 181.
87 [SECRET] *Report on Pakistani Organized Subversion*, op. cit., 18.
88 Ibid.
89 Ibid., 22.
90 Ibid., 161–164.
91 Ibid., 164.
92 Yoginder Sikand, "Changing Course of Kashmiri Struggle", *Economic and Political Weekly* (Mumbai), January 20, 2001. Online http://www.epw.org.in/showArticles.php?root=2001&leaf=01&filename=2.
93 Ibid. Sikand's quotations are from the head of the Markaz Dawa wal'Irshad (later re-named the Jamaat-ud-Dawa), Hafiz Muhammad Sayeed.
94 [SECRET] *Report on Pakistani Organized Subversion*, op. cit., 22–23.
95 [SECRET] Ibid., 19.
96 [SECRET] Ibid., 23–24.
97 Schofield, op. cit., 91.
98 K. Mubarak Shah (ed.), *The Kashmir Conspiracy Case Report 1* (Srinagar: Legal Defense Committee, 1964), 8.
99 Mullik, op. cit., 87.
100 Gundevia, op. cit., 118.
101 [SECRET] *Report on Pakistani Organized Subversion,* op. cit., 26.
102 Gundevia, op. cit., 117–118.
103 [SECRET] *Report on Pakistani Organized Subversion,* op. cit., 26.
104 Ibid., 57.
105 Ibid., 24.

106 Kh. Mubarak Shah (ed.), *The Kashmir Conspiracy Case Report VII* (Srinagar: Legal Defense Committee, 1964), 24–25.
107 Ibid., 37.
108 Karan Singh, *Autobiography* (New Delhi: Oxford University Press, 1994), 265.
109 Mullik, op. cit., 87.
110 Schofield, op. cit., 97.
111 Gundevia, op. cit., 134.
112 Behera, op. cit., 116.
113 Cited in Sheikh Mohammad Abdullah, *Flames of the Chinar*, translated from the Urdu by Khushwant Singh (New Delhi: Viking, 1960), 147.
114 Karan Singh, op. cit., 267–268.
115 Behera, op. cit., 116.
116 Bazaz, op. cit., 18.
117 Gundevia, op. cit., 118.
118 Contrary to the assertion of one scholar, Sheikh Abdullah was not actually exonerated of the charges against him, at least in a legal sense (Schofield, op. cit., 91). The state merely dropped the prosecution against him, leaving the question of his guilt unresolved. Apart from [SECRET] *Report on Pakistani Organized Subversion,* op. cit., 27–28, also see Gundevia, op. cit., 118.
119 Cited in Sheikh Mohammad Abdullah, op. cit., 144.
120 [SECRET] *Report on Pakistani Organized Subversion,* op. cit., 28.
121 Akbar Khan, op. cit., 52.
122 Ibid., 156.
123 Eric Hobsbawm, *Revolutionaries* (New York: The New Press, 2001), 198.
124 Akbar Khan, op. cit., 191–192.
125 Ibid., 191.
126 AG Noorani, *The Kashmir Question* (Bombay: Manaktalas, 1964), 60.
127 Cited in Ibid., 60.

3 THE MASTER CELL

1 [SECRET] *Report on Pakistani Organized Subversion, Sabotage and Infiltration in Jammu and Kashmir* (Jammu and Kashmir Government: Criminal Investigation Department, 1966). Appendix IV is a photograph of the original communication which was obtained by Indian intelligence, with a copy marked to Brigadier AAK Niazi, overall in command of 1 Sector during the 1965 India–Pakistan war. This is, to my knowledge, the only surviving original mimeograph of this document.
2 Ibid., 161–162.
3 Ibid., 50.
4 Ibid., 13.
5 Ibid., 50–51.
6 Ibid., 13–14.
7 [RESTRICTED] P.B. Sinha and A.A. Athale *History of the Conflict With China, 1962* (New Delhi: Ministry of Defence, 1962). XXXIV.
8 Victoria Schofield, *Kashmir in the Crossfire* (London: IB Tauris, 1996), 193–194.
9 Ibid., 197.
10 [SECRET] *Report on Pakistani Organized Subversion,* op. cit., 69.
11 Ibid., 69.
12 Ibid., 70.
13 Ibid., 71–72.
14 Y.D. Gundevia, "Sheikh Abdullah: A Monograph", in *The Testament of Sheikh Abdullah* [no author] (New Delhi: Palit and Palit, 1974), 123.

15 Sarvepalli Gopal, *Jawaharlal Nehru: A Biography* (New Delhi: Oxford University Press, 2004), 264.
16 Cited in Victoria Schofield, *Kashmir In Conflict: India, Pakistan and the Unending War* (Minneapolis and London: IB Tauris, 2003), 104–105.
17 Mohammad Ayub Khan, *Friends, Not Masters: A Political Autobiography* (London: Oxford University Press, 1967), 128.
18 Altaf Gauhar, *Ayub Khan: Pakistan's First Military Ruler* (Karachi: Oxford University Press, 1996), 265.
19 Gundevia says, speaking for India's foreign-affairs establishment, that the rejection had been "definitely expected by us" (Gundevia, op. cit., 130).
20 Stanley Wolpert, *Zulfi Bhutto of Pakistan: His Life and Times* (New York: Oxford University Press, 1993), 78.
21 Balraj Puri, *Jammu and Kashmir: Triumph and Tragedy of Indian Federalisation* (New Delhi: Sterling, 1983), 93.
22 A.G. Noorani, *The Kashmir Question* (Bombay: Manaktalas, 1964), 63.
23 Rauf Klasra, "Kashmiris 'misled' Ayub in 1965 war: Gohar", *The News* (Karachi), June 2, 2005. Online at http://www.jang-group.com/thenews/jun2005-daily/02-06-2005/main/main2.htm.
24 Gul Hassan Khan, *The Memoirs of Lt. Gen. Gul Hassan Khan* (Karachi: Oxford University Press, 1994), 182–183.
25 Ibid., 114.
26 Ibid., 178.
27 Ibid., 116.
28 [SECRET] *Report on Pakistani Organized Subversion*, op. cit., 81.
29 Ibid., 81.
30 Sarabhai's ardent advocacy of Sheikh Abdullah's cause incensed many in the Indian establishment. Gundevia, a friend of the prominent political figure, writes that "at least a fifth of the 'sins' of Sheikh Abdullah are the 'sins' of Mirdula Sarabhai; and although she has never been 'charged' with conspiracy and sedition, she has shared part of the honours with him of being imprisoned for several months without trial over the Kashmir issue". It is unclear, though interesting to contemplate, just why Sarabhai did not face criminal charges. Gundevia gives some insight when he notes that "I had been warned to be 'very careful' because she was reputed to be a 'dangerous woman.' Why? Because 'she has direct access to the Prime Minister!'" Gundevia was at the time engaged in negotiations on the repatriation of Indian women – including Jammu and Kashmir residents – kidnapped and taken to Pakistan at the time of Partition.

> Mridula was setting about this business of the abducted women like a tigress-with-cubs and I had to work with her. . . . After we had played about with a hash-of-a-draft Mridula had worked out, we went to Lahore, and came back with a complete agreement with Pakistan on the question of the recovery of the abducted women including the Kashmiri girls. (Gundevia, op. cit., 92–93)

31 [SECRET] *Report on Pakistani Organized Subversion,* op. cit., 83.
32 'BA Kitchloo laid to rest,' *The Daily Excelsior* (Jammu), August 28, 2001.
33 [SECRET] *Report on Pakistani Organized Subversion*, op. cit., 86–87.
34 Kara [also spelt Karra] eventually fell out with Sheikh Abdullah, and emerged as one of his most bitter critics. In 1977, he founded the Janata Party's state unit in Srinagar. On April 4 that year,

> a crowd of 7000 people turned up at the historic Lal Chowk to hear Karra when he launched the unit. In his speech, Karra made a blistering attack on

the policies of Abdullah saying he had only succeeded in giving the State a "corrupt, inefficient and ruthless administration".

See Prem Nath Bazaz, *Democracy Through Terror and Intimidation* (New Delhi: Heritage Publishers, 1978), 44–45.

35 [SECRET] *Report on Pakistani Organized Subversion*, op. cit., 103.
36 [SECRET] *Report on Pakistani Organized Subversion*, op. cit., 110.
37 [SECRET] *Report on Pakistani Organized Subversion,* op. cit., Appendix VI, 179–182.
38 Sumit Ganguly, *The Crisis in Kashmir* (Washington, DC: Woodrow Wilson Centre Press and Cambridge University Press, 1997), 31–37.
39 Ibid., 74.
40 [RESTRICTED] B.C. Chakravorty, *History of the Indo-Pak War, 1965* (New Delhi: History Division, Ministry of Defence, Government of India, 1992), 48–51.
41 Schofield (1996), op. cit., 202.
42 [SECRET] *Report on Pakistani Organized Subversion,* op. cit., 31.
43 Ibid., op. cit., 32–33.
44 Ibid., op. cit., 43.
45 [RESTRICTED] Chakravorty, op. cit., 59, 66.
46 Wolpert, op. cit., 90.
47 Victoria Schofield (1996), op. cit., 107–108.
48 Brian Cloughley, *A History of Pakistan Army: Wars and Insurrections* (Karachi: Oxford University Press, 2000), 70–71.
49 [RESTRICTED] Chakravorty, op. cit., 185.
50 [RESTRICTED] Chakravorty, op. cit., 242.
51 Schofield (1996), op. cit., 203–204.
52 Gul Hassan Khan, op. cit., 183–185.
53 Altaf Gauhar, op. cit., 352–353.
54 Poster issued by the Inqilabi Council, reproduced in [SECRET] *Report on Pakistani Organized Subversion,* op. cit., 109.
55 Ibid., 113–114.
56 At several points in these pages of the [SECRET] *Report on Pakistani Organized Subversion*, op. cit., the month of September is recorded as October, evidently a typographical error, 115–116.
57 Ibid., 143.
58 Ibid.
59 Pradeep Thakur, *Militant Monologues* (New Delhi: Parity Publisher, 2003), 99.
60 Ibid., 166 [Urdu; my translation].
61 Ibid., 167 [Urdu; my translation].
62 Ibid., 72–73.
63 [SECRET] *Report on Pakistani Organized Subversion,* op. cit., 73. The *Report* refers to Butt being released on May 1, 1965. This should be May 1, 1956.
64 Ibid., 75.
65 Ibid., 79.
66 From Samuel M. Burke, *Mainsprings of Indian and Pakistan Foreign Policies*, cited in Schofield (2003), op. cit., 106.
67 Gundevia, op. cit., 133.
68 Ibid., 132–133.
69 Schofield (2003), op. cit., 106.
70 [SECRET] *Report on Pakistani Organized Subversion,* op. cit., 29–30.
71 Some accounts that the National Conference, at least, saw these brawls in distinctly military terms – its street volunteer force's leader, Bakshi Ghulam Mohammad, used the military title *Salar-i-Azam* or Commander-in-Chief. See F.M. Hassnain, *Freedom Struggle in Kashmir* (New Delhi: Rima Publishing House, 1988), 97.

72 Bazaz writes that Abdullah's supporters would "terrorize and cow down into silence everyone who disagreed, criticised or opposed Sher-i-Kashmir or his caucus. In the late thirties and forties, Sheikh himself was seen moving around in Srinagar with a stick in his hand beating men whom he suspected to harbour ill will towards him." See Bazaz, op. cit., 10.
73 Navnita Chadha Behera, *State, Identity and Violence: Jammu, Kashmir and Ladakh* (New Delhi: Manohar, 2000), 122.
74 Ibid., 125.
75 Decisive Islamic battle where the prophet Mohammad and a small group of followers decisively defeated superior forces. For a brief account of the battle, see http://www.geocities.com/badr_313/battle.htm.
76 [SECRET] *Report on Pakistani Organized Subversion,* op. cit., 111–112.
77 Ibid., 139.
78 Owen Bennett Jones, *Pakistan: eye of the storm* (New Haven and London: Yale University Press, 2003), 77.
79 [SECRET] *Report on Pakistani Organized Subversion,* op. cit., 119.
80 Thakur, op. cit., 99.
81 [SECRET] *Report on Pakistani Organized Subversion,* op. cit., 119–126.
82 Ibid., 106(b).
83 Ibid., 106.
84 Ibid., 134–136.
85 Ibid., 133.
86 Ibid., 147.
87 Ibid., 147–148.
88 Ibid., 148, 150.
89 Ibid., 140–141.
90 Ibid.
91 Major-General Akbar Khan, *Raiders in Kashmir* (Islamabad: National Book Foundation, 1975), 191–192.
92 "Proclamation of War of Liberation" by the Revolutionary Council of Kashmir, reproduced in [SECRET] *Report on Pakistani Organized Subversion,* op. cit. page facing 167.
93 Wolpert, op. cit., 94.

4 AL-FATAH

1 [SECRET] *Report on Pakistani Organised Subversion, Sabotage and Infiltration in Jammu and Kashmir* (Jammu and Kashmir: Criminal Investigation Department, 1966), 4–5.
2 "As PM arrives, Army begins to pull out," *Kashmir Images* (Srinagar), November 17, 2004.
3 "PM promises new Kashmir, announces Rs. 24,000 cr revival plan," *The Daily Excelsior* (Jammu), November 18, 2004.
4 Ibid.
5 Ahmed Ali Fayyaz, "Two fidayeen killed near Manmohan's rally venue," *The Daily Excelsior* (Jammu), November 18, 2004.
6 [SECRET] *Report on Pakistani-Organised Subversion,* op. cit., 181.
7 [SECRET] Ibid., 182.
8 "Politicians used to send gifts to militants: Beig," *Kashmir Times* (Jammu), December 11, 2004.
9 Praveen Swami, "A Campaign With Guns," *Frontline* (Chennai), May 7, 2004. Online at http://www.frontlineonnet.com/fl2109/stories/20040507003803600.htm.

10 "BA Kitchloo laid to rest," *The Daily Excelsior* (Jammu), August 28, 2001.
11 [SECRET] *Report on Underground Organisation Known As Al-Fatah*, (Jammu and Kashmir: Criminal Investigation Department, 1961), 4.
12 Owen Bennett Jones, *Pakistan: eye of the storm* (New Haven and London: Yale University Press, 2003), 79.
13 Text of Tashkent Agreement, available online at the South Asia Terrorism Portal, http://www.satp.org/satporgtp/countries/india/document/papers/tashkent_declaration.htm.
14 [RESTRICTED] B.C. Charavorty, *History of the Indo-Pak War, 1965* (New Delhi: Ministry of Defence, Government of India, 1992), 320–321.
15 Victoria Schofield, *Kashmir in Conflict* (New York: IB Tauris, 2003), 112.
16 Cited in ibid., 112.
17 [SECRET] *Report on Underground Organisation*, op. cit., 7–8.
18 Anon., *Sultan-ul-Arifeen Sheikh Hamza Makhdum.* Online at Google.com's cache of http://www.myasa.net/sufindia/sultan-ul-arifeenhazrat sheikh hamza makhdum.html as retrieved on Dec 13, 2004 17:47:58 GMT.
19 [RESTRICTED] P.B. Sinha and A.A. Athale, *History of the Conflict With China, 1962* (New Delhi: Ministry of Defence, 1992), xxiv.
20 [RESTRICTED] Ibid., xxi.
21 "Tense Vigil On The Nathu-La Pass," *The Illustrated Weekly* (New Delhi), October 31, 1965. Online at http://www.bharat-rakshak.com/LAND-FORCES/Army/History/1960s/Nathu.html.
22 [RESTRICTED] Sinha and Athale, op. cit., xxi.
23 Praveen Swami, "Open Doors For Mole Recruitment," *The Hindu* (Chennai), June 14, 2004. Online at http://www.hindu.com/2004/06/14/stories/2004061408020100.htm.
24 Jones, op. cit., 158.
25 Ghulam Rasool Zahgir, fragments of handwritten notes maintained from November 1968 to January 1969 [hereafter *Notes*], 1.
26 *Notes*, 1.
27 *Notes*, 1.
28 *Notes*, 2.
29 *Notes*, 2.
30 *Notes*, 3.
31 *Notes*, 4.
32 Stephen P. Cohen, *The Idea of Pakistan* (New Delhi: Oxford University Press, 2005), 104.
33 *Notes*, 5.
34 *Notes*, 6.
35 *Notes*, 16.
36 *Notes*, 16.
37 Karen Armstrong, *Holy Wars: The Crusades and their Impact on Today's World* (New York: Anchor, 2001), 38.
38 Derived from *Statistical Report On General Election, 1967, To The Legislative Assembly of Jammu and Kashmir* (New Delhi: Election Commission of India, undated), 85–92.
39 Y.D. Gundevia, "On Sheikh Abdullah" in *The Testament of Sheikh Abdullah* [no author], (New Delhi: Palit and Palit, 1974), 138.
40 Ibid., 138.
41 A.G. Noorani, *The Kashmir Question* (Bombay: Manaktalas, 1964), 118.
42 Gundevia, op. cit., 139.
43 Balraj Puri, "Jan Sangh's Decline Creates Political Vacuum," *Economic and Political Weekly* (Mumbai), September 1, 1973, 1593–4, cited in Navnita Chadha Behera, *State, Identity and Violence: Jammu, Kashmir and Ladakh* (New Delhi: Manohar, 2000), 119.

44 Gundevia, op. cit., 139.
45 Cited in Behera, op. cit., 127.
46 For an account of the fabrication of such explosives, and for an instance of their continued use in subsequent decades, see N.B. Bardhan, A. Dey, and S.P. Khanna, *Characteristics of New Improvised Explosive Device* (New Delhi: Central Bureau of Investigations, undated). Online at http://cbi.nic.in/cfsl/bardhan.htm.
47 Facsimile in [SECRET] *Report on Underground Organisation*, op. cit., 19–20.
48 [SECRET] *Report on Underground Organisation*, op. cit., 33–34.
49 Facsimile in ibid., 60–61.
50 For a comprehensive account of the history of codes and code-breaking, see Simon Singh, *The Code Book: The Science of Secrecy From Ancient Egypt to Quantum Cryptography* (New York: Anchor Books, 2000).
51 Facsimile in [SECRET] *Report on Underground Organisation*, op. cit., 59–60.
52 [SECRET] *Report on Underground Organisation*, op. cit., 57.
53 Anil Maheshwari, *Crescent Over Kashmir: Politics of Mullaism* (E-Book published online at http://ikashmir.org/Crescent/Chapter1.html).
54 Pradeep Thakur, *Militant Monologues* (New Delhi: Parity Paperbacks, 2003), 100.
55 Ibid. 100.
56 Sayyid Mir Qasim, *My life and times* (New Delhi: Allied Publishers, 1992), 129.
57 Sati Sahni, *Kashmir Underground* (New Delhi: Har-Anand, 1999), 171–172.
58 Thakur, op. cit., 100.

5 YEARS OF RETREAT AND REVIVAL

1 Sati Sahni, *Kashmir Underground* (New Delhi: Har Anand, 1999), 56.
2 Ibid., 57.
3 Several scholarly and media accounts describe Amar Chand as an intelligence official. Sati Sahni's work on the period, based on first-hand knowledge of the affair, affirms otherwise. See ibid., 57.
4 Ibid.
5 Anwer Abbas, "February 11 & Kashmir Struggle," *The Pakistan Times* (Internet), February 14, 2004. Online at http://pakistantimes.net/2004/02/14/guest2.htm.
6 [SECRET] *Report on Pakistani Organised Subversion, Sabotage and Infiltration in Jammu and Kashmir* (Jammu and Kashmir: Criminal Investigation Department, 1966), 150.
7 Jammu Kashmir Liberation Front, "Maqbool Butt anniversary signature campaign." Online at http://www.geocities.com/jklf-kashmir/maqboolstory.html.
8 Rafique A Khan, "Maqbool Butt: His Life, Letters and Legacy" (Los Angeles: Kashmir Human Rights Foundation, undated). Online at http://www.kanger.crosswinds.net/maqbool.html.
9 Mohammad Sayeed Asad, *Shaur-e-Farda* [Urdu] (Mirpur: National Institute of Kashmir Studies, 2001), 54.
10 St. Joseph's College seems to have been a magnet for political fracas – although its pre-Independence brush with notoriety was tinged with humour. In 1925, the institution was charged with having engineered the conversion to Christianity of a Kashmiri Pandit, Dinanath, and having secured his marriage with a Christian woman, with arranging his surreptitious disappearance from the town. The affair sparked off a furor, and was an important phase in the communal consolidation of Hindus in the Kashmir valley. For an account, see JN Kachroo, "National Schools of Kashmir" in *Milchar* (Mumbai), September 2003. Online at http://www.milchar.com/Sept2003/11.html.
11 Rafique A Khan, op. cit.

12 Jammu Kashmir Liberation Front, "Maqbool Butt anniversary signature campaign," op. cit.
13 Rafique A Khan, op. cit.
14 Sahni, op. cit., 55.
15 Ibid., 55–56.
16 AG Noorani, "Contours of Militancy," *Frontline* (Chennai), October 13, 2000. Online at http://www.flonnet.com/fl1720/17200800.htm.
17 Sahni, op. cit., 57.
18 Jammu Kashmir Liberation Front, op. cit.
19 Ajit Bhattacharjea, *Kashmir: The Wounded Valley* (New Delhi: UBSPD, 1994), 246.
20 Kashmir Human Rights Site, "Statement of Maqbool Bhat [Butt] before a special court in Pakistan." Online at http://www.kashmir-hr.net/mainfile.php/v1n4/228/.
21 Pradeep Thakur, *Militant Monologues* (New Delhi: Parity Paperbacks, 2003), 30. The quote from Khan appears inaccurate either in its telling or recounting, as the LoC did not exist at the time. There is no reason, however, to disbelieve Khan's general recollection of the judgement.
22 Ibid., 31.
23 Ibid.
24 Ibid. The circumstances described by Altaf Khan of his arrest are, perhaps, a little exaggerated. May is generally a warm month through Kashmir, although snowfall or rain do on occasion hit mid-altitude passes like the one used by Altaf Khan and his associates. It is more likely that Altaf Khan and his group were under-equipped or ill-trained.
25 Ibid., 30.
26 Siddiq Salik, *Witness to Surrender* (Karachi: Oxford University Press, 1977), 3.
27 Owen Bennett Jones, *Pakistan: eye of the storm* (New Haven and London: Yale University Press, 2003), 153.
28 Ibid., 157.
29 Hassan Abbas, *Pakistan's Drift Into Extremism: Allah, the Army, and America's War on Terror* (Armonk and London: ME Sharpe, 2005), 57.
30 Ibid., 58.
31 Ibid., 59.
32 Jones, op. cit., 160.
33 Ibid., 160–161.
34 G.W. Chowdhury, *The Last Days of United Pakistan* (Karachi: Oxford Univeristy Press, 1993), 15.
35 Jones, op. cit., 162.
36 Ibid., 162.
37 Abbas, op. cit., 63.
38 Jones, op.cit., 168.
39 Sahni, op. cit., 58.
40 For a brief account of the PFLP hijackings, see "Dawson's Field Hijackings" in Wikipedia, http://en.wikipedia.org/wiki/Dawson%27s_Field_hijackings.
41 Interview with Ashraf Qureshi in *The Frontier Post* (Peshawar), March 1992, cited in Hashim Qureshi, *The Unveiling of Truth: A Political Analysis of Kashmir* (Lahore: Jeddojudh Publications, 1999), 308.
42 Qureshi, op. cit., 309.
43 Alistair Lamb, *Kashmir: A Disputed Legacy, 1846–1990* (Hertingfordbury: Roxford Books, 1991), 288.
44 Ibid., 288.
45 Gul Hasan Khan, *Memoirs of Lt. Gen. Gul Hasan Khan* (Karachi: Oxford University Press, 1993), 260.
46 Lamb, op. cit., 288.

47 Qureshi, op. cit., 287–288.
48 Lamb, op. cit., 288.
49 Gul Hasan Khan, op. cit., 260.
50 Richard Sisson and Leo E. Rose, *War and Secession*, cited in Ardeshir Cowasjee, "Gareeb no koi nahi," *Dawn* (Karachi), January 13, 2002. Online at http://www.dawn.com/weekly/cowas/20020113.htm.
51 Lamb, op. cit., 289.
52 Abbas, op. cit., 61.
53 Sahni, op. cit., 58.
54 Hassan Abbas, op. cit., 61.
55 Anil Maheshwari, *Crescent Over Kashmir: Politics of Mullaism* (e-book published online at http://ikashmir.org/Crescent/Chapter1.html). I have been unable to locate the original UNI dispatch, if indeed it exists.
56 BM Sinha, *The Samba Spying Case* (New Delhi: Vikas, 1981), 21–23.
57 Ibid., 22.
58 Lamb, op. cit., 293.
59 Ibid., 290.
60 Ibid., 291.
61 Qureshi, op. cit., 293–294.
62 Lamb, op. cit., 287, and Sinha, op. cit., 21–22.
63 Lamb, op. cit., 292.
64 Qureshi, op. cit., 249.
65 [RESTRICTED] SN Prasad, ed., *History of the Bangladesh War, 1971*(New Delhi: Ministry of Defence, 1992), 68.
66 Gul Hasan Khan, op. cit., 261.
67 Katherine Frank, *Indira: The Life of Indira Nehru Gandhi* (Boston and New York: Houghton Mifflin, 2002), 329.
68 Kenneth Conboy and James Morrison, *The CIA's Secret War in Tibet* (Lawrence: University Press of Kansas, 2002), 256.
69 [RESTRICTED] Prasad, op. cit., 174.
70 [RESTRICTED] Ibid., 164.
71 [RESTRICTED] Ibid., 165–66.
72 [RESTRICTED] Ibid., 169.
73 "Special Frontier Force," online at http://bharat-rakshak.com/LAND-FORCES/Special-Forces/SFF.html.
74 Claude Arpi, "Tibetan Faujis in Bluestar." Originally published on Rediff.com; article now cached at http://vigilonline.com/reference/columns/vicharmala_view.asp?col_id=126.
75 "Special Frontier Force," op. cit.
76 Conboy Morrison, op. cit., 257.
77 Ibid., 210–211.
78 Ibid., 233–234.
79 Ibid., 235.
80 "Special Frontier Force," op. cit.
81 Shelford Bidwell and Ray Bonds, eds., *The Encyclopedia of Land Warfare in the 20th Century* (New York: Crowell, 1976), 166.
82 Sumit Ganguly, *Conflict Unending: India–Pakistan Tensions* (Columbia University Press and Woodrow Wilson Center Press: New York and Washington D.C., 2001), 68.
83 Frank, op. cit., 342.
84 Jones, op. cit., 146.
85 Ibid., 185.
86 Ibid., 185–186.
87 Ibid., 186.

88 Ibid., 184.
89 Thakur, op. cit., 32.
90 Ibid., 32.
91 [SECRET] *Report on Pakistani-Organised Subversion*, op. cit., Also see Chapter 4.
92 Thakur, op. cit., 33.
93 For the text of the Shimla Agreement 1972, see the South Asia Terrorism Portal, http://www.satp.org/satporgtp/countries/india/document/papers/shimla_agreement_1972.htm.
94 Ganguly, op. cit., 71.
95 Victoria Schofield, *Kashmir in Conflict: India, Pakistan and the Unending War* (London: IB Tauris, 2003), 122.
96 *Statistical Report on General Elections, 1972, to the Legislative Assembly of Jammu & Kashmir* (New Delhi: Election Commission of India, undated), 6.
97 Schofield, op. cit., 122.
98 Cited in Ajit Bhattacharjee, *Kashmir: The Wounded Valley* (New Delhi: UBSPD, 1994), 234.
99 Frank, op. cit., 365.
100 Yoginder Sikand, "The Emergence and Development of the Jamaat-e-Islami of Jammu and Kashmir," *Modern Asian Studies* 36, 3 (Cambridge Univeristy Press: Cambridge, 2002), 733.
101 Ibid., 733–34.
102 Schofield, op. cit., 125.
103 Katherine Frank, *Indira: The Life of Indira Nehru Gandhi* (Boston and New York: Houghton Mifflin, 2002), 366.
104 Mirza Mohammad Abdullah Beg and G. Parthasarathi, *Agreed Conclusions* (New Delhi, November 13, 1974). Available online at http://www.satp.org/satporgtp/countries/india/states/jandk/documents/papers/sheikh_indira_accord_1975.htm.
105 Schofiled, op. cit., 124.
106 Cited in Navnita Chadha Behera, *State, Identity and Violence: Jammu, Kashmir and Ladakh* (New Delhi: Manohar, 2000), 143.
107 Sikand, op. cit., 736.
108 Behera, op. cit., 143.
109 Ibid., 143.
110 Schofield, op. cit., 125.
111 Behera, op. cit., 140.
112 *Key Highlights of General Election, 1977, to the Legislative Assembly of Jammu and Kashmir* (New Delhi: Election Commission of India, undated), 8.
113 Ibid., 4–5.
114 Schofield, op. cit., 122.
115 MJ Akbar, *Kashmir: Behind the Vale* (New Delhi and New York: Viking, 1991), 188.
116 *The Pakistan Times*, July 19, 1972, cited in Schofield, op. cit., 120.
117 Asad, op. cit., 59.
118 Abbas, op. cit., 79.
119 Ibid., 81.
120 Ibid., 81–82.
121 Seyyed Vali Reza Nasr, *The Vanguard of the Islamic Revolution: the Jama'at-ı Islami of Pakistan* (Berkley and Los Angeles: University of California Press, 1994), 182.
122 Thakur, op. cit., 33.
123 Syed Nazir Gilani, "Ravindra Mhatre to Akhtar Bano – Blood on Our Hands," *Kashmir Images* (Srinagar), November 20, 2003.
124 Qureshi, op. cit., 68.
125 Qureshi, op. cit., 30.
126 Sahni, op. cit., 58.

127 Roger Ballard, "Remittances and their economic impact in areas of mass overseas migration: two contrasting South Asian case studies" (London: Department for International Development SLSO-ASREP Lunchtime Seminar, March 4, 2003).
128 "Kashmiri expatriate Ayub Thakur dies," *Daily Times* (Karachi), March 9, 2005.
129 "Ayub Thakur admits sending money to Kashmir," *The Hindu* (Chennai), June 19, 2002.
130 Sahni, op. cit., 59.
131 Akbar, op. cit., 197.
132 Schofield, op. cit., 126.
133 Sikand, op. cit., 737.
134 Behera, op. cit., 144.
135 Ibid., 144.
136 Qureshi, op. cit., 43.
137 Ibid., 44.
138 Ibid.
139 Ibid., 45.
140 Ibid., 46.
141 "Man charged with diplomat murder," BBC News, November 4, 2004. Online at http://news.bbc.co.uk/go/pr/fr/-/2/hi/uk_news/england/west_midlands/3981411.stm.
142 Nabanita Sircar, "British Muslim charged with Mhatre's murder," *The Hindustan Times* (New Delhi), November 6, 2004.
143 Sahni, op. cit., 59.
144 T.V. Parasuram, "Indian envoy's murder: Suspect arrested in US," Rediff.com (Online), March 6, 2004. Online at http://www.rediff.com//news/2004mar/06uk.htm.
145 Gilani, op. cit. In European folklore, girls could practice certain divinatory rituals on The Eve of St. Agnes before going to bed in order to have a vision of their future husbands. See Meredith Ringel, "The Theme of 'The Eve of St. Agnes' in the Pre-Raphaelite Movement" in *The Victorian Web,* http://www.victorianweb.org/painting/prb/ringel12.html.
146 Ibid.
147 Thakur, op. cit., 31.
148 Asad, op. cit., 129.
149 Ibid., 80.
150 Sikand, op. cit., 746.

6 THE WAR OF MANY FRONTS

1 In Sunil Fotedar (ed.), *Kashmir: Articles & Opinions* (Austin: Kashmir News Network, 2002), 153.
2 Manoj Joshi, *The Lost Rebellion: Kashmir in the Nineties*. (New Delhi: Penguin, 1991), 23.
3 Government of Jammu and Kashmir, "Major Events, 1975–1990." Online at http://jammukashmir.nic.in/profile/majev2.htm.
4 Hassan Abbas, *Pakistan's Drift Into Extremism: Allah, the Army and America's War on Terror* (Armonk and London: ME Sharpe, 2005), 86.
5 Ibid., 90–91.
6 Ibid., 91.
7 Ibid., 97.
8 Seyyed Vali Reza Nasr, *The Vanguard of the Islamic Revolution: The Jamaat-e-Islami of Pakistan* (Berkley and Los Angeles: University of California Press, 1994), 182.
9 Stephen P. Cohen, *The Idea of Pakistan* (New Delhi: Oxford University Press, 2005), 221.

10 Ibid., 220.
11 S. Akbar Zaidi, *Pakistan's Economic & Social Development* (New Delhi: Rupa, 2004), 20.
12 Hamza Alavi, "Class and State" in Hassan Gardezi and Jamil Rashid, ed., *Pakistan: The Roots of Dictatorship* (London: Zed Press, 1983), 51.
13 Abbas, op. cit., 87.
14 Ibid., 87.
15 "Interview with Naseem Hassan Shah," *The News* (Karachi), April 23, 1994.
16 Nasr, op. cit., 182.
17 Cohen, op. cit., 119.
18 Abbas, op. cit., 96.
19 Steve Coll, *Ghost Wars* (London: Penguin Books, 2005), 42.
20 Giles Dorronsoro, *Revolution Unending: Afghanistan, 1979 to the Present* (New York: Colombia University Press, 2005), 133.
21 Coll, op. cit., 51.
22 Dorronsoro, op. cit., 81–83.
23 Ibid., 83.
24 Abbas, op. cit., 111.
25 Dennis Kux, *The United States and Pakistan: Disenchanted Allies* (Washington, DC: Woodrow Wilson Center Press, 2000), 257–262.
26 Mohammad Yousaf and Mark Adkin, *Afghanistan: The Bear Trap* (Havertown, PA: Casemate, 2001), 111.
27 Ibid., 98.
28 Ibid., 20.
29 For a sample of the scholarly work on the genesis and course of the Khalistan movement, see D.H. Butani, *The Third Sikh War? Towards or Away From Khalistan?* (New Delhi: Promilla & Co., Publishers, 1986); Satyapal Dang, *Genesis of Terrorism: An Analytical Study of Punjab Terrorists* (New Delhi: Patriot Publishers, 1988); K.P.S. Gill, *Punjab: The Knights of Falsehood* (New Delhi: Har Anand Publishers, 1997); Sharda Jain, *Politics of Terrorism in India: The Case of Punjab* (New Delhi: Deep and Deep Publications, 1995); Manoj Joshi, *Combating Terrorism in Punjab: Indian Democracy in Crisis* (London: Research Institute for the Study of Conflict, 1993); Harkishan Singh Surjeet, *Deepening Punjab Crisis: A Democratic Solution* (New Delhi: Patriot Publishers, 1992); Mark Tully and Satish Jacob, *Amritsar: Mrs. Gandhi's Last Battle* (London: Jonathan Cape, 1986). I am indebted to Razili Datta for her thorough review of the literature on the Khalistan movement, and her analytic insights on the issue.
30 Katherine Frank, *Indira: The Life of Indira Nehru Gandhi* (London: Harper Collins, 2002), 455.
31 For an insider account of Operation Bluestar, see Lieutenant-General KS Brar, *Operation Bluestar: The True Story* (New Delhi: UBSPD, 1993). For a brief narration of the course of the fighting, see http://www.bharat-rakshak.com/LAND-FORCES/Army/History/1970s/Bluestar.html.
32 Maqsood Ahmed Masood and Peter Stockdale, *The Khalistan Riddle* (Islamabad: Modern Book Depot, 1998), 192.
33 Paul Wallace and Surendra Chopra, *Political Dynamics and Crisis in Punjab* (Amritsar: Department of Political Science, Guru Nanak Dev University, 1988), 475.
34 Ibid., 477.
35 Praveen Swami, "Failed Threats and Flawed Fences: India's Military Responses to Pakistan's Proxy War", *India Review* 3, no. 2 (2004), 149.
36 Ibid., 150.
37 Jain, op. cit., 236.

38 Tully and Jacob, op. cit., 212; *Human Rights in India: Punjab in Crisis* (New York: Human Rights Watch, 1991), 174.
39 Manoj Joshi, *Combating Terrorism in Punjab,* op. cit.
40 Ibid., 3.
41 Tara Kartha, "Early Warning and Light Weapons in South Asia" in *Arms Watching: Integrating Small Arms and Light Weapons Into the Early Warning of Violent Conflict*, ed. Edward J. Laurence, (International Alert, 1999), 34.
42 Shereen Ilahi, "The Radcliffe Boundary Commission and the Fate of Kashmir" *India Review* Vol. 2, No. 1, (London: Frank Cass, January 2003), 96.
43 See, generally, *Human Rights* op. cit. and Dang, op. cit.
44 "CSIS mole 'had hand' in Kanishka bombing," *The Tribune* (Chandigarh), June 6, 2003.
45 Swami, "Failed Threats and Flawed Fences," op. cit., 150–151.
46 "I was arrested, says Zafarwal," *The Tribune* (Chandigarh), April 13, 2001.
47 Dang, op. cit., 90.
48 Tara Kartha, "Early Warning and Light Weapons." The South Asia Terrorism Portal, http://www.satp.org/satporgtp/publication/faultlines/volume4/fault4-tarakf1.htm#10.
49 Owen Bennett Jones, *Pakistan: eye of the storm* (New Haven and London: Yale University Press), 2003, 187.
50 Ibid., 196.
51 Shahid-ur-Rehman, *Long Road to Chagai: Untold Story of Pakistan's Nuclear Quest* (Islamabad: Print Wise Publications), 1999, 51.
52 Abbas, op. cit., 83.
53 Cohen, op. cit., 140, 170.
54 Jones, op. cit., 202.
55 For a succinct account of these activities, see "Weapons of Mass Destruction" and related hyperlinks (Washington DC: Global Security, 2004). Online at http://www.globalsecurity.org/wmd/world/pakistan/khan.htm.
56 *From Surprise to Reckoning: Kargil Review Committee Report* (New Delhi: Government of India Press, 1999), 159.
57 *Time Magazine*, March 30, 1987, cited in *From Surprise to Reckoning,* op. cit., 177 footnote 39.
58 David Albright, Frans Berkhout, William Walker, *Plutonium and Highly Enriched Uranium 1996: World Inventories, Capabilities, and Policies.* (Stockholm: Stockholm International Peace Research Institute, 1997), 273.
59 Pravin Sawhney, *The Defence Makeover* (New Delhi: Sage Publications, 2002), 346.
60 Devin Hagerty, "Nuclear Deterrence in South Asia." *International Security,* Winter 1995–1996, Volume 20, No. 3 cited in Pravin Sawhney, *The Defence Makeover*, (New Delhi: Sage Publications, 2002), 350.
61 Sumit Ganguly and R. Harrison Wagner, "India and Pakistan: Bargaining in the Shadow of Nuclear War," *The Journal of Strategic Studies*, Vol. 27, No. 3, September 2004 (London: Francis and Taylor), 479–507.
62 A.G. Noorani, "Contours of Militancy" in *Frontline* (Chennai), October 13, 2000. Available online at http://www.flonnet.com/fl1720/17200800.htm.
63 Dang, op. cit., 91.
64 Ibid., 77.
65 B. Raman, "Musharraf's speech was an anti-climax," *Rediff.com*, May 28, 2002. Online at http://www.rediff.com/news/2002/may/28guest.htm.
66 Praveen Swami "Open doors for mole recruitment," *The Hindu* (Chennai), June 14, 2004. Online at http://www.hindu.com/2004/06/14/stories/2004061408020100.htm.
67 "Apex court rejects Sarabjit petition," *The News* (Lahore), September 28, 2005.
68 For a more detailed discussion of this network, see Swami, "Failed Threats and Flawed Fences", op. cit., 157–159.

69 Joshi, *Combating Terrorism* op. cit., 23.
70 Ibid., 3, and Kartha, "Early Warning and Light Weapons in South Asia", op. cit., 40.
71 Kartha, "Early Warning and Light Weapons in South Asia" op. cit., 40.
72 M.J. Akbar, *Kashmir: Behind the Vale* (New Delhi and New York: Viking, 1991), 197.
73 The Jammu and Kashmir Resettlement Bill continues to be a cause of controversy. In 2005, after visitors from Pakistan-administered Kashmir laid claim to property in Indian-administered Kashmir, controversy flared up an account. For an account, see "NC Locks Horns With Government Over Resettlement Act," *The Tribune* (Chandigarh), May 3, 2005.
74 *Key Highlights of General Elections, 1983, to the Legislative Assembly of Jammu and Kashmir* (Election Commission of India, New Delhi, undated), 6.
75 Navnita Chadha Behera, *State, Identity and Violence: Jammu, Kashmir and Ladakh* (New Delhi: Manohar, 2000), 150–151.
76 Basharat Peer, "The trek to Muzaffarabad, sans passports," *Tehelka* (New Delhi) April 16, 2005).
77 Farooq Abdullah, *My Dismissal* (New Delhi: Privately Published, 1985), 32.
78 Victoria Schofield, *Kashmir in Conflict: India, Pakistan and the Unending War* (London and New York: IB Tauris, 2003), 133.
79 Ibid., 134.
80 Balraj Puri, *Kashmir Towards Insurgency* (New Delhi: Orient Longman, 1993), 34.
81 Behera, op. cit., 154.
82 Puri, op. cit., 35.
83 Behera, op. cit., 155.
84 Ibid., 156.
85 P.S. Verma, "Muslim United Front" in S. Bhatnagar and Pradeep Kumar, ed., *Regional Political Parties in India* (New Delhi: Ess Ess Publications, 1988), 191.
86 Behera, op. cit., 158.
87 P.S. Verma, *Jammu and Kashmir at the Political Crossroad* (New Delhi: Vikas, 1994), 159.
88 Mohammad Ishaq Khan, "Six Centuries of Islamisation in Kashmir: Retrospect and Prospects" in Balraj Puri, ed. *5000 Years of Kashmir* (New Delhi: Ajanta, 1997).
89 Verma, "Muslim United Front" op. cit., 195.
90 "We are not militants, we're victims J&K Jamaat," *Asian Age* (New Delhi), November 14, 1998.
91 Behera, op. cit., 159.
92 Schofield, op. cit., 137.
93 PS Verma, "Muslim United Front" op. cit., 197.
94 Manoj Joshi, *The Lost Rebellion,* op. cit., 16–18.
95 Pradeep Thakur, *Militant Monologues* (New Delhi: Parity Paperbacks, 2003), 34.
96 Muhammad Amir Rana (trans. Saba Ansari), *A–Z of Jihadi Organizations in Pakistan* (Lahore: Mashal Books, 2004), 86.
97 Thakur, op. cit., 35.
98 Thakur, op. cit., 100–101.
99 Joshi, op. cit., 50.
100 "JKLF: Our Ideology, Aims and Objectives", http://shell.comsats.net.pk/~jklf/i2.htm.
101 Newstrack Video Magazine. Transcript at http://www.ikashmir.org/Video/BittaKarate_tr.html.
102 Manoj Joshi, *The Lost Rebellion,* op. cit., 26.
103 Ibid., 28.
104 "After burqa, militants order men, women apart on buses." *The Indian Express*, New Delhi, September 7, 2002. http://www.expressindia.com/fullstory.php?newsid=1980.
105 Rashid Ahmed, "Valley to Stage New Morality Play", *The Hindustan Times,* March 8, 1999. http://www.jammu-kashmir.com/archives/archives1999/99march08b.html.

106 For a comprehensive list of abuses targeting women, see *Atrocities on Women and Children* (New Delhi: Institute for Conflict Management, 2003).
107 Praveen Swami, "Taliban-type terror in Rajouri." *Frontline,* Chennai, January 17, 2003.
108 Hashim Qureshi, *The Unveiling of Truth: A Political Analysis of Kashmir* (Lahore: Jeddojudh Publications, 1999), 136–137.
109 Suchita Vemuri, "Kashmir Drenched in Women's Tears" Women in Action 1–2: 2000, http://www.isiswomen.org/wia/wia100/soc00011.html.
110 Joshi, *The Lost Rebellion* op. cit., 1.
111 [SECRET] Jammu and Kashmir Police Criminal Investigation Department, *Interrogation Report of Mushtaq Ahmad Zargar* @ [alias] *Latram.* Srinagar, June 1992. Unsigned.
112 Maj. Gen. Afsir Karim, *Kashmir: The Troubled Frontiers* (New Delhi: Lancer Publishers, 1994), 311–312.
113 Ibid., 254.
114 Joshi, *The Lost Rebellion,* op. cit., 73–74.
115 Ibid., 43.
116 Cohen, op. cit., 104.
117 Ibid., 104–105.
118 Dorronsoro, op. cit., 187–188.
119 Yusaf and Adkin, op. cit., 42.
120 Ibid., 96.

7 THE NUCLEAR JIHAD

1 S.K. Malik, *The Quranic Concept of War* (New Delhi: Himalayan Books, 1986, in arrangement with Wajid Alis Limited, Lahore), 59.
2 Sumit Ganguly and Devin T Hagerty, *Fearful Symmetry: India–Pakistan Crises in the Shadow of Nuclear Weapons* (New Delhi: Oxford University Press, 2005), 88.
3 George Perkovich, *India's Nuclear Bomb: The Impact on Global Proliferation* (Berkley: University of California Press, 1999), 307.
4 Raja Asghar, "Bhutto Predicts Victory for Kashmir Independence Campaign," *Reuters Library Report*, March 13, 1990.
5 Moses Manoharan, "Indian Leader Tells Pakistan to Stay Out of Kashmir Uprising," *Reuter Library Report*, March 13, 1990.
6 Steve Coll, "Assault on Pakistan Gains Favor in India," *The Washington Post* (Washington DC), April 15, 1990.
7 Ganguly and Hagerty, op. cit., 104.
8 Pradeep Thakur, *Militant Monologues* (New Delhi: Parity Publishers, 2003), 115.
9 Manoj Joshi, *The Lost Rebellion: Kashmir in the Nineties* (New Delhi: Penguin Books, 1999), 402–403.
10 Ghulam Rasool Zahgir, fragments of handwritten notes maintained from November 1968 to January 1969, 1.
11 Navnita Chadha Behera, *State, Identity and Violence: Jammu, Kashmir and Ladakh* (New Delhi: Manohar, 2000), 172.
12 Thakur, op. cit., 64–66.
13 Ibid., 146.
14 Muhammad Amir Rana (trans. Saba Ansari), *A–Z of Jihadi Organizations in Pakistan* (Lahore: Mashal, 2004), 86–87.
15 Joshi, op. cit., 105–107.
16 Mohammad Yousaf and Mark Adkin, *Afghanistan: The Bear Trap* (Havertown: Casemate, 2001), 50.

17 Ibid., 105–106.
18 Ganguly and Hagerty, op. cit., 47.
19 Praveen Swami, "A beleaguered force," *Frontline* (Chennai), February 12, 1999.
20 Eric Hobsbawm, *Bandits* (New York: The New Press, 2000), 129.
21 Rana, op. cit., 95.
22 U.S. Department of State, *1995 Patterns of Global Terrorism* (Washington DC: U.S. Department of State, Office of the Coordinator for Counterterrorism, April 1996).
23 B. Raman, "Harkat-ul-Mujahideen: An Update" (Chennai: South Asia Analysis Group, Paper 42, March 20, 1999).
24 B. Raman, "The Lashkar-e-Taiba" (Chennai: South Asia Analysis Group Paper 374, December 15, 2001). Online at http://www.saag.org/papers4/paper374.html.
25 Ibid.
26 Ibid.
27 Rana, op. cit., 137–138.
28 Ibid., 94.
29 Sikand, "The Jihad in Kashmir and India," op. cit.
30 Pankaj Pachauri and Zafar Meraj, "Kashmir Valley: Drifting Dangerously," *India Today* (New Delhi), January 15, 1990.
31 Behera, op. cit., 173.
32 Praveen Swami, "The Nadimarg Outrage," *Frontline* (Chennai), April 25, 2003. Online at http://www.frontlineonnet.com/fl2008/stories/20030425002903800.htm.
33 Yoginder Sikand, "For Islam and Kashmir: The Prison Diaries of Sayyed 'Ali Gilani of the Jama'at-i-Islami of Jammu and Kashmir," op. cit.
34 www.dawacenter.com/kashmir/kashmir.html, cited in Yoginder Sikand, "The Jihad in Kashmir and India," op. cit.
35 Hafiz Muhammad Saeed, "No More Dialogue on Kashmir," http://www.dawacenter.com/magazines/voiceofislam/sept99/editorial.html, cited in Sikand, Ibid.
36 Sikand, Ibid.
37 Quoted in the Pakistan-based Urdu paper, *Ausaf*, December 27, 2000, cited on the Indian Army's official website, www.armyinkashmir.org/articles/jihad_machine.html.
38 Praveen Swami, "Unknown heroes of Batalik," *Frontline* (Chennai), July 30, 1999. Online at http://www.frontlineonnet.com/fl1615/16150070.htm.
39 Major-General Ashok Kalyan Verma, *Kargil: Blood on the Snow* (New Delhi: Manohar, 2002), 95–96.
40 Hassan Abbas, *Pakistan's Drift Into Extremism: Allah, the Army and America's War on Terror* (Armonk and London: ME Sharpe, 2005), 170.
41 For one account of coercion, see Ajith Pillai, "Vote Marshaled," *Outlook* (New Delhi), June 5, 1996. Also see Shiraz Sidhva, "Guns and Votes" in *Frontline* (Chennai), June 14, 1996.
42 Praveen Swami, "A Surprise in Kashmir: The Myth of Coercion at Large," *Frontline* (Chennai), June 14, 1996. Also see Aarti Dhar, "A Vote for Peace," *The Hindu* (Chennai), June 9, 1996.
43 *Statistical Report on General Election, 1996, to the Legislative Assembly of Jammu & Kashmir* (New Delhi: Election Commission of India, 1997), 10.
44 T. Jayaraman, "Deterrence and other myths," *Frontline* (Chennai) May 21, 1999. Online at http://www.frontlineonnet.com/fl1610/16101150.htm.
45 Sumit Ganguly and Devin T Hagerty, *Fearful Symmetry: India–Pakistan Crises in the Shadow of Nuclear Weapons* (New Delhi: Oxford University Press, 2005), 151–152.
46 For example, see Ayaz Amir, "The ghost that won't go away," *Dawn* (Karachi), July 23, 2004. Online at http://www.dawn.com/weekly/ayaz/20040723.htm.
47 For a thorough discussion of the war, including Indian military failures, see Lieutenant-General Y.M. Bammi, *Kargil: The Impregnable Conquered* (NOIDA: Gorkha

Publishers, 2002). Also see Praveen Swami, *The Kargil War* (New Delhi: Left Word, 1999) for a critical account of Indian military and political decision-making prior to the war.

48 Ganguly and Hagerty, op. cit., 155–156.

49 Ashley J Tellis, C. Christine Fair, and Jamison Jo Medby, *Limited Conflict Under The Nuclear Umbrella: Indian and Pakistani Lessons From the Kargil Crisis* (Washington DC: Rand Corporation, 2001), 7.

50 Pervez Musharraf, "From the President's Desk," on the official website of the President of Pakistan, http://www.presidentofpakistan.gov.pk/FromThePresidentsDesk.aspx.

51 CSDS Team with Suhas Palshikar, "Split factor decides the outcome", *Frontline* (Chennai), November 19, 1999.

52 Praveen Swami, "A dangerous game," *Frontline* (Chennai), September 29, 2000. Online at http://www.frontlineonnet.com/fl1719/17190260.htm.

53 "Hizbul Declares Unilateral Ceasefire For Three Months," *The Asian Age*, Mumbai: July 24, 1997.

54 "Hizbul expelled from jehad council," *The Tribune* (New Delhi), July 27, 2000.

55 "8 militants killed, blasts in Srinagar," *The Tribune* (New Delhi), July 27, 2000.

56 "100 killed in Kashmir's night of terror," *The Times of India* (New Delhi), August 3, 2000.

57 Praveen Swami, "Disturbed Doda," *Frontline* (Chennai), August 31, 2001.

58 "Ceasefire decision hasty: Hurriyat," *The Tribune* (New Delhi), July 28, 2000.

59 Amit Baruah, "Militant chiefs warn Musharraf," *The Hindu* (Chennai), February 6, 2005.

60 "75 pc of Militants in J&K Foreigners," *The Times of India* (New Delhi), August 8, 2003.

61 "Hizb commdr Bilal-e-Habshi killed in Budgam," *The Daily Excelsior* (Jammu), January 1, 2004.

62 "Killer of 23 Pandits shot dead," *The Tribune* (New Delhi), March 15, 2002.

63 "Lashkar, Harkat issue fresh threats", *The Times of India* (New Delhi), December 12, 1999.

64 Associated Press of India dispatch of April 12, 1948, cited in Lieutenant-General LP Sen, *Slender Was The Thread.* Orient Longman (New Delhi), 1994. Page 13.

65 Lt. Gen. Gul Hasan Khan, *Memoirs.* Oxford University Press (Karachi), 1993. Page 230–231.

66 Cited Rana op. cit., 49.

67 Praveen Swami, "The 'liberation' of Hyderabad," *Frontline* (Chennai), May 26, 2000.

68 [SECRET] *A Note of Background Information of Interrogation*, (Jammu and Kashmir Police, Joint Interrogation Centre, Kot Bhawal, signed by the Deputy Superintendent of Police, Joint Interrogation Centre, Kot Bhalwal), 11.

69 The Dar-ul-Uloom seminary at Deoband denies any association with terrorists who draw inspiration from their teachings. For example, see Sankarshan Thakur, "What Taliban have done is horrifying, un-Islamic but you need to ask why," *The Indian Express* (New Delhi), March 14, 2001.

70 Praveen Swami, "Extending Terror", *Frontline* (Chennai), March 26, 1999.

71 Praveen Swami, "From terror to politics", *Frontline* (Chennai), March 1, 2002.

72 "Hizbul against US mediation in Kashmir," *The Indian Express* (New Delhi), December 10, 1998.

73 "Lashkar, Harkat issue fresh threats," *The Times of India* (New Delhi), December 12, 1999.

74 B. Muralidhar Reddy, "Lashkar threatens more Red Fort type incidents," *The Hindu* (Chennai), November 21, 2001.

75 T.V. Parsuraman, "Islamic extremists are biggest threat in J&K: Lone," Rediff.com (Internet: December 7, 2004). Online at http://www.rediff.com/news/2000/dec/07jk4.htm.

76 Lawrence Lifschultz, "Voice of Peace Abdul Ghani Lone 1932–2002," *Newsline* (Karachi), August 2002.
77 "Rift in Hurriyat Reappears," *The Tribune* (Chandigarh), April 26, 2002.
78 Umar Javed, "For a real cease fire," *Greater Kashmir* (Srinagar), May 1, 2002.
79 Ahmed Ali Fayyaz, "Hizb expels Majeed Dar, Zaffar, Yazdani," *The Daily Excelsior* (Jammu), May 4, 2002.
80 Owais Tohid and Khawar M. Rizvi, "Vale of uncertainty," *Newsline* (Karachi), July 2003.
81 B. Raman, "Before and after Agra," South Asia Analysis Group (Chennai), Paper 280, July 18, 2001. Online at http://www.saag.org/papers3/paper280.html.
82 Glenn Snyder, "The Balance of Power and the Balance of Terror" in Paul Seabury (ed.) *The Balance of Power* (San Francisco: Chandler, 1965).
83 Sumit Ganguly, "Indo-Pakistani Nuclear Issues and the Stability/Instability Paradox" *Studies in Conflict and Terrorism*, Vol. 18, No. 4, 1995, 325–334.
84 Eric Hobsbawm, *Revolutionaries* (New York: The New Press, 2001), 197.

8 TOWARDS PEACE?

1 "Crowds Applaud As PM Recites Kashmiri Poem", *The Hindustan Times* (New Delhi), April 18, 2003. My translation.
2 "All options are open: PM", *The Hindu* (Chennai), December 19, 2001.
3 "India's missile system 'in position': Fernandes", *The Hindu* (Chennai), December 26, 2001.
4 "Army chief warns Pak against N-strike", *The Tribune (Chandigarh)*, January 12, 2002. Online at http://www.tribuneindia.com/2002/20020112/main1.htm.
5 Mukhtar Ahmad, "Geelani abetted Abdul Gani Lone's assassination: Sajjad Lone", *Rediff.com*, July 8, 2003. http://www.rediff.com/news/2003/jul/08jk.htm.
6 Thom Shanker and Elizabeth Bumiller, "Bush Presses Pakistan on Kashmir and Orders Rumsfeld to Region", *The New York Times* (New York), May 31, 2002.
7 Sumit Ganguly and Devin Hagerty, *Fearful Symmetry: India–Pakistan Crisis in the Shadow of Nuclear Weapons* (New Delhi: Oxford University Press, 2005), 180.
8 Pervez Musharraf, "From the President's Desk", on the official website of the President of Pakistan, http://www.presidentofpakistan.gov.pk/FromThePresidentsDesk.aspx.
9 Hassan Abbas, *Pakistan's Drift into Extremism: Allah, the Army and America's War on Terror* (Armonk and London: ME Sharpe, 2005), 221.
10 Musharraf, op. cit.
11 Pavlos Xanthoulis, "Pakistan sponsoring terrorism in Kashmir", Politics, October 7, 2002, reproduced by the Ministry of External Affairs (New Delhi), http://mea.gov.in/interview/2002/10/07i02.htm.
12 Praveen Swami, "Lashkar still has its way in Pakistan", *The Hindu* (Chennai), April 1, 2004.
13 "Musharraf escapes assassination bid", *The Tribune* (Chandigarh), December 26, 2004.
14 Stephen Philip Cohen, *The Idea of Pakistan* (New Delhi: Oxford University Press, 2005), 129.
15 Ibid., 124.
16 Abbas, op. cit.
17 S. Akbar Zaidi, *Pakistan's Economic & Social Development* (New Delhi: Rupa, 2004), 54–55.
18 B. Raman, South Asia Analysis Group (Chennai), "Before and After Agra", South Asia Analysis Group (Chennai), Paper 280, July 18, 2001. Online at http://www.saag.org/papers3/paper280.html.

19 *Statistical Report on General Election, 2002, to the Legislative Assembly of Jammu and Kashmir* (New Delhi: Election Commission of India, 2002), 8.
20 Ibid., 5–6.
21 Praveen Swami, "The peace meltdown", *Frontline* (Chennai), September 26, 2003.
22 Praveen Swami, "Fissures in the Hizb", *Frontline* (Chennai), May 9, 2003.
23 S.P. Sharma, "Akshardham attack planned at minister's house: Lashkar activist", *The Tribune* (Chandigarh), September 8, 2003.
24 Praveen Swami, "Vote in the time of terror", *Frontline* (Chennai), October 25, 2002.
25 Praveen Swami, "Public funds for terrorism", *Frontline* (Chennai), July 30, 2004.
26 Praveen Swami, "The Nadimarg Outrage", *Frontline* (Chennai), April 25, 2003. Also see "Hizb commdr Bilal-e-Habshi killed in Budgam", *The Daily Excelsior* (Jammu), January 1, 2005.
27 Eric Hobsbawm, *Bandits* (New York: The New Press, 1999), 99.
28 Harish Dugh, "Hizb calls for renewed jihad in Kashmir", *The Indian Express* (New Delhi), June 23, 2005.
29 Praveen Swami and Mohammad Shehzad, "Lashkar raising Islamist brigades for Iraq", *The Hindu* (Chennai), June 13, 2004.
30 Abdus Sattar Ghazali, "Four Pakistanis held in US for running al-Qaeda cell", *The Dawn* (Karachi), June 9, 2005.
31 N.C. Aizenman, "Karzai asks Musharraf for Border Help", *The Washington Post* (Washington, DC), June 22, 2005.
32 Air Commodore Tariq M Ashraf, "Doctrinal Reawakening of the Indian Armed Forces", in *The Combined Armed Center Military Review* (Fort Leavenworth: US Army Combined Arms Center, November–December 2004), 58.
33 Sheikh Mohammad Abdullah [trans. Khushwant Singh], *Flames of the Chinar: An Autobiography* (New York: Viking, 1993), 90.

BIBLIOGRAPHY

Newspapers and magazines

BBC.co.uk (Online)
Bharat-Rakshak.com (Online)
Frontline (Chennai)
Globalsecurity.org (Online)
India Today (New Delhi)
Government of Jammu and Kashmir, http://www.jammukashmir.nic.in (Online)
Kashmir Images (Srinagar)
Kashmir Times (Srinagar)
Newsline (Karachi)
Newstrack (New Delhi)
Outlook (New Delhi)
Rediff.com (Online)
Tehelka (New Delhi)
The Daily Excelsior (Jammu)
The Daily Times (Lahore)
The Dawn (Karachi)
The Hindu (Chennai)
The Hindustan Times (New Delhi)
The Indian Express (New Delhi)
The New York Times (New York)
The News (Lahore)
The official website of the President of Pakistan, http://www.presidentofpakistan.gov.pk/FromThePresidentsDesk.aspx
The Pakistan Times Online
The South Asia Terrorism Portal (Online)
The South Asian Analysis Group (Online)
The Times of India (New Delhi)
The Tribune (Chandigarh)
The Washington Post (Washington, DC)
Time (New York)
Wikipedia.com (Online)

Official documents and resources

[RESTRICTED] B.C. Chakravorty, *History of the Indo-Pak War, 1965* (New Delhi: History Division, Ministry of Defence, Government of India, 1992).

[RESTRICTED] P.B. Sinha and A.A. Athale *History of the Conflict With China, 1962* (New Delhi: Ministry of Defence, 1962).

[RESTRICTED] S.N. Prasad, ed., *History of the Bangladesh War, 1971*(New Delhi: Ministry of Defence, 1992).

[SECRET] *A Note of Background Information of Interrogation*, (Jammu and Kashmir Police, Joint Interrogation Centre, Kot Bhawal, signed by the Deputy Superintendent of Police, Joint Interrogation Centre, Kot Bhalwal).

[SECRET] Jammu and Kashmir Police Criminal Investigation Department, *Interrogation Report of Mushtaq Ahmad Zargar* @ [alias] *Latram*. Srinagar, June 1992. Unsigned.

[SECRET] *Report on Pakistani Organized Subversion, Sabotage and Infiltration in Jammu and Kashmir* (Jammu and Kashmir: Criminal Investigation Department, 1966).

[SECRET] *Report on Underground Organisation Known As Al-Fatah* (Jammu and Kashmir: Criminal Investigation Department, 1961).

Atrocities on Women and Children (New Delhi: Institute for Conflict Management, 2003).

Statement of Maqbool Bhat [Butt] before a special court in Pakistan.

Key Highlights of General Election, 1977, to the Legislative Assembly of Jammu and Kashmir (New Delhi: Election Commission of India, undated).

Key Highlights of General Election, 1983, to the Legislative Assembly of Jammu and Kashmir (Election Commission of India, New Delhi, undated).

Mirza Mohammad Abdullah Beg and G. Parthasarathi, *Agreed Conclusions*

Statistical Report On General Election, 1967, To The Legislative Assembly of Jammu and Kashmir (New Delhi: Election Commission of India, undated).

Statistical Report on General Election, 1996, to the Legislative Assembly of Jammu & Kashmir (New Delhi: Election Commission of India, 1997).

Statistical Report on General Election, 2002, to the Legislative Assembly of Jammu and Kashmir (New Delhi: Election Commission of India, 2002).

Statistical Report on General Election, 1972, to the Legislative Assembly of Jammu & Kashmir (New Delhi: Election Commission of India, undated).

Statistical Report on General Election, 1987, to the Legislative Assembly of Jammu & Kashmir (New Delhi: Election Commission of India, undated).

The Shimla Agreement, 1972

Books and articles

A.G. Noorani, *The Kashmir Question* (Bombay: Manaktalas, 1964).

Ajit Bhattacharjee, *Kashmir: The Wounded Valley* (New Delhi: UBSPD, 1994).

Akbar Khan, *Raiders in Kashmir* (Islamabad: National Book Foundation, 1970).

Alistair Lamb, *Kashmir: A Disputed Legacy, 1856–1990* (Hertingfordbury: Roxford Books, 1991).

Altaf Gauhar, *Ayub Khan: Pakistan's First Military Ruler* (Karachi: Oxford University Press, 1996).

Altaf Hussain, *Shabbir Shah: A Living Legend in Kashmir History*, (Srinagar: Noble Publishing House, not dated).

Anil Maheshwari, *Crescent Over Kashmir: Politics of Mullaism* (e-book, published online at http://ikashmir.org/Crescent/Chapter1.html

Anon., 'Sultan-ul-Arifeen, Sheikh Hamza Makhdum, online at Google.com's cache of http://www.myasa.net/sufindia/sultan-ul-arifeen hazrat sheikh hamza makhdum.html as retrieved on Dec. 13, 2004 17:47:58 GMT.

Ashley J Tellis, C. Christine Fair, and Jamison Jo Medby, *Limited Conflict Under The Nuclear Umbrella: Indian and Pakistani Lessons From the Kargil Crisis* (Washington DC: Rand Corporation, 2001).

Balraj Puri, "Jammu and Kashmir" in Myron Weiner, ed., *State Politics in India* (Princeton: Princeton University Press, 1966).

Balraj Puri, *Jammu and Kashmir: Triumph and Tragedy of Indian Federalisation* (New Delhi: Sterling, 1983).

B.M. Sinha, *The Samba Spying Case* (New Delhi: Vikas, 1981).

B.N. Mullik, *Chinese Betrayal: My Years With Nehru* (Bombay: Allied Publishers, 1971).

Brian Cloughley, *A History of Pakistan Army: Wars and Insurrections* (Karachi: Oxford University Press, 2000).

C.A. Bayly, *Empire and Information: Intelligence Gathering and Social Communication in India, 1780–1870* (Cambridge: Cambridge University Press, 1999).

Chandrashekhar Dasgupta, *War and Diplomacy in Kashmir: 1947–1948* (New Delhi: Sage Publications, 2002).

D.H. Butani, *The Third Sikh War? Towards or Away From Khalistan?* (New Delhi: Promilla & Co., Publishers, 1986).

David Albright, Frans Berkhout, William Walker, *Plutonium and Highly Enriched Uranium 1996: World Inventories, Capabilities, and Policies*. (Stockholm: Stockholm International Peace Research Institute), 1997.

Dennis Kux, *The United States and Pakistan: Disenchanted Allies* (Washington, DC: Woodrow Wilson Center Press, 2000).

Devendra Swarup and Sushil Aggarwal, *The Roots of the Kashmir Problem: The Continuing Battle Between Secularism and Communal Separatism* (New Delhi: Manthan Prakashan, 1992).

Devin Hagerty, "Nuclear Deterrence in South Asia." *International Security,* Winter 1995–1996, Volume 20, No. 3 cited in Pravin Sawhney, *The Defence Makeover* (New Delhi: Sage Publications, 2002).

Eric Hobsbawm, *Bandits* (New York: The New Press, 1999).

Eric Hobsbawm, *Revolutionaries* (New York: The New Press, 2001).

F.M. Hassnain, *Freedom Struggle in Kashmir* (New Delhi: Rima Publishing House, 1988).

Farooq Abdullah, *My Dismissal* (New Delhi: Privately Published, 1985).

Francis Tucker, *India's Partition and Human Debasement, Vol. 1* (New Delhi: Akashdeep Publishing House, 1988).

Geoffrey Roberts, *Victory at Stalingrad* (Edinburgh and London: Pearson, 2002).

Giles Dorronsoro, *Revolution Unending: Afghanistan, 1979 to the Present* (New York: Colombia University Press, 2005).

Glenn Snyder, "The Balance of Power and the Balance of Terror", in Paul Seabury (ed.) *The Balance of Power* (San Francisco: Chandler, 1965).

Gul Hasan Khan, *Memoirs* (Karachi: Oxford Univeristy Press, 1993).

G.W. Chowdhury, *The Last Days of United Pakistan* (Karachi: Oxford Univeristy Press, 1993).

Hamza Alavi, "Class and State", in Hassan Gardezi and Jamil Rashid (eds), *Pakistan: The Roots of Dictatorship* (London: Zed Press, 1983).

Hari Om, *Mulism of Jammu & Kashmir: A Study of the Spread of Education and Consciousness, 1857–1925* (New Delhi: Archives Publishers, 1982).

Harkishan Singh Surjeet, *Deepening Punjab Crisis: A Democratic Solution* (New Delhi: Patriot Publishers, 1992)

Hashim Qureshi, *The Unveiling of Truth: A Political Analysis of Kashmir* (Lahore: Jeddojudh Publications, 1999).

Hassan Gardezi, "The Resurgence of Islam: Islamic Ideology and Encounters With Imperialism", in Hassan Gardezi and Jamil Rashid (eds), *Pakistan: The Roots of Dictatorship* (London: Zed Press, 1983).

Human Rights in India: Punjab in Crisis (New York: Human Rights Watch, 1991)

Jack Weatherford, *Genghis Khan and the Making of the Modern World* (New York: Three Rivers Press, 2005).

J.N. Kachroo, "National Schools of Kashmir" in *Milchar* (Mumbai), September 2003.

Joseph Korbel, *Danger in Kashmir* (Princeton: Princeton University Press, 1966).

K.P.S. Gill, *Punjab: The Knights of Falsehood* (New Delhi: Har Anand Publishers, 1997).

Kanchan L., "The Media in Terror: Analysing Reportage From Theatres of Conflict", *Faultlines*, Volume 8 (New Delhi: Institute for Conflict Management, April 2001).

Karan Singh, *Autobiography* (New Delhi: Oxford University Press, 1994).

Karen Armstrong, *Holy Wars: The Crusades and their Impact on Today's World* (New York: Anchor, 2001).

Katherine Frank, *Indira: The Life of Indira Nehru Gandhi* (Boston and New York: Houghton Mifflin, 2002).

Kenneth Conboy and James Morrison, *The CIA's Secret War in Tibet* (Lawrence: University Press of Kansas, 2002).

Kh. Mubarak Shah (ed.), *The Kashmir Conspiracy Case Report 1* (Srinagar: Legal Defense Committee, 1964).

Kh. Mubarak Shah (ed.), *The Kashmir Conspiracy Case Report VII* (Srinagar: Legal Defense Committee, 1964).

Lieutenant-General K.S. Brar, *Operation Bluestar: The True Story* (New Delhi: UBSPD, 1993).

Lieutenant-General Y.M. Bammi, *Kargil: The Impregnable Conquered* (NOIDA: Gorkha Publishers, 2002).

Lieutenant-General L.P. Sen, *Slender Was The Thread* (New Delhi: Orient Longman, 1994).

Major-General Afsir Karim, *Kashmir: The Troubled Frontiers* (New Delhi: Lancer Publishers, 1994).

Major-General Ashok Kalyan Verma, *Kargil: Blood on the Snow* (New Delhi: Manohar, 2002).

Manoj Joshi, *The Lost Rebellion: Kashmir in the Nineties* (New Delhi: Penguin Books, 1991).

Manoj Joshi, *Combating Terrorism in Punjab: Indian Democracy in Crisis* (London: Research Institute for the Study of Conflict, 1993)

Maqsood Ahmed Masood and Peter Stockdale, *The Khalistan Riddle* (Islamabad: Modern Book Depot, 1998).

Margaret Bourke-White, *Halfway To Freedom* (Mumbai: Asia Publishing House, 1950).

Mark Tully and Satish Jacob, *Amritsar: Mrs. Gandhi's Last Battle* (London: Jonathan Cape, 1986)

M.J. Akbar, Kashmir: Behind the Vale (New Delhi: Viking, 1991).

Mohammad Ayub Khan, *Friends, Not Masters: A Political Autobiography* (London: Oxford University Press, 1967).

Mohammad Ishaq Khan, "Six Centuries of Islamisation in Kashmir: Retrospect and Prospects" in Balraj Puri (ed.), *5000 Years of Kashmir* (New Delhi: Ajanta, 1997).

Mohammad Sayeed Asad, *Shaur-e-Farda* [Urdu] (Mirpur: National Institute of Kashmir Studies, 2001).

Mridu Rai, *Hindu Rulers, Muslim Subjects: Islam, Rights and the History of Kashmir* (Princeton University Press: Princeton, 2004).

Muhammad Amir Rana [trans. Saba Ansari], *A–Z of Jehadi Organizations in Pakistan* (Lahore: Mashal Books, 2004).

Navnita Chadha Behera, *State, Identity and Violence: Jammu, Kashmir and Ladakh* (New Delhi: Manohar Books, 2000).

N.B. Bardhan, A. Dey, and S.P. Khanna, *Characteristics of New Improvised Explosive Device* (New Delhi: Central Bureau of Investigations, undated).

Owen Bennett Jones, *Pakistan: Eye of the Storm* (New Haven and London: Yale University Press, 2003).

Paul Wallace and Surendra Chopra, *Political Dynamics and Crisis in Punjab* (Amritsar: Department of Political Science, Guru Nanak Dev University, 1988).

Peter Hopkirk, *The Great Game: The Struggle for Empire in Central Asia* (Tokyo and New York: Kodansha International, 1994).

Peter Hopkirk, *Trespassers on the Roof of the World: The Secret Exploration of Tibet* (Tokyo and New York: Kodansha International, 1995).

P.L. Lakhanpal, *Essential Documents and Notes on Kashmir Dispute* (New Delhi: International Books, 1965).

Pradeep Thakur, *Militant Monologues* (New Delhi: Parity Publisher, 2003).

Praveen Swami, "Failed Threats and Flawed Fences: India's Military Responses to Pakistan's Proxy War," *India Review* 3, No. 2 (2004).

Praveen Swami, *The Kargil War* (New Delhi: Left Word, 1999)

Pravin Sawhney, *The Defence Makeover* (New Delhi: Sage Publications, 2002).

Prem Nath Bazaz, *Democracy Through Intimidation and Terror* (New Delhi: Heritage, 1978).

Prem Shankar Jha, *Kashmir, 1947: rival versions of history* (New Delhi: Oxford University Press, 1996).

Prithvi Nath Kaul Bamzai, *Kashmir and Power Politics From Lake Success To Tashkent* (Metropolitan Book Company: New Delhi), 1966.

P.S. Verma, "Muslim United Front" in S. Bhatnagar and Pradeep Kumar (ed.), *Regional Political Parties in India* (New Delhi: Ess Ess Publications, 1988).

P.S. Verma, *Jammu and Kashmir at the Political Crossroad* (New Delhi: Vikas, 1994).

Rafique A. Khan, "Maqbool Butt: His Life, Letters and Legacy" (Los Angeles: Kashmir Human Rights Foundation, undated).

S. Akbar Zaidi, *Pakistan's Economic & Social Development* (New Delhi: Rupa, 2004).

Salman Rushdie, *Midnight's Children* (New York: Penguin, 1991).

Sarvepalli Gopal, *Jawaharlal Nehru: A Biography* (New Delhi: Oxford University Press, 2004).

Satyapal Dang, *Genesis of Terrorism: An Analytical Study of Punjab Terrorists* (New Delhi: Patriot Publishers, 1988);

Sayyid Mir Qasim, *My life and Times* (New Delhi: Allied Publishers, 1992).

Seyyed Vali Reza Nasr, *The Vanguard of the Islamic Revolution: The Jama'at-i Islami of Pakistan* (Berkley and Los Angeles: University of California Press, 1994).

Shahid-ur-Rehman, *Long Road to Chagai: Untold Story of Pakistan's Nuclear Quest* (Islamabad: Print Wise Publications), 1999.

Sharda Jain, *Politics of Terrorism in India: The Case of Punjab* (New Delhi: Deep and Deep Publications, 1995)

Sheikh Mohammad Abdullah [trans. Khushwant Singh], *Flames of the Chinar: An Autobiography* (New York: Viking, 1993).

Shelford Bidwell and Ray Bonds, eds, *The Encyclopedia of Land Warfare in the 20th Century* (New York: Crowell, 1976) .

Shereen Ilahi, "The Radcliffe Boundary Commission and the Fate of Kashmir", *India Review* Vol. 2. No. 1 (London: Frank Cass, January 2003).

Siddiq Salik, *Witness to Surrender* (Karachi: Oxford University Press, 1977).

Simon Singh, *The Code Book: The Science of Secrecy From Ancient Egypt to Quantum Cryptography* (New York: Anchor Books, 2000).

Stanley Wolpert, *Zulfi Bhutto of Pakistan: His Life and Times* (New York: Oxford University Press, 1993).

Steve Coll, *Ghost Wars* (London: Penguin Books, 2005).

Steven P. Cohen, *The Idea of Pakistan* (New Delhi: Oxford University Press, 2005).

Suchita Vemuri, "Kashmir Drenched in Women's Tears", Women in Action 1–2: 2000, online at http://www.isiswomen.org/wia/wia100/soc00011.html).

Sumit Ganguly and Devin T. Hagerty, *Fearful Symmetry: India-Pakistan Crises in the Shadow of Nuclear Weapons* (New Delhi: Oxford University Press, 2005).

Sumit Ganguly and R. Harrison Wagner, "India and Pakistan: Bargaining in the Shadow of Nuclear War", *The Journal of Strategic Studies*, Vol. 27, No. 3, September 2004 (London: Francis and Taylor).

Sumit Ganguly, "Indo-Pakistani Nuclear Issues and the Stability/Instability Paradox", *Studies in Conflict and Terrorism,* Vol. 18, No. 4, 1995.

Sumit Ganguly, *The Crisis in Kashmir: Portents of War, Prospects of Peace* (Cambridge and Washington, DC: Cambridge University Press and Woodrow Wilson Center Press, 1997).

Sunil Fotedar, ed., *Kashmir: Articles & Opinions* (Austin: Kashmir News Network, 2002).

Tara Kartha, "Early Warning and Light Weapons in South Asia", in *Arms Watching: Integrating Small Arms and Light Weapons Into the Early Warning of Violent Conflict,* ed. Edward J. Laurence, (International Alert, 1999).

Tariq M. Ashraf, "Doctrinal Reawakening of the Indian Armed Forces", in *The Combined Armed Center Military Review* (Fort Leavenworth: US Army Combined Arms Center, November-December 2004).

Victoria Schofield, *Kashmir in Conflict: India, Pakistan and the Unending War* (London: I.B. Tauris, 2003).

Victoria Schofield, *Kashmir in the Crossfire* (London: IB Tauris, 1996).

Y.D. Gundevia, "On Sheikh Abdullah", in *The Testament of Sheikh Abdullah* [no author] (New Delhi: Palit and Palit, 1974).

Yoginder Sikand, "Changing Course of Kashmiri Struggle", *Economic and Political Weekly* (Mumbai), January 20, 2001.

Yoginder Sikand, "The Emergence and Development of the Jamaat-e-Islami of Jammu and Kashmir," *Modern Asian Studies* Vol. 36, No. 3 (Cambridge Univeristy Press: Cambridge, 2002).

INDEX

eBooks

eBooks – at www.eBookstore.tandf.co.uk

A library at your fingertips!

eBooks are electronic versions of printed books. You can store them on your PC/laptop or browse them online.

They have advantages for anyone needing rapid access to a wide variety of published, copyright information.

eBooks can help your research by enabling you to bookmark chapters, annotate text and use instant searches to find specific words or phrases. Several eBook files would fit on even a small laptop or PDA.

NEW: Save money by eSubscribing: cheap, online access to any eBook for as long as you need it.

Annual subscription packages

We now offer special low-cost bulk subscriptions to packages of eBooks in certain subject areas. These are available to libraries or to individuals.

For more information please contact webmaster.ebooks@tandf.co.uk

We're continually developing the eBook concept, so keep up to date by visiting the website.

www.eBookstore.tandf.co.uk